T0133966

SPATIO-TEMPORAL STATISTICS WITH R

Chapman & Hall/CRC
The R Series

For more information about this series, please visit: https://www.crcpress.com/go/the-r-series

SPATIO-TEMPORAL STATISTICS WITH R

CHRISTOPHER K. WIKLE
ANDREW ZAMMIT-MANGION
NOEL CRESSIE

CRC Press
Taylor & Francis Group
Boca Raton London New York

CRC Press is an imprint of the
Taylor & Francis Group, an **informa** business

A CHAPMAN & HALL BOOK

Cover Illustration: Julinu (Julian Mallia)
www.julinu.com

CRC Press
Taylor & Francis Group
6000 Broken Sound Parkway NW, Suite 300
Boca Raton, FL 33487-2742

© 2019 by Taylor & Francis Group, LLC
CRC Press is an imprint of Taylor & Francis Group, an Informa business

No claim to original U.S. Government works

ISBN 13: 978-1-138-71113-6 (hbk)

Library of Congress Cataloging-in-Publication Data

Names: Wikle, Christopher K., 1963- author. | Zammit-Mangion, Andrew, author. | Cressie, Noel A. C., author.
Title: Spatio-temporal statistics with R / Christopher K. Wikle, Andrew Zammit-Mangion, Noel Cressie.
Description: Boca Raton, Florida : CRC Press, [2019] | Includes bibliographical references and index.
Identifiers: LCCN 2018048440| ISBN 9781138711136 (hardback : alk. paper) | ISBN 9781351769723 (e-book : alk. paper)
Subjects: LCSH: Spatial analysis (Statistics) | Statistics. | R (Computer program language)
Classification: LCC QA278.2 .W55 2019 | DDC 519.5/37--dc23
LC record available at https://lccn.loc.gov/2018048440

Visit the Taylor & Francis Web site at
http://www.taylorandfrancis.com

and the CRC Press Web site at
http://www.crcpress.com

Contents

Acknowledgements

When Noel and I finished the multi-year project that became *Statistics for Spatio-Temporal Data* in 2010, I'm pretty sure I didn't think that I would be writing another book on this topic! But, it's eight years later and here we are .... It has been a great pleasure to work with Andrew and Noel on this project and I thank them deeply for all of the stimulating discussion, idea-sharing, advice, and hard work they put into this project. I learned a great deal and it could never have happened without them! In particular, Andrew has worked magic to make the R Labs integrate into the methodological content, and this is the feature of the book that makes it unique. I want to thank my spatio-temporal colleagues at Mizzou (Scott Holan, Sakis Micheas, and Erin Schliep) as well as students and postdocs who have continued to make this an exciting and fun topic in which to work. My eternal thanks to Olivia, Nathan, and Andrea for their support of this project and all it entailed and for enriching my life always! Last, and most importantly, I would like to thank Carolyn, who is on the "front lines" of dealing with the effects of these sorts of projects, and always provides tremendous support, sanity, and encouragement along the way. I could not do what I do if it were not for her!

C.K.W.

More than ten years have passed since the day when I was sitting opposite my honors thesis supervisor, Simon Fabri, at the University of Malta with a scholarship offer from the University of Sheffield in my hand, and a pen in the other. "What is spatio-temporal modeling, and is there any future in it?" I mumbled inquisitively. It is largely thanks to his reply and my PhD supervisor Visakan Kadirkamanathan that I took an interest in spatio-temporal modeling, and in the field of statistics in general. Since then, I have had other mentors from numerous disciplines, from statistics to computer science and geography, and I would like to thank them all for their advice and for the opportunities they have provided me with; they include Guido Sanguinetti, Jonathan Rougier, Jonathan Bamber, and more recently Noel Cressie.

In the last ten years I have had the privilege to work and have discussions with several other colleagues with similar interests. Some of these have inspired my work in several ways; they include Tara Baldacchino, Parham Aram, Michael Dewar, Kenneth Scerri, Sean

Anderson, Botond Cseke, Finn Lindgren, Bohai Zhang, and Thomas Suesse. Above all, they have made my time in this field of research enjoyable, intriguing, and rewarding.

Chris and Noel were seen as the pioneers of dynamic spatio-temporal models by our research group in Sheffield, and much of our early work was based on theirs. I therefore feel very privileged and honored to have had the opportunity to work with them. I would like to thank them for all they have taught me in the last few years during the writing of this book.

Finally, I would like to thank my family: my parents, Patricia and Louis, for all the opportunities they have given me; my dear wife, Anaïd, who was extremely supportive and always there throughout the writing of the book; and my son, Benjamin, who was born in the last stages of the book and had to make figures of spatio-temporal data his favorite toys for much of his early months. Thank you!

A.Z.-M.

Books are like children, and this is my fourth (book). You love them differently because they are unique, and so it's impossible to prefer one over another. I won't tell you about the others, but you can get some idea about them by reading this one; all of them share some of the same genetic material. This book started with discussions between Chris and me just after I moved to the University of Wollongong (UOW); we were talking about doing a second edition of our 2011 book, *Statistics for Spatio-Temporal Data,* until we realized a great need for something else. Speaking for myself, I felt that my research wasn't having the impact in the sciences I was hoping for. It became clear to me that I was having a one-way conversation, but I also knew that software can be a powerful medium of communication. This book is a very exciting development because spatio-temporal statistical modeling has found another voice, one that talks with scientists through software as well as methodology. Chris and Andrew share this view and have been instrumental in making our two-way conversation with others happen. Chris and I have been hanging out for a long time and I always learn from him, something that happened in spades on this project. Andrew was a gift to me and UOW from half-way around the world, and in the four years since he came we have shared a number of papers and now a book. Back to genes, my parents Ray and Rene gave me so much from so little. My children Amie and Sean are interwoven in all that I do, and I hope they sense their mathematical talent in what follows. Elisabeth is my muse, and words cannot express how important she is in my life.

N.C.

Our sincere thanks go to a number of people who have contributed to the completion of this project. Material from the book was trialed at various short courses, including ones sponsored by ASA, AIMS, ACEMS, IBS-AR, and NIASRA. Valuable feedback came from

Russel Yost and Michael Kantar of the University of Hawaii, Giri Gopalan of the University of Iceland, Petra Kuhnert of CSIRO, Nathan Wikle of Pennsylvania State University, the Space-Time Reading Group at the University of Missouri (Chris Hassett, Alex Oard, Toryn Schafer, Erin Schliep, Matt Simpson), and Mevin Hooten of the USGS and Colorado State University. We are very grateful to Simon Wood, Finn Lindgren, Johan Lindström, and Clint Shumack for test-driving and commenting on many of the Labs and to Patrick McDermott who contributed functions to the book's R package **STRbook** for the ESN implementation in Appendix F. Karin Karr LATEXed the first and epilogical chapters and helped compile the author index, Bohai Zhang was a resource for Karin, and Clint Shumack produced Figures 1.2 and 2.13 and implemented the book's website: many thanks for their assistance. CKW and NC wish to acknowledge travel support from the US National Science Foundation and the US Census Bureau under National Science Foundation grant SES-1132031, funded through the National Science Foundation Census Research Network (NCRN) program. AZM was partially supported by an Australian Research Council (ARC) Discovery Early Career Research Award, DE180100203. Rob Calver at Chapman & Hall/CRC has been our rock as we've navigated aspects of publishing new to us. A bound hard-cover copy of the book can be purchased (with a stunning cover produced by Julian Mallia) from our publisher, Chapman & Hall/CRC, at http://www.crcpress.com/9781138711136, or it is free for download from our interactive website, https://spacetimewithr.org. Finally, we would like to express our appreciation to the whole R community, upon whose shoulders we stand!

Preface

We live in a complex world, and clever people are continually coming up with new ways to observe and record increasingly large parts of it so we can comprehend it better (warts and all!). We are squarely in the midst of a "big data" era, and it seems that every day new methodologies and algorithms emerge that are designed to deal with the ever-increasing size of these data streams.

It so happens that the "big data" available to us are often *spatio-temporal data*. That is, they can be indexed by spatial locations and time stamps. The space might be geographic space, or socio-economic space, or more generally network space, and the time scales might range from microseconds to millennia. Although scientists have long been interested in spatio-temporal data (e.g., Kepler's studies based on planetary observations several centuries ago), it is only relatively recently that statisticians have taken a keen interest in the topic. At the risk of two of us being found guilty of self-promotion, we believe that the book *Statistics for Spatio-Temporal Data* by Cressie and Wikle (2011) was perhaps the first dedicated and comprehensive statistical monograph on the topic. In the decade (almost) since the publication of that book, there has been an exponential increase in the number of papers dealing with spatio-temporal data analysis – not only in statistics, but also in many other branches of science. Although Cressie and Wikle (2011) is still extremely relevant, it was intended for a fairly advanced, technically trained audience, and it did not include software or coding examples. In contrast, the present book provides a more accessible introduction, with hands-on applications of the methods through the use of R Labs at the end of each chapter. At the time of writing, this unique aspect of the book fills a void in the literature that can provide a bridge for students and researchers alike who wish to learn the basics of spatio-temporal statistics.

What level is expected of readers of this book? First, although each chapter is fairly self-contained and they can be read in any order, we ordered the book deliberately to "ease" the reader into more technical material in later chapters. Spatio-temporal data can be complex, and their representations in terms of mathematical and statistical models can be complex as well. They require a number of indices (e.g., for space, for time, for multiple variables). In addition, being able to account for dependent random processes requires a bit of statistical sophistication that cannot be completely avoided, even in an applications-based introductory book. We believe that a reader who has taken a class or two in calculus-based probability and inference, and who is comfortable with basic matrix-algebra representations of statistical models (e.g., a multiple regression or a multivariate time-series representation), could comfortably get through this book. For those who would like a brief refresher on matrix algebra, we provide an overview of the components that we use in an appendix. To make this a bit easier on readers with just a few statistics courses on their transcript, we have interspersed "technical notes" throughout the book that provide short, gentle reviews of methods and ideas from the broader statistical literature.

Chapter 1 is the place to start, to get you intrigued and perhaps even excited about what is to come. We organized the rest of the book to follow what we believe to be good statistical practice. First, look at your data and do exploratory analyses (Chapter 2), then fit simple statistical models to the data to indicate possible patterns and see if assumptions are violated (Chapter 3), and then use what you learned in these analyses to build a spatio-temporal model that allows valid inferences (Chapters 4 and 5). The end of the cycle is to evaluate your model formally to find areas of improvement and to help choose the best model possible (Chapter 6). Then, if needed, repeat with a better-informed spatio-temporal model.

The bulk of the material on spatio-temporal modeling appears in Chapters 4 and 5. Chapter 4 covers descriptive (*marginal*) models formed by characterizing the spatio-temporal dependence structure (mainly through spatio-temporal covariances), which in turn leads to models that are analogous to the ubiquitous geostatistical models used in kriging. Chapter 5 focuses on dynamic (*conditional*) models that characterize the dynamic evolution of spatial processes through time, analogous to multivariate time-series models. Like Cressie and Wikle (2011), both Chapters 4 and 5 are firmly rooted in the notion of *hierarchical thinking* (i.e., hierarchical statistical modeling), which makes a clear distinction between the data and the underlying latent process of interest. This is based on the very practical notion that "[w]hat you see (data) is not always what you want to get (process)" (Cressie and Wikle, 2011, p. xvi).

Spatio-temporal statistics is such a vast field and this modestly sized book is necessarily not comprehensive. For example, we focus primarily on data whose spatial reference is a point, and we do not explore issues related to the "change-of-support" problem, nor do we deal with spatio-temporal point processes. Further, we mostly limit our discussion to models and methodologies that are relatively mature, understood, and widely used. Some of the applications our readers are confronted with will undoubtedly require cutting-edge methods beyond the scope of this book. In that regard, the book provides a down-to-earth introduction. We hope you find that the path is wide and the slope is gentle, ultimately giving you the confidence to explore the literature for new developments. For this reason, we have named our epilogical chapter *Pergimus*, Latin for "let us continue to progress."

A substantial portion of this book is devoted to "Labs," which enable the reader to put his or her understanding into practice using the programming language R. There are several reasons why we chose R: it is one of the most versatile languages designed for statistics; it is open source; it enjoys a vibrant online community whose members post solutions to virtually any problem you will encounter when coding; and, most importantly, a large number of packages that can be used for spatio-temporal modeling, exploratory data analysis, and statistical inference (estimation, prediction, uncertainty quantification, and so forth) are written in R. The last point is crucial, as it was our aim right from the beginning to make use of as much tried-and-tested code as possible to reduce the analyst's barrier to entry. Indeed, it is fair to say that this book would not have been possible without the excellent work, openness, and generosity of the R community as a whole.

In presenting the Labs, we intentionally use a "code-after-methodology" approach, since we firmly believe that the reader should have an understanding of the statistical methods being used before delving into the computational details. To facilitate the connections between methodology and computation, we have added "R Tips" where needed. The Labs themselves assume some prior knowledge of R and, in particular, of the *tidyverse*, which is built on an underlying philosophy of how to deal with data and graphics. Readers who would like to know more can consult the excellent book by Wickham and Grolemund (2016) for background reading (freely available online).

Finally, our goal when we started this project was to help as many people as we could to start analyzing spatio-temporal data. Consequently, with the generous support of our editors at Chapman & Hall/CRC, we have made the .pdf file of this book and the accompanying R package, **STRbook**, freely available for download from the website listed below. In addition, this website is a place where users can post *errata*, comment on the code examples, post their own code for different problems, their own spatio-temporal data sets, and articles on spatio-temporal statistics. You are invited to go to:

```
https://spacetimewithr.org
```

We hope you find this book useful for your endeavors as you begin to explore the complexities of the spatio-temporal world around us – and within us! Let's get started . . .

Christopher K. Wikle
Columbia, Missouri, USA

Andrew Zammit-Mangion
Wollongong, NSW, Australia

Noel Cressie
Sydney, NSW, Australia

Chapter 1

Introduction to Spatio-Temporal Statistics

> "I feel all things as dynamic events, being, changing, and interacting with each other in space and time even as I photograph them." (Wynn Bullock, 1902–1975, American photographer)

Wynn Bullock was an early pioneer of modern photography, and this quote captures the essence of what we are trying to get across in our book – except in our case the "photographs" are fuzzy and the pictures are incomplete! The top panel of Figure 1.1 shows the July 2014 launch of the US National Aeronautics and Space Administration (NASA) *Orbiting Carbon Observatory-2* (OCO-2) satellite, and the bottom panel shows the "photographer" in action. OCO-2 reached orbit successfully and, at the time of writing, is taking pictures of the dynamic world below. They are taken every fraction of a second, and each "photograph" is made up of measurements of the sun's energy in selected spectral bands, reflected from Earth's surface.

After NASA processes these measurements, an estimate is obtained of the fraction of carbon dioxide (CO_2) molecules in an atmospheric column between Earth's surface and the OCO-2 satellite. The top panel of Figure 1.2 shows these estimates in the boreal winter at locations determined by the geometry of the satellite's 16-day repeat cycle (the time interval after which the satellite retraces its orbital path). (They are color-coded according to their value in units of parts per million, or ppm.) Plainly, there are gaps caused by OCO-2's orbit geometry, and notice that the higher northern latitudes have very few data (caused by the sun's low angle at that time of the year). The bottom panel of Figure 1.2 shows 16 days of OCO-2 data obtained six months later, in the boreal summer, where the same comments about coverage apply, except that now the higher southern latitudes have very few data. Data incompleteness here is a moving target in both space and time. Furthermore, any color-coded "dot" on the map represents a datum that should not be totally believed, since

Figure 1.1: Top: Launch of NASA's OCO-2 satellite, on 02 July 2014 (credit: NASA/JPL). Bottom: An artist's impression of the OCO-2 satellite in orbit (credit: NASA/JPL).

it is an estimate obtained from measurements made through 700 km of atmosphere with clouds, water vapor, and dust getting in the way. That is, there is "noise" in the data.

There is a "+" on the global maps shown in Figure 1.2, which is at the location of the Mauna Loa volcano, Hawaii. Near the top of this volcano, at an altitude of 4.17 km, is the US National Oceanic and Atmospheric Administration (NOAA) Mauna Loa Observatory that has been taking monthly measurements of CO_2 since the late 1950s. The data are shown as a time series in Figure 1.3. Now, for the moment, put aside issues associated with measurements being taken with different instruments, on different parcels of air, at

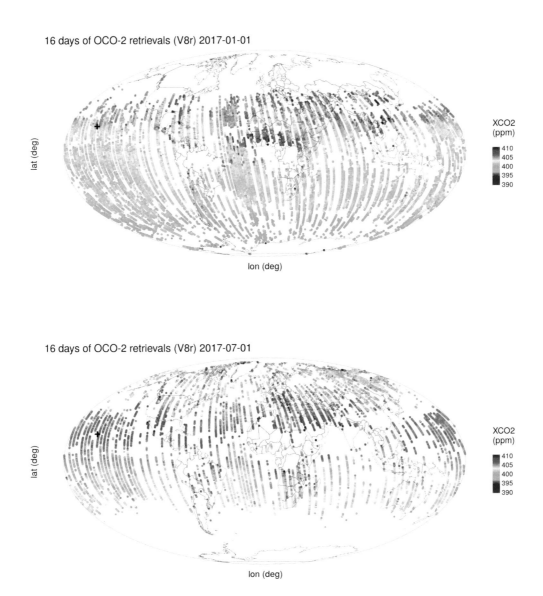

Figure 1.2: Sixteen days of CO_2 data from the OCO-2 satellite. Top: Data from 25 December 2016 to 09 January 2017 (boreal winter). Bottom: Data from 24 June 2017 to 09 July 2017 (boreal summer). The panel titles identify the eighth day of the 16-day window.

different locations, and for different blocks of time; these can be dealt with using quite advanced spatio-temporal statistical methodology found in, for example, Cressie and Wikle

(2011). What is fundamental here is that underlying these imperfect observations is a spatio-temporal process that itself is not perfectly understood, and we propose to capture this uncertainty in the process with a spatio-temporal statistical model.

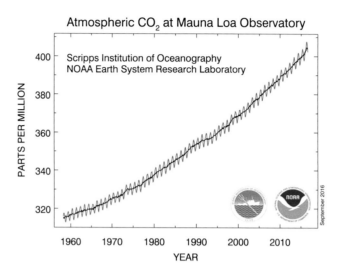

Figure 1.3: Monthly mean atmospheric CO_2 (ppm) at the NOAA Mauna Loa Observatory, Hawaii. The smooth line represents seasonally corrected data (Credit: Scripps Institution of Oceanography and NOAA Earth System Research Laboratory).

The atmospheric CO_2 process varies in space and in time, but the extent of its spatio-temporal domain means that exhaustive measurement of it is not possible; and even if it were possible, it would not be a good use of resources (a conclusion you should find evident after reading our book). Figure 1.2 shows two spatial views during short time periods that are six months apart; that is, it gives two spatial "snapshots." Figure 1.3 shows a temporal view at one particular location as it varies monthly over a 50-year time period; that is, it gives a temporal "profile." This is a generic problem in spatio-temporal statistics, namely our noisy data traverse different paths through the "space-time cube," but we want to gain knowledge about unobserved (and even observed) parts of it. We shall address this problem in the chapters, the Labs, and the technical notes that follow, drawing on a number of data sets introduced in Chapter 2.

Humans have a longing to understand their place (temporally and spatially) in the universe. In an Einsteinian universe, space and time interact in a special, "curved" way; however, in this book our methodology and applications are for a Newtonian world. Rick Delmonico, author of the book, *The Philosophy of Fractals* (Delmonico, 2017), has been quoted elsewhere as saying that "light is time at maximum compression and matter is space at maximum compression." Our Newtonian world is definitely more relaxed than this! Nevertheless, it is fascinating that images of electron motion at a scale of 10^{-11} meters look very much like images of the cosmos at a scale of 10^{17} meters (Morrison and Morrison, 1982).

Trying to understand spatio-temporal data and how (and ultimately why) they vary in space and time is not new – just consider trying to describe the growth and decline of populations, the territorial expansion and contraction of empires, the spread of world religions, species (including human) migrations, the dynamics of epidemics, and so on. Indeed, history and geography are inseparable. From this "big picture" point of view, there is a complex system of interacting physical, biological, and social processes across a range of spatial/temporal scales.

How does one do spatio-temporal statistics? Well, it is not enough to consider just spatial snapshots of a process at a given time, nor just time-series profiles at a given spatial location – the behavior at spatial locations at one time point will almost certainly affect the behavior at nearby spatial locations at the next time point. Only by considering time and space together can we address how spatially coherent entities change over time or, in some cases, why they change. It turns out that a big part of the *how* and *why* of such change is due to *interactions* across space and time, and across multiple processes.

For example, consider an influenza epidemic, which is generally in the winter season. Individuals in the population at risk can be classified as susceptible (S), infected (I), or recovered (R), and a well-known class of multivariate temporal models, called SIR models, capture the transition of susceptibles to infecteds to recovereds and then possibly back to susceptibles. At a micro level, infection occurs in the household, in the workplace, and in public places due to the interaction (contact) between infected and susceptible individuals. At a macro level, infection and recovery rates can be tracked and fitted to an SIR model that might also account for the weather, demographics, and vaccination rates. Now suppose we can disaggregate the total-population SIR rates into health-district SIR rates. This creates a spatio-temporal data set, albeit at a coarse spatial scale, and the SIR rates can be visualized dynamically on a map of the health districts. Spatio-temporal interactions may then become apparent, and the first steps of spatio-temporal modeling can be taken.

Spatio-temporal interactions are not limited to similar types of processes nor to spatial and temporal scales of variability that seem obvious. For example, El Niño and La Niña phenomena in the tropical Pacific Ocean correspond to periods of warmer-than-normal and colder-than-normal sea surface temperatures (SST), respectively. These SST "events" occur every two to seven years, although the exact timing of their appearance and their end is not regular. But it is well known that they have a tremendous impact on the weather across the globe, and weather affects a great number of things! For example, the El Niño and La Niña events can affect the temperature and rainfall over the midwest USA, which can affect, say, the soil moisture in the state of Iowa, which would likely affect corn production and could lead to a stressed USA agro-economy during that period. Simultaneously, these El Niño and La Niña events can also affect the probability of tornado outbreaks in the famed "tornado alley" region of the central USA, and they can even affect the breeding populations of waterfowl in the USA.

Doing some clever smoothing and sharp visualizations of the spatial, temporal, and spatio-temporal variability in the data is a great start. But the information we glean from these data analyses needs to be organized, and this is done through models. In the next section, we make the case for spatio-temporal models that are *statistical*.

1.1   Why Should Spatio-Temporal Models Be *Statistical?*

In the physical world, phenomena evolve in space and time following deterministic, perhaps "chaotic," physical rules (except at the quantum level), so why do we need to consider randomness and uncertainty? The primary reason comes from the uncertainty resulting from incomplete knowledge of the science and of the mechanisms driving a spatio-temporal phenomenon. In particular, *statistical* spatio-temporal models give us the ability to model components in a physical system that appear to be random and, even if they are not, the models are useful if they result in accurate and precise predictions. Such models introduce the notion of uncertainty, but they are able to do so without obscuring the salient trends or regularities of the underlying process (that are typically of primary interest).

Take, for instance, the raindrops falling on a surface; to predict exactly where and when each drop will fall would require an inconceivably complex, deterministic, meteorological model, incorporating air pressure, wind speed, water-droplet formation, and so on. A model of this sort at a large spatial scale is not only infeasible but also unnecessary for many purposes. By studying the temporal intensity of drops on a regular spatial grid, one can test for spatio-temporal interaction or look for dynamic changes in spatial intensity (given in units of "per area") for each cell of the grid. The way in which the intensity evolves over time may reveal something about the driving mechanisms (e.g., wind vectors) and be useful for prediction, even though the exact location and time of each incident raindrop is uncertain.

Spatio-temporal statistical models are *not* at odds with deterministic ones. Indeed, the most powerful (in terms of predictive performance) spatio-temporal statistical models are those that are constructed based on an understanding of the biological or physical mechanisms that give rise to spatio-temporal variability and interactions. Hence, we sometimes refer to them as *physical-statistical models* (see the editorial by Kuhnert, 2014), or generally as *mechanistically motivated statistical models*. To this understanding, we add the reality that observations may have large gaps between them (in space and in time), they are observed with error, our understanding of the physical mechanisms is incomplete, we have limited knowledge about model parameters, and so on. Then it becomes clear that incorporating statistical distributions into the model is a very natural way to solve complex problems. Answers to the problems come as estimates or predictions along with a quantification of their uncertainties. These physical-statistical models, in the temporal domain, the spatial domain, and the spatio-temporal domain, have immense use in everything from anthropology to zoology and all the "ologies" in-between.

1.2   Goals of Spatio-Temporal Statistics

What are we trying to accomplish with spatio-temporal data analysis and statistical modeling? Sometimes we are just trying to gain more understanding of our data. We might be interested in looking for relationships between two spatio-temporally varying processes, such as temperature and rainfall. This can be as simple as visualizing the data or exploring them through various summaries (Chapter 2). Augmenting these data with scientific theories and statistical methodologies allows valid inferences to be made (Chapter 3). For example, successive reports from the United Nations Intergovernmental Panel on Climate Change have concluded from theory and data that a build-up of atmospheric CO_2 leads to a greenhouse effect that results in global warming. Models can then be built to answer more focused questions. For example, the CO_2 data shown in Figure 1.2 are a manifestation of Earth's carbon cycle: can we find precisely the spatio-temporal "places" on Earth's surface where carbon moves in and out of the atmosphere? Or, how might this warming affect our ability to predict whether an El Niño event will occur within 6 months?

Broadly speaking, there are three main goals that one might pursue with a spatio-temporal statistical model: (1) prediction in space and time (filtering and smoothing); (2) inference on parameters; and (3) forecasting in time. More specific goals might include data assimilation, computer-model emulation, and design of spatio-temporal monitoring networks. These are all related through the presence of a spatio-temporal statistical model, but they have their own nuances and may require different methodologies (Chapters 4 and 5).

1.2.1   The Two Ds of Spatio-Temporal Statistical Modeling

There have been two approaches to spatio-temporal statistical modeling that address the goals listed above. These are the "two Ds" referred to in the title of this subsection, namely the *descriptive* approach and the *dynamic* approach. Both are trying to capture statistical dependencies in spatio-temporal phenomena, but they go about it in quite different ways.

Probably the simplest example of this is in time-series modeling. Suppose that the dependence between any two data at different time points is modeled with a stationary first-order autoregressive process (AR(1)). *Dynamically*, the model says that the value at the current time is equal to a "propagation factor" (or "transition factor") times the value at the previous time, plus an independent "innovation error." This is a mechanistic way of presenting the model that is easy to simulate and easy to interpret.

Descriptively, the same probability structure can be obtained by defining the correlation between two values at any two given time points to be an exponentially decreasing function of the lag between the two time points. (The rate of decrease depends on the AR(1) propagation factor.) Viewing the model this way, it is not immediately obvious how to simulate from it nor what the behavior of the correlation function means physically.

The "take-home" message here is that, while there is a single underlying probability model common to the two specifications, the dynamic approach has some attractive interpretable features that the descriptive approach does not have. Nevertheless, in the absence of knowledge of the dynamics, it can be the descriptive approach that is more "fit for purpose." With mean and covariance functions that are sufficiently flexible, a good fit to the data can be obtained and, consequently, the spatio-temporal variability can be well described.

1.2.2   Descriptive Modeling

The descriptive approach typically seeks to characterize the spatio-temporal process in terms of its mean function and its covariance function. When these are sufficient to describe the process, we can use "optimal prediction" theory to obtain predictions and, crucially, their associated prediction uncertainties. This approach has a distinguished history in spatial statistics and is the foundation of the famed *kriging* methodology. (Cressie, 1990, presents the early history of kriging.) In a spatio-temporal setting, the descriptive approach is most useful when we do not have a strong understanding of the mechanisms that drive the spatio-temporal phenomenon being modeled. Or perhaps we are more interested in studying how covariates in a regression are influencing the phenomenon, but we also recognize that the errors that occur when fitting that relationship are statistically dependent in space and time. That is, the standard assumption given in Chapter 3, that errors are independent and identically distributed (iid), is *not* tenable. In this case, knowing spatio-temporal covariances between the data is enough for statistically efficient inferences (via generalized least squares) on regression coefficients (see Chapter 4). But, as you might suspect, it can be quite difficult to specify all possible covariances for complex spatio-temporal phenomena (and, for nonlinear processes, covariances are not sufficient to describe the spatio-temporal statistical dependence within the process).

Sometimes we can describe spatio-temporal dependence in a phenomenon by including in our model covariates that capture spatio-temporal "trends." This large-scale spatio-temporal variability leaves behind smaller-scale variability that can be modeled statistically with spatio-temporal covariances. The descriptive approach often relies on an important statistical characteristic of dependent data, namely that nearby (in space and time) observations tend to be more alike than those far apart. In spatial modeling, this is often referred to as "Tobler's first law of geography" (Tobler, 1970), and it is often a good guiding principle. It is fair to point out, though, that there are exceptions: there might be "competition" (e.g., only smaller trees are likely to grow close to or under bigger trees as they compete over time for light and nutrients), or things may be more alike on two distant mountain peaks at the same elevation than they are on the same mountain peak at different elevations.

It is important to take a look back at the writings of the pioneers in statistics and ask why spatio-temporal statistical dependencies were not present in early statistical models if they are so ubiquitous in real-world data. Well, we know that some people definitely were aware of these issues. For example, in his ground-breaking treatise on the design of experiments in agriculture, R. A. Fisher (1935, p. 66) wrote: "After choosing the area we usually have no guidance beyond the widely verified fact that patches in close proximity are commonly more alike, as judged by the yield of crops, than those which are further apart." In this case, the spatial variability between plots is primarily due to the fact that the soil properties vary relatively smoothly across space at the field level. Unfortunately, Fisher could not implement complex error models that included spatial statistical dependence due to modeling and computational limitations at that time. So he came up with the brilliant solution of introducing randomization into the experimental design in order to avoid confounding plot effects and treatment effects (but note, only at the plot scale). This was one of the most important innovations in twentieth-century science, and it revolutionized experimentation, not only in agriculture but also in industrial and medical applications. Readers interested in more details behind the development of spatial and spatio-temporal statistics could consult Chapter 1 of Cressie (1993) and Chapter 1 of Cressie and Wikle (2011), respectively.

1.2.3   Dynamic Modeling

Dynamic modeling in the context of spatio-temporal data is simply the notion that we build statistical models that posit (either probabilistically or mechanistically) how a spatial process changes through time. It is inherently a conditional approach, in that we condition on knowing the past, and then we model how the past statistically evolves into the present. If the spatio-temporal phenomenon is what we call "stationary," we could take what we know about it in the present (and the past) and forecast what it will look like in the future.

Building spatio-temporal models using the dynamic approach is closer to how scientists think about the etiology of processes they study – that is, most spatio-temporal data *really do* correspond to a mechanistic real-world process that can be thought of as a spatial process evolving through time. This connection to the mechanism of the process allows spatio-temporal dynamic models a better chance to establish answers to the "why" questions (causality) – is this not the ultimate goal of science? Yet, there is no free lunch – the power of these models comes from established knowledge about the process's behavior, which may not be available for the problem at hand. In that case, one might specify more flexible classes of dynamic models that can adapt to various types of evolution, or turn to the descriptive approach and fit flexible mean and covariance functions to the data.

From a statistical perspective, dynamic models are closer to the kinds of statistical models studied in time series than to those studied in spatial statistics. Yet, there are two fundamental differences between spatio-temporal statistical models that are dynamic, and the usual multivariate time-series models. The first is that dynamic spatio-temporal models have to represent realistically the kinds of spatio-temporal interactions that take place in the phenomenon being studied – not all relationships that one might put into a multivariate time-series model make physical (or biological or economic or ...) sense. The second reason has to do with dimensionality. It is very often the case in spatio-temporal applications that the dimensionality of the spatial component of the model prohibits standard inferential methods. That is, there would be too much "multi" if one chose a multivariate time-series representation of the phenomenon. Special care has to be taken as to how the model is parameterized in order to obtain realistic yet parsimonious dynamics. As discussed in Chapter 5, this has been facilitated to a large extent by the development of basis function expansions within hierarchical statistical models.

Irrespective of which "D" is used to model a spatio-temporal data set, its sheer size can overwhelm computations. Model formulations that use basis functions are a powerful way to leap-frog the computational bottleneck caused by inverting a very large covariance matrix of the data. The general idea is to represent a spatio-temporal process as a mixed linear model with known covariates whose coefficients are unknown and non-random, together with known basis functions whose coefficients are unknown and *random* (Chapters 4 and 5). Usually the basis functions are functions of space and their coefficients define a multivariate time series of dependent random vectors. Depending on the type of basis functions considered, this formulation gives computational advantages due to reduced dimensions and/or sparse covariance/precision matrices that facilitate or eliminate the need for matrix inversions.

There are many classes of basis functions to choose from (e.g., Fourier, wavelets, bisquares) and many are multi-resolutional, although physically based functions (e.g., elevation) can easily be added to the class. If the basis functions are spatial and their random coefficients depend only on time, then the temporal dependence of the coefficients can capture complex spatio-temporal interactions. These include phenomena for which fine spatial scales affect coarse spatial scales and, importantly, vice versa.

1.3   Hierarchical Statistical Models

We believe that we are seeing the end of the era of constructing marginal-probability-based models for complex data. Such models are typically based on the specification of likelihoods from which unknown parameters are estimated. However, these likelihoods can be extremely difficult (or impossible) to compute when there are complex dependencies, and they cannot easily deal with the reality that the data are noisy versions of an underlying real-world process that we care about.

An alternative way to introduce statistical uncertainty into a model is to think conditionally and build complexity through a series of conditional-probability models. For example, if most of the complex dependencies in the data are due to the underlying process of interest, then one should model the distribution of the data *conditioned* on that process (data model), followed by a model of the process' behavior and its uncertainties (process model).

There will typically be unknown parameters present, in both the statistical model for the data (conditioned on the process) and the statistical model for the process.

When a dynamic model of one or several variables is placed within a hierarchical model formulation (see below), one obtains what has been historically called a *state-space model* in the time-series literature. That is, one has data that are collected sequentially in time (i.e., a time series), and they are modeled as "noisy" observations of an underlying *state process* evolving (statistically) through time. These models are at the core of a number of engineering applications (e.g., space missions), and the challenge is to find efficient approaches to perform inference on the underlying state process of interest while accounting for the noise.

In general, there are three such situations of interest when considering state-space models: *smoothing, filtering*, and *forecasting*. *Smoothing* refers to inference on the hidden state process during a fixed time period in which we have observations throughout the time period. (The reader might note that this is the temporal analog of spatial prediction on a bounded spatial domain.) Now consider a time period that always includes the most current time, at which the latest observation is available. *Filtering* refers to inference on the hidden state value at the most current time based on the current and all past data. The most famous example of filtering in this setting is a methodology known widely as the Kalman filter (Kalman, 1960). Finally, *forecasting* refers to inference on the hidden state value at any time point beyond the current time, where data are either not available or not considered in the forecast. In this book, instead of modeling the evolution of a single variable or several variables, we model entire spatial processes evolving through time, which often adds an extra layer of modeling complexity and computational difficulty. Chapter 5 discusses how basis-function representations can deal with these difficulties.

In addition to uncertainty associated with the data and the underlying spatio-temporal process, there might be uncertainties in the parameters. These uncertainties could be accounted for statistically by putting a prior distribution on the parameters. To make sense of all this, we use *hierarchical (statistical) models* (HMs), and follow the terminology of Berliner (1996), who defined an HM to include a *data model*, a *process model*, and a *parameter model*. Technical Note 1.1 gives the conditional-probability structure that ties these models together into a coherent joint probability model of all the uncertainties. The key to the Berliner HM framework is that, at any level of a spatio-temporal HM, it is a good strategy to put as much of the dependence structure as possible in the conditional-mean specification in order to simplify the conditional-covariance specification.

When the parameters are given prior distributions (i.e., a parameter model is posited) at the bottom level of the hierarchy, then we say that the model is a *Bayesian hierarchical model* (BHM). A BHM is often necessary for complex-modeling situations, because the parameters themselves may exhibit quite complex (e.g., spatial or temporal) structure. Or they may depend on other covariates and hence could be considered as processes in their own right. In simpler models, an alternative approach is to estimate the parameters present in the top two levels in some way using the data or other sources of data; then we like to say that the hierarchical model is an *empirical hierarchical model* (EHM). When applicable, an EHM may be preferred if the modeler is reluctant to put prior distributions on parameters about which little is known, or if computational efficiencies can be gained.

It is clear that the BHM approach allows very complex processes to be modeled by going deeper and deeper in the hierarchy, but at each level the conditional-probability model can be quite simple. Machine learning uses a similar approach with its *deep models*. A cascade of levels, where the processing of output from the previous level is relatively simple, results in a class of machine-learning algorithms known as *deep learning*. A potential advantage of the BHM approach over deep learning is that it provides a unified probabilistic framework that allows one to account for uncertainty in data, model, and parameters.

A very important advantage of the data–process–parameter modeling paradigm in an HM is that, while marginal-dependence structures are difficult to model directly, conditional-dependence structures usually come naturally. For example, it is often reasonable to assume that the *data covariance matrix* (given the corresponding values of the hidden process) is simply a diagonal matrix of measurement-error variances. This frees up the *process covariance matrix* to capture the "pure" spatio-temporal dependence, ideally (but, not necessarily) from physical or mechanistic knowledge. Armed with these two covariance matrices, the seemingly complex *marginal covariance matrix* of the data can be simply obtained. This same idea is used in mixed-effects modeling (e.g., in longitudinal data analysis), and it is apparent in the spatio-temporal statistical models described in Chapters 4 and 5.

The product of the conditional-probability components of the HM gives the joint probability model for all random quantities (i.e., all "unknowns"). The HM could be either a BHM or an EHM, depending on whether, respectively, a prior distribution is put on the parameters (i.e., a parameter model is posited) or the parameters are estimated. (A hybrid situation arises when some but not all parameters are estimated and the remaining have a prior distribution put on them.) In this book, we are primarily interested in obtaining the (finite-dimensional) distribution of the hidden (discretized) spatio-temporal process given the data, which we call the *predictive distribution*. The BHM also allows one to obtain the posterior distribution of the parameters given the data, whereas the EHM requires an estimate of the parameters. Predictive and posterior distributions are obtained using *Bayes' Rule* (Technical Note 1.1).

Since predictive and posterior distributions must have total probability mass equal to 1, there is a critical normalizing constant to worry about. Generally, it cannot be calculated in closed form, in which case we rely on computational methods to deal with it. Important advances in the last 30 years have alleviated this problem by making use of Monte Carlo samplers from a Markov chain whose stationary distribution is the predictive (or the posterior) distribution of interest. These *Markov chain Monte Carlo* (MCMC) methods have revolutionized the use of HMs for complex modeling applications, such as those found in spatio-temporal statistics.

Technical Note 1.1: Berliner's Bayesian Hierarchical Model (BHM) paradigm

First, the fundamental notion of the *law of total probability* allows one to decompose a joint distribution into a series of conditional distributions: $[A, B, C] = [A \mid B, C][B \mid C][C]$, where the "bracket notation" is used to denote probability distributions; for example, $[A, B, C]$ is the *joint distribution* of random variables A, B, and C, and $[A \mid B, C]$ is the *conditional distribution* of A given B and C.

Mark Berliner's insight (Berliner, 1996) was that one should use this simple decomposition as a way to formulate models for complex dependent processes. That is, the joint distribution, [data, process, parameters], can be factored into three levels.

At the top level is the *data model*, which is a probability model that specifies the distribution of the data given an underlying "true" process (sometimes called the hidden or latent process) and given some parameters that are needed to specify this distribution. At the next level is the *process model*, which is a probability model that describes the hidden process (and, thus, its uncertainty) given some parameters. Note that at this level the model does not need to account for measurement uncertainty. The process model can then use science-based theoretical or empirical knowledge, which is often physical or mechanistic. At the bottom level is the parameter model, where uncertainty about the parameters is modeled. From top to bottom, the levels of a BHM are:

1. Data model: [data | process, parameters]
2. Process model: [process | parameters]
3. Parameter model: [parameters]

Importantly, each of these levels could have sub-levels, for which conditional-probability models could be given.

Ultimately, we are interested in the posterior distribution, [process, parameters | data] which, conveniently, is proportional to the product of the levels of the BHM given above:

$$
\begin{aligned}
[\text{process, parameters} \mid \text{data}] \propto \quad & [\text{data} \mid \text{process, parameters}] \\
& \times [\text{process} \mid \text{parameters}] \\
& \times [\text{parameters}],
\end{aligned}
$$

where "$\propto$" means "is proportional to." (Dividing the right-hand side by the normalizing constant, [data], makes it equal to the left-hand side.) Note that this result comes from application of Bayes' Rule, applied to the hierarchical model. Inference based on complex models typically requires numerical evaluation of the posterior (e.g., MCMC methods), because the normalizing constant cannot generally be calculated in closed form.

An empirical hierarchical model (EHM) uses just the first two levels, from which the predictive distribution is

$$[\text{process} \mid \text{data, parameters}] \propto \quad [\text{data} \mid \text{process, parameters}]$$
$$\times \, [\text{process} \mid \text{parameters}],$$

where *parameter estimates* are substituted in for "parameters." Numerical evaluation of this (empirical) predictive distribution is also typically needed, since the EHM's normalizing constant cannot generally be calculated in closed form.

1.4   Structure of the Book

The remaining chapters in this book are arranged in the way that we often approach statistical modeling in general and spatio-temporal modeling in particular. That is, we begin by exploring our data. So, Chapter 2 gives ways to do this through visualization and through various summaries of the data. We note that both of these types of exploration can be tricky with spatio-temporal data, because we have one or more dimensions in space and one in time. It can be difficult to visualize information in more than two dimensions, so it often helps to slice through or aggregate over a dimension, or use color, or build animations through time. Similarly, when looking at numerical summaries of the data, we have to come up with innovative ways to help reduce the inherent dimensionality and to examine dependence structures and potential relationships in time and space.

After having explored our data, it is often the case that we would like to fit some fairly simple models – sometimes to help us do an initial filling-in of missing observations that will assist with further exploration, or sometimes just to see if we have enough covariates to adequately explain the important dependencies in the data. This is the spirit of Chapter 3, which presents some ways to do spatial prediction that are not based on a statistical model or are based on very basic statistical models that do not explicitly account for spatio-temporal structure (e.g., linear regression, generalized linear models, and generalized additive models).

If the standard models presented in Chapter 3 are not sufficient to accomplish the goals we gave in Section 1.2, what are we to do? This is when we start to consider the descriptive and dynamic approaches to spatio-temporal modeling discussed above. The descriptive approach has been the "workhorse" of spatio-temporal statistical modeling for most of the history of the discipline, and these methods (e.g., kriging) are described in Chapter 4. But, as mentioned above, when we have strong mechanistic knowledge about the underlying process and/or are interested in complex prediction or forecasting scenarios, we often benefit from the dynamic approach described in Chapter 5. Take note that Chapters 4 and 5 will require a bit more patience to go through, because process models that incorporate statistical dependence require more mathematical machinery. Hence, in these two chapters, the notation and motivation will be somewhat more technical than for the models presented in Chapter 3. It should be kept in mind, though, that the aim here is not to make you an expert, rather it is to introduce you (via the text, the Labs, and the technical notes) to the motivations, main concepts, and practicalities behind spatio-temporal statistical modeling.

After building a model, we would like to know how good it is. There are probably as many ways to evaluate models as there are models! So, it is safe to say that there is no standard way to evaluate a spatio-temporal statistical model. However, there are some common approaches that have been used in the past to carry out model evaluation and model comparison, some of which apply to spatio-temporal models (see Chapter 6). We note that the aim there is not to show you how to obtain the "best" model (as there isn't one!). Rather, it is to show you how a model or a set of models can be found that does a reasonable job with regard to the goals outlined in Section 1.2.

Last, but certainly not least, each of Chapters 2–6 contain Lab vignettes that go through the implementation of many of the important methods presented in each chapter using the R programming language. This book represents the first time such a comprehensive collection of R examples for spatio-temporal data have been collected in one place. We believe that it is essential to "get your hands dirty" with data, but we recognize that quite a few of the methods and approaches used in spatio-temporal statistics can be complicated and that it can be daunting to program them yourself from scratch. Therefore, we have tried to identify some useful (and stable) R functions from existing R packages (see the list following the appendices) that can be used to implement the methods discussed in Chapters 2–6. We have also put a few functions of our own, along with the data sets that we have used, in the R package, **STRbook**, associated with this book (instructions for obtaining this package are available at https://spacetimewithr.org). We note that there are many other R packages that implement various spatio-temporal methods, whose approaches could arrive at the same result with more or less effort, depending on familiarity. As is often the case with R, one gets used to doing things a certain way, and so most of our choices are representative of this.

Chapter 2

Exploring Spatio-Temporal Data

Exploration into territory unknown, or little known, requires both curiosity and survival skills. You need to know where you are, what you are looking at, and how it relates to what you have seen already. The aim of this chapter is to teach you those skills for exploring spatio-temporal data sets. The curiosity will come from you!

Spatio-temporal data are everywhere in science, engineering, business, and industry. This is driven to a large extent by various automated data acquisition instruments and software. In this chapter, after a brief introduction to the data sets considered in this book, we describe some basic components of spatio-temporal data structures in R, followed by spatio-temporal visualization and exploratory tools. The chapter concludes with fairly extensive Labs that provide examples of R commands for data wrangling, visualization, and exploratory data analysis.

When you discover the peaks and valleys, trends and seasonality, and changing landscapes in your data set, what then? Are they real or illusory? Are they important? Chapters 3–6 will give you the inferential and modeling skills required to answer these questions.

2.1   Spatio-Temporal Data

Time-series analysts consider univariate or multivariate sequential data as a random process observed at regular or irregular intervals, where the process can be defined in continuous time, discrete time, or where the temporal event is itself the random event (i.e., a *point process*). Spatial statisticians consider spatial data as either temporal aggregations or temporally frozen states ("snapshots") of a spatio-temporal process. Spatial data are traditionally thought of as random according to either *geostatistical*, *areal* or *lattice*, or *point process* (and sometimes *random set*) behavior. We think of geostatistical data as the kind where we could have observations of some variable or variables of interest (e.g., temperature and wind speed) at continuous locations over a given spatial domain, and where we seek to predict those variables at unknown locations in space (e.g., using interpolation methodology such as *kriging*). Lattice processes are defined on a finite or countable subset in space (e.g., grid nodes, pixels, polygons, small areas), such as the process defined by work-force indicators on a specific political geography (e.g., counties in the USA) over a specific period of time. A spatial point process is a stochastic process in which the locations of the points (sometimes called *events*) are random over the spatial domain, where these events can have attributes given in terms of *marks* (e.g., locations of trees in a forest are random events, with the diameter at breast height being the mark). Given the proliferation of various data sources and geographical information system (GIS) software, it is important to broaden the perspective of spatial data to include not only points and polygons, but also *lines*, *trajectories*, and *objects*. It is also important to note that there can be significant differences in the abundance of spatial information versus temporal information.

R tip: Space-time data are usually provided in comma-separated value (CSV) files, which can be read into R using `read.csv` or `read.table`; shapefiles, which can be read into R using functions from **rgdal** and **maptools**; NetCDF files, which can be read into R using a variety of packages, such as **ncdf4** and **RNetCDF**; and HDF5 files, which can be read into R using the package **h5**.

It should not be surprising that data from spatio-temporal processes can be considered from either a time-series perspective or a spatial-random-process perspective, as described in the previous paragraph. In this book, we shall primarily consider spatio-temporal data that can be described by processes that are discrete in time and either geostatistical or on a lattice in space. For a discussion of a broader collection of spatio-temporal processes, see Cressie and Wikle (2011), particularly Chapters 5–9.

Throughout this book, we consider the following data sets:

- *NOAA daily weather data.* These daily data originated from the US National Oceanic and Atmospheric Administration (NOAA) National Climatic Data Center and can be obtained from the IRI/LDEO Climate Data Library at Columbia University.[1] The data set we consider consists of four variables: daily maximum temperature (`Tmax`) in degrees Fahrenheit (°F), minimum temperature (`Tmin`) in °F, dew point temperature (`TDP`) in °F, and precipitation (`Precip`) in inches at 138 weather stations in the central USA (between 32°N–46°N and 80°W–100°W), recorded between the years 1990 and 1993 (inclusive). These data are considered to be discrete and regular in time (daily) and geostatistical and irregular in space. However, the data are not complete, in that there are missing measurements at various stations and at various time points, and the stations themselves are obviously not located everywhere in the central USA. We will refer to these data as the "NOAA data set." Three days of `Tmax` measurements from the NOAA data set are shown in Figure 2.1.

---

[1] `http://iridl.ldeo.columbia.edu/SOURCES/.NOAA/.NCDC/.DAILY/.FSOD/`

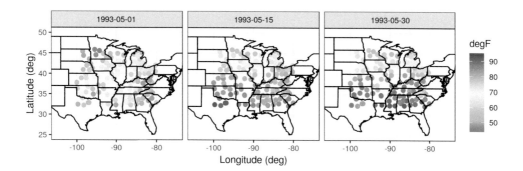

Figure 2.1: Maximum temperature (Tmax) in °F from the NOAA data set on 01, 15, and 30 May 1993.

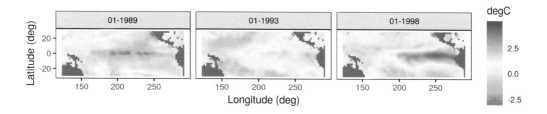

Figure 2.2: Sea-surface temperature anomalies in °C for the month of January in the years 1989, 1993, and 1998. The year 1989 experienced a La Niña event (colder than normal temperatures) while the year 1998 experienced an El Niño event (warmer than normal temperatures).

- *Sea-surface temperature anomalies.* These sea-surface temperature (SST) anomaly data are from the NOAA Climate Prediction Center as obtained from the IRI/LDEO Climate Data Library at Columbia University.[2] The data are gridded at a 2° by 2° resolution from 124°E–70°W and 30°S–30°N, and they represent monthly anomalies from a January 1970–December 2003 climatology (averaged over time). We refer to this data set as the "SST data set." Three individual months from the SST data set are shown in Figure 2.2.

- *Breeding Bird Survey (BBS) counts.* These data are from the North American Breeding Bird Survey.[3] In particular, we consider yearly counts of the house finch (*Carpodacus mexicanus*) at BBS routes for the period 1966–2000 and the Carolina wren

---

[2]http://iridl.ldeo.columbia.edu/SOURCES/.CAC/
[3]K. L. Pardieck, D. J. Ziolkowski Jr., M. Lutmerding, and M.-A. R. Hudson, US Geological Survey, Patux-

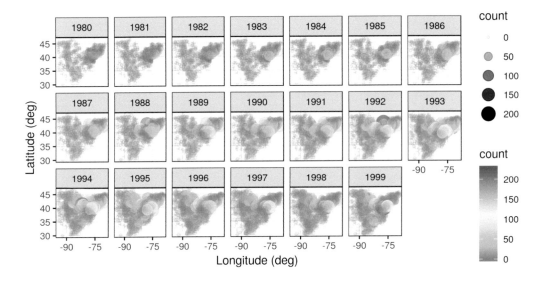

Figure 2.3: Counts of house finches between 1980 and 1999. The size of the points is proportional the number of observed birds, while transparency is used to draw attention to regions of high sampling density or high observed counts.

(*Thryothorus ludovicianus*) for the period 1967–2014. The BBS sampling unit is a roadside route of length approximately 39.2 km. In each sampling unit, volunteer observers make 50 stops and count birds for a period of 3 minutes when they run their routes (typically in June). There are over 4000 routes in the North American survey, but not all routes are available every year. For the purposes of the analyses in this book, we consider the total route counts to occur yearly (during the breeding season) and define the spatial location of each route to be the route's centroid. Thus, we consider the data to be discrete in time, geostatistical and irregular in space, and non-Gaussian in the sense that they are counts. We refer to this data set as the "BBS data set." Counts of house finches for the period 1980–1999 are shown in Figure 2.3.

- *Per capita personal income.* We consider yearly per capita personal income (in dollars) data from the US Bureau of Economic Analysis (BEA).[4] These data have areal spatial support corresponding to USA counties in the state of Missouri, and they cover the period 1969–2014. We refer to this data set as the "BEA income data set." Figure 2.4 shows these data, on a log scale, for the individual years 1970, 1980, and

ent Wildlife Research Center (`https://www.pwrc.usgs.gov/bbs/RawData/`). Note that we used the archived 2016.0 version of the data set, doi: 10.5066/F7W0944J, which is accessible through the data archive link on the BBS website (`ftp://ftpext.usgs.gov/pub/er/md/laurel/BBS/Archivefiles/Version2016v0/`).

[4]`http://www.bea.gov/regional/downloadzip.cfm`

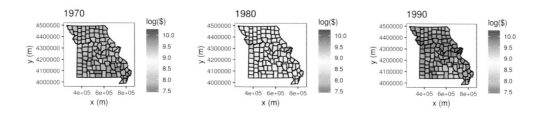

Figure 2.4: Per capita personal income (in dollars) by county for residents in Missouri in the years 1970, 1980, and 1990, plotted on a log scale. The data have been adjusted for inflation. Note how both the overall level of income as well as the spatial variation change with time.

1990; note that these data have been adjusted for inflation.

- *Sydney radar reflectivity.* These data are a subset of consecutive weather radar reflectivity images considered in the World Weather Research Programme (WWRP) Sydney 2000 Forecast Demonstration Project. There are 12 images at 10-minute intervals starting at 08:25 UTC on 03 November, 2000 (i.e., 08:25–10:15 UTC). The data were originally mapped to a 45×45 grid of 2.5 km pixels centered on the radar location. The data used in this book are for a region of dimension 28×40, corresponding to a 70 km by 100 km domain. All reflectivities are given in "decibels relative to Z" (dBZ, a dimensionless logarithmic unit used for weather radar reflectivities). We refer to this data set as the "Sydney radar data set." For more details on these data, shown in Figure 2.5, see Xu et al. (2005).

- *Mediterranean winds.* These data are east–west (u) and north–south (v) wind-component observations over the Mediterranean region (from 6.5°W–16.5°E and 33.5°N–45.5°N) for 28 time periods (every 6 hours) from 00:00 UTC on 29 January 2005 to 18:00 UTC on 04 February 2005. There are two data sources: satellite wind observations from the QuikSCAT scatterometer, and surface winds and pressures from an analysis by the European Center for Medium Range Weather Forecasting (ECMWF). The ECMWF-analysis winds and pressures are given on a $0.5° \times 0.5°$ spatial grid (corresponding to 47 longitude locations and 25 latitude locations), and they are available at each time period for all locations. The QuikSCAT observations are only available intermittently in space, due to the polar orbit of the satellite, but at much higher spatial resolution (25 km) than the ECMWF data when they are available. The QuikSCAT observations given for each time period correspond to all observations available in the spatial domain within 3 hours of time periods stated above. There are no QuikSCAT observations available at 00:00 UTC and 12:00 UTC in the spatial domain and time periods considered here. We refer to this data set as the "Mediterranean winds data set." Figure 2.6 shows the wind vectors ("quivers")

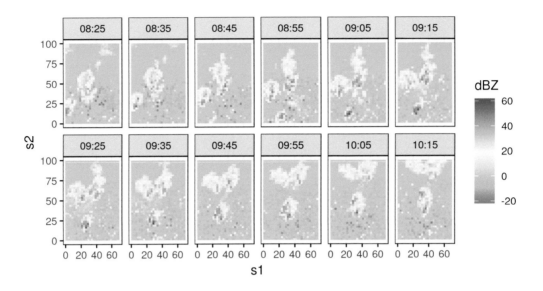

Figure 2.5: Weather radar reflectivities in dBZ for Sydney, Australia, on 03 November 2000. The images correspond to consecutive 10-minute time intervals from 08:25 UTC to 10:15 UTC.

for the ECMWF data at 06:00 UTC on 01 February 2005. These data are a subset of the data described in Cressie and Wikle (2011, Chapter 9) and Milliff et al. (2011).

2.2   Representation of Spatio-Temporal Data in R

Although there are many ways to represent spatial data and time-series data in R, there are relatively few ways to represent spatio-temporal data. In this book we use the class definitions defined in the R package **spacetime**. These classes extend those used for spatial data in **sp** and time-series data in **xts**. For details, we refer the interested reader to the package documentation and vignettes in Pebesma (2012). Here, we just provide a brief introduction to some of the concepts that facilitate thinking about spatio-temporal data structures.

Although spatio-temporal data can come in quite sophisticated relational forms, they most often come in the form of fairly simple "tables." Pebesma (2012) classifies these simple tables into three classes:

- *time-wide*, where columns correspond to different time points;

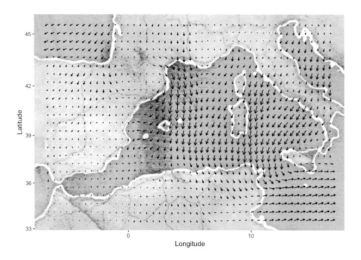

Figure 2.6: ECMWF wind vector observations over the Mediterranean region for 06:00 UTC on 01 February 2005.

- *space-wide*, where columns correspond to different spatial features (e.g., locations, regions, grid points, pixels);

- *long formats*, where each record corresponds to a specific time and space coordinate.

> **R tip:** Data in long format are space inefficient, as spatial coordinates and time attributes are required for each data point, whether or not data are on a lattice. However, it is easy to subset and manipulate data in long format. Powerful "data wrangling" tools in packages such as **dplyr** and **tidyr**, and visualization tools in **ggplot2**, are designed for data in long format.

Tables are very useful elementary data objects. However, an object from the **spacetime** package contains additional information, such as the map projection and the time zone. Polygon objects may further contain the individual areas of the polygons as well as the individual bounding boxes. These objects have elaborate, but consistent, class definitions that greatly aid the geographical (e.g., spatial) component of the analysis.

Pebesma (2012) considers four classes of space-time data:

- *full grid* (STF), a combination of any **sp** object and any **xts** object to represent all possible locations on the implied space-time lattice;

- *sparse grid* (STS), as STF, but contains only the non-missing space-time combinations on a space-time lattice;

- *irregular* (STI), an irregular space-time data structure, where each point is allocated a spatial coordinate and a time stamp;

- *simple trajectories* (STT), a sequence of space-time points that form trajectories.

Note that the "grid" in the first two classes corresponds to a *space-time lattice* – but the spatial locations may or may not be on a lattice! The sparse grid is most effective when there are missing observations, or when there are a relatively few spatial locations that have different time stamps, or when there are a relatively small number of times that have differing spatial locations.

It is important to note that the class objects that make up the **spacetime** package are not used to store data; this is accomplished through the use of the R data frame. As illustrated in Lab 2.1 at the end of this chapter and in Pebesma (2012), there are several important methods in **sp** and **spacetime** that help with the construction and manipulation of these spatio-temporal data sets. In particular, there are methods to construct an object, replace/select data or various spatial or temporal subsets, coerce spatio-temporal objects to other classes, overlay spatio-temporal observations, and aggregate over space, time, or space-time.

> **R tip:** When spatio-temporal data have non-trivial support (i.e., a spatio-temporal region over which a datum is defined), and if the geometry allows it, use **SpatialPixels** and not **SpatialPolygons** as the underlying **sp** object. This results in faster geometric manipulations such as when finding the overlap between points and polygons using the function `over`.

2.3 Visualization of Spatio-Temporal Data

A picture – or a video – can be worth a thousand tables. Use of maps, color, and animation is a very powerful way to provide insight that suggests exploratory data analysis that then leads to spatio-temporal models (Chapters 3–5). Although there are distinct challenges in visualizing spatio-temporal data due to the fact that several dimensions often have to be considered simultaneously (e.g., two or three spatial dimensions and time), there are some fairly common tools that can help explore such data visually. For the most part, we are somewhat selective in what we present here as we want to convey fairly simple methods that have consistently proven useful in our own work and in the broader literature. These can be as simple as static spatial maps and time-series plots, or they can be interactive explorations of the data (e.g., Lamigueiro, 2018). In addition, because of the special dynamic component of many spatio-temporal processes, where spatial processes evolve through time, it is often quite useful to try to visualize this evolution. This can be done in the context of one-dimensional space through a space-time (*Hovmöller*) plot, or more generally through *animations*. We conclude by discussing an increasingly popular approach to help with visualization of very high-dimensional data.

> **R tip:** Spatio-temporal visualization in R generally proceeds using one of two methods: the trellis graph or the grammar of graphics. The command `plot` invokes the trellis graph when **sp** or **spacetime** objects are supplied as arguments. The commands associated with the package **ggplot2** invoke the grammar of graphics. The data objects frequently need to be converted into a data frame in long format for use with **ggplot2**, which we often use throughout this book.

2.3.1 Spatial Plots

Snapshots of spatial processes for a given time period can be plotted in numerous ways. If the observations are irregular in space, then it is often useful to plot a symbol at the data location and give it a different color and/or size to reflect the value of the observation. For example, consider `Tmax` for 01 May 1993 from the NOAA data set plotted in the left panel of Figure 2.1. In this case, the circle center corresponds to the measurement location and the color of the filled-in circle corresponds to the value of the maximum temperature. Notice the clear visual trend of decreasing temperatures from the southeast to the northwest over this region of the USA.

Spatial plots of gridded data are often presented as contour plots, so-called "image" plots, or surface plots. For example, Figure 2.2 shows image representations for three individual months of the Pacific SST data set. Note the La Niña signal (cooler than normal SSTs) in 1989 and the El Niño signal (warmer than normal SSTs) in 1998 in the tropical Pacific Ocean. Figure 2.7 shows contour and surface representations of the SST anomalies in January 1998, corresponding to the right panel (i.e., the El Niño event) in Figure 2.2.

It is often useful to plot a sequence of spatial maps for consecutive times to gain greater insight into the *changes in* spatial patterns through time. Figure 2.8 shows a sequence of SST spatial maps for the months January–June 1989. Note how the initially strong La Niña event dissipates by June 1989.

> **R tip:** Multiple time-indexed spatial maps can be plotted from one long-format table using the functions `facet_grid` or `facet_wrap` in **ggplot2** with time as a grouping variable.

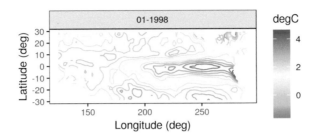

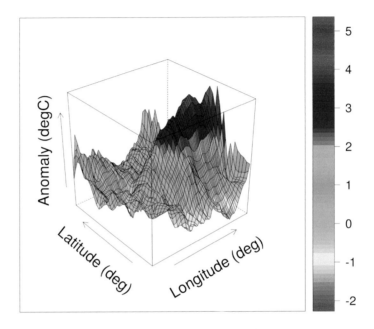

Figure 2.7: Sea-surface temperature anomalies (in °C) for January 1998 as a contour plot (top) and as a surface plot (bottom).

2.3.2   Time-Series Plots

It can be instructive to plot time series corresponding to an observation location, an aggregation of observations, or multiple locations simultaneously. For example, Figure 2.9 shows time-series plots of daily `Tmax` for 10 of the NOAA stations (chosen randomly from the 139 stations) for the time period 01 May 1993–30 September 1993. The time-series plots are quite noisy, as is to be expected from the variability inherent in mid-latitude weather systems. However, there is an overall temporal trend corresponding to the annual seasonal

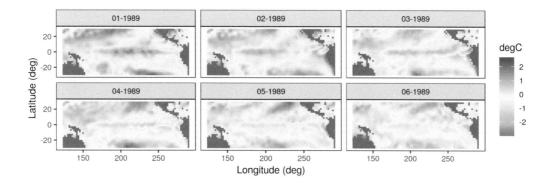

Figure 2.8: Sea-surface temperature anomalies (in °C) for January–June 1989.

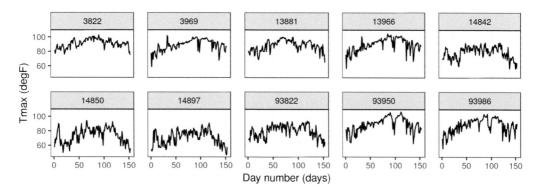

Figure 2.9: Maximum temperature (°F) for ten stations chosen from the NOAA data set at random, as a function of the day number, with the first day denoting 01 May 1993 and the last day denoting 30 September 1993. The number in the grey heading of each plot denotes the station ID.

cycle. That is, all of the time series appear to peak somewhat towards the center of the time horizon, which corresponds to the month of July. In this case, since we are using only five months of data, this trend appears to be roughly quadratic in time. Periodic functions are often used when considering a whole year or multiple years of data, especially with weather and economic data. Although all of these temperature series contain a seasonal component, some appear shifted on the vertical axis (Tmax) relative to one another (e.g., station 13881 has higher temperatures than station 14897). This is due to the latitudinal trend apparent in Figure 2.1.

2.3.3 Hovmöller Plots

A Hovmöller plot (Hovmöller, 1949) is a two-dimensional space-time visualization in which space is collapsed (projected or averaged) onto one dimension and where the second dimension denotes time. These plots have traditionally been considered in the atmospheric-science and ocean-science communities to visualize propagating features. For example, the left panel of Figure 2.10 shows monthly SST anomalies averaged from 1°S–1°N and plotted such that longitude (over the Pacific Ocean) is on the x-axis and time (from 1996 to 2003) is on the y-axis (increasing from top to bottom). The darker red colors correspond to warmer than normal temperatures (i.e., El Niño events) and the darker blue colors correspond to colder than normal temperatures (i.e., La Niña events). Propagation through time is evident if a coherent color feature is "slanted." In this plot, one can see several cases of propagating features along the longitudinal axis (e.g., both of the major La Niña events show propagation from the eastern longitudes towards the western longitudes.)

Hovmöller plots are straightforward to generate with regular spatio-temporal data, but they can also be generated for irregular spatio-temporal data after suitable interpolation to a regular space-time grid. For example, in Figure 2.11, we show Hovmöller plots for the `Tmax` variable in the NOAA data set between 01 May 1993 and 30 September 1993. We see that the temporal trend is fairly constant with *longitude* (left panel), but it decreases considerably with increasing *latitude* (right panel) as expected, since overall maximum temperature decreases with increasing latitude in the conterminous USA. Such displays may affect modeling decisions of the trend (e.g., a time–latitude interaction might become evident in such plots).

2.3.4 Interactive Plots

Programming tools for interactive visualization are becoming increasingly accessible. These tools typically allow for a more data-immersive experience, and they allow one to explore the data without having to resort to scripting. In the simplest of cases, one can "hover" a cursor over a figure, and some information related to the data corresponding to the current location of the cursor is conveyed to the user. For example, in Figure 2.12 we show the interaction of the user with a spatial plot of SST using the package **plotly**. This package works in combination with a web portal for more advanced exploration methods (e.g., the exploration of three-dimensional data).

There are several interactive plots that may aid with the visualization of spatio-temporal data. One of the most useful plots builds on *linked brushing*, with the link acting between time and space. Here, one hovers a cursor over a spatial observation or highlights a spatial area, and then the time series corresponding to that point or area is visualized; see Figure 2.12. This allows one to explore the time series corresponding to known geographic areas with minimal effort. Code for generating a linked brush is available from the book's website (`https://spacetimewithr.org`).

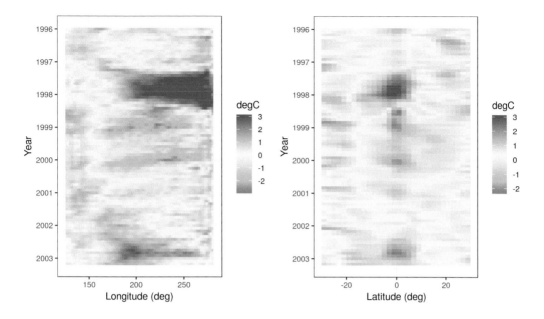

Figure 2.10: Hovmöller plots for both the longitude (left) and latitude (right) coordinates for the SST data set. The color denotes the temperature anomaly in °C.

2.3.5 Animations

Everyone loves a movie. Animation captures our attention and can suggest structure in a way that a sequence of still frames cannot. Good movies should be watched again and again, and that is our intention here for understanding why the spatio-temporal data behave the way they do.

An animation is typically constructed by plotting spatial data frame-by-frame, and then stringing them together in sequence. When doing so, it is important to ensure that all spatial axes and color scales remain constant across all frames. In situations with missing or unequally spaced observations, one may sometimes improve the utility of an animation by performing a simple interpolation (in space and/or time) before constructing the sequence. Animations in R can be conveniently produced using the package **animation**. We provide an example using this package in Lab 2.2.

2.3.6 Trelliscope: Visualizing Large Spatio-Temporal Data Sets

Most spatio-temporal statistical analyses to date have been carried out on manageable data sets that can fit into a computer's memory which, at the time of writing, was in the order of a few tens or a couple of hundreds of gigabytes in size. Being able to visualize these data is important and useful in many respects. Proceeding with modeling and prediction where

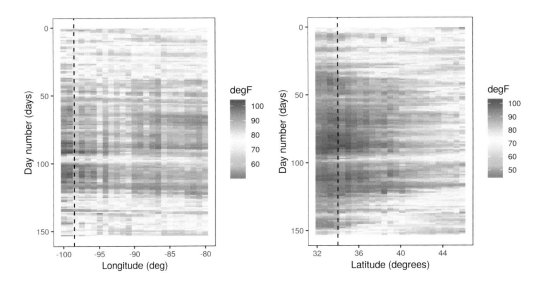

Figure 2.11: Hovmöller plots for both the longitude (left) and latitude (right) coordinates for the `Tmax` variable in the NOAA data set between 01 May 1993 and 30 September 1993, where the data are interpolated as described in Lab 2.2. The color denotes the maximum temperature in °F. The dashed lines correspond to the longitude and latitude coordinates of station 13966 (compare to Figure 2.9).

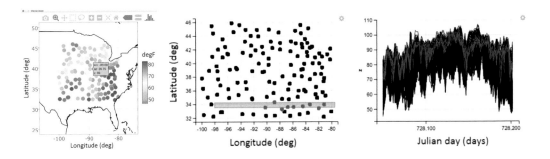

Figure 2.12: Interactively exploring maximum temperatures on 01 May 1993 using the NOAA data set. The "hover" feature can be added to **ggplot2** objects by using `ggplotly` from the package **plotly** (left). A linked brush can be used to explore the time series (right) corresponding to a user-chosen set of spatial locations (middle) with the package **ggvis**.

not all the data can be processed in a single place (known as parallel-data algorithms) is an active area of research and will not be discussed here.

The Trelliscope system, available with the package **trelliscope**, helps users visualize massive data sets. The first advantage of **trelliscope** is that it facilitates exploration when, due to their size, the data may only be visualized using hundreds or thousands of plots (or panels). When this is the case, the Trelliscope system can calculate subset summaries (known as *cognostics*) that are then used for filtering and sorting the panels. For example, consider the SST data set. If a grouping is made by month, then there are over 300 spatial maps that can be visualized between, say, 1970 and 2003. Alternatively, one may decide to visualize only those months in which the SST exceeded a certain maximum or minimum threshold. One can formulate a cognostic using the monthly spatial mean values of SST averaged over their spatial domain and visualize them in a quantile plot (see Figure 2.13). The analyst can use this approach to quickly view the strongest El Niño and La Niña events in this time period.

The second advantage is that the **trelliscope** package is designed to visualize data that are on a distributed file system that may be residing on more than one node. The data are processed in a *divide and recombine* fashion; that is, the data are divided and processed by group in parallel fashion and then recombined. In **trelliscope**, this can be useful for generating both the cognostics and the viewing panels efficiently. Therefore, the Trelliscope system provides a way to visualize terabytes of space-time data but, as quoted in its package manual, it "can also be very useful for small data sets."

> **R tip:** Processing and visualizing large data sets residing on a distributed file system using divide and recombine may seem like a daunting task. The R package **datadr**, which can be used together with **trelliscope**, provides an easy-to-use front-end for data residing on distributed file systems. More importantly, it reduces the barrier to entry by allowing the same, or very similar, code to be used for data residing in memory and data residing on a distributed file system such as Hadoop.

2.3.7 Visualizing Uncertainty

One of the main things that separates statistics from other areas of data science is the focus on uncertainty quantification. Uncertainties could be associated with data (e.g., measurement error in satellite observations or sampling error in a survey), estimates (e.g., uncertainty in regression parameter estimates), or predictions (e.g., uncertainties in a forecast of SST anomalies). Taking a Bayesian point of view, uncertainties could also be associated with the parameters themselves. In the case where these uncertainties are indexed in time, space, or space-time, one can use any of the methods discussed in this section to produce visualizations of these uncertainties. It is increasingly the case that one seeks methods to visualize both the values of interest and their uncertainty simultaneously. This is challenging given the difficulties in visualizing information in multiple dimensions, and it is an active area of research both in geography and statistics (see, for example, the discussion of

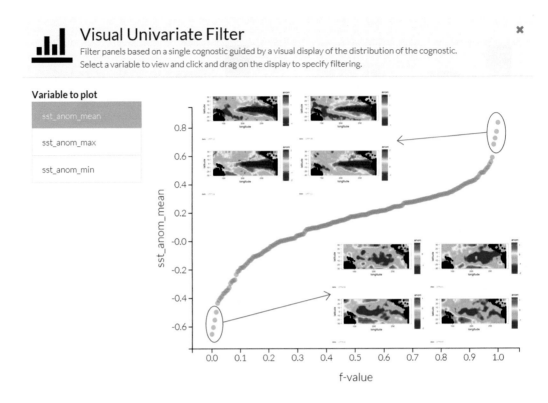

Figure 2.13: Exploring a large spatio-temporal data set with Trelliscope. Quantile plot of monthly averages of sea-surface temperature from the SST data set; the insets are what would be displayed if the user highlighted the circle points, corresponding to El Niño and La Niña events.

"visuanimation" in Genton et al., 2015). For a recent overview in the case of areal data, and an accompanying R vignette, see Lucchesi and Wikle (2017) and the R package **Vizumap**.[5]

2.4 Exploratory Analysis of Spatio-Temporal Data

Visualization of data is certainly an important and necessary component of exploratory data analysis. In addition, we often wish to explore spatio-temporal data in terms of summaries of first-order and second-order characteristics. Here we consider visualizations of empirical means and empirical covariances, spatio-temporal covariograms and semivariograms, the use of empirical orthogonal functions and their associated principal-component time series, and spatio-temporal canonical correlation analysis. To do this, we have to start using some mathematical symbols and formulas. Mathematics is the language of science (and of

---

[5]https://doi.org/10.5281/zenodo.1479951

statistical science), and we introduce this language along the way to help readers who are a bit less fluent. For reference, we present some fundamental definitions of vectors and matrices and their manipulation in Appendix A. Readers who are not familiar with the symbols and basic manipulation of vectors and matrices would benefit from looking at this material before proceeding.

2.4.1 Empirical Spatial Means and Covariances

It can be useful to explore spatio-temporal data by examining the empirical means and empirical covariances. Assume for the moment that we have observations $\{Z(\mathbf{s}_i; t_j)\}$ for spatial locations $\{\mathbf{s}_i : i = 1, \ldots, m\}$ and times $\{t_j : j = 1, \ldots, T\}$. The empirical spatial mean for location $\mathbf{s}_i$, $\widehat{\mu}_{z,s}(\mathbf{s}_i)$, is then found by averaging over time:

$$\widehat{\mu}_{z,s}(\mathbf{s}_i) \equiv \frac{1}{T} \sum_{j=1}^{T} Z(\mathbf{s}_i; t_j).$$

If we consider the means for all spatial data locations and assume that we have T observations at each location, then we can write down the spatial mean as an m-dimensional vector, $\widehat{\boldsymbol{\mu}}_{z,s}$, where

$$\widehat{\boldsymbol{\mu}}_{z,s} \equiv \begin{bmatrix} \widehat{\mu}_{z,s}(\mathbf{s}_1) \\ \vdots \\ \widehat{\mu}_{z,s}(\mathbf{s}_m) \end{bmatrix} = \begin{bmatrix} \frac{1}{T} \sum_{j=1}^{T} Z(\mathbf{s}_1; t_j) \\ \vdots \\ \frac{1}{T} \sum_{j=1}^{T} Z(\mathbf{s}_m; t_j) \end{bmatrix} = \frac{1}{T} \sum_{j=1}^{T} \mathbf{Z}_{t_j}, \tag{2.1}$$

and $\mathbf{Z}_{t_j} \equiv (Z(\mathbf{s}_1; t_j), \ldots, Z(\mathbf{s}_m; t_j))'$.

This mean vector is a spatial quantity whose elements are indexed by their location. Therefore, it can be plotted on a map, as in the case of the maximum temperature in the NOAA data set (see Figure 2.1), or as a function of the spatial coordinates (e.g., longitude or latitude) as in Figure 2.14. From these plots one can see that there is a clear trend in the empirical spatial mean of maximum temperature with latitude, but not so much with longitude. Note that one may not have the same number of observations at each location to calculate the average, in which case each location in space must be calculated separately (e.g., $\widehat{\mu}_{z,s}(\mathbf{s}_i) = (1/T_i) \sum_{j=1}^{T_i} Z(\mathbf{s}_i; t_j)$, where T_i is the number of time points at which there are data at location $\mathbf{s}_i$).

Additionally, one can average across space and plot the associated time series. The empirical temporal mean for time t_j, $\widehat{\mu}_{z,t}(t_j)$, is given by

$$\widehat{\mu}_{z,t}(t_j) \equiv \frac{1}{m} \sum_{i=1}^{m} Z(\mathbf{s}_i; t_j). \tag{2.2}$$

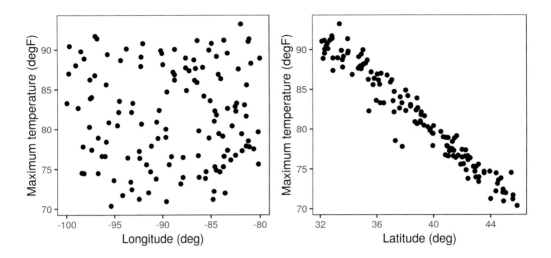

Figure 2.14: Empirical spatial mean, $\widehat{\mu}_{z,s}(\cdot)$, of `Tmax` (in °F) as a function of station longitude (left) and station latitude (right).

For example, Figure 2.15 shows the time series of `Tmax` for the NOAA temperature data set averaged across all of the spatial locations. This plot of the empirical temporal means shows the seasonal nature of the mid-latitude temperature over the central USA, but it also shows variations in that seasonal pattern due to specific large-scale weather systems.

> **R tip:** Computing empirical means is quick and easy using functions in the package **dplyr**. For example, to find a temporal average, the data in a long-format data frame can first be grouped by spatial location using the function `group_by`. A mean can then be computed for every spatial location using the function `summarise`. See Lab 2.1 for more details on these functions.

It is often useful to consider the empirical spatial covariability in the spatio-temporal data set. This covariability can be used to determine to what extent data points in the data set covary (behave similarly) as a function of space and/or time. In the context of the data described above, the empirical lag-τ covariance between spatial locations $\mathbf{s}_i$ and $\mathbf{s}_k$ is given by

$$\widehat{C}_z^{(\tau)}(\mathbf{s}_i, \mathbf{s}_k) \equiv \frac{1}{T - \tau} \sum_{j=\tau+1}^{T} (Z(\mathbf{s}_i; t_j) - \widehat{\mu}_{z,s}(\mathbf{s}_i))(Z(\mathbf{s}_k; t_j - \tau) - \widehat{\mu}_{z,s}(\mathbf{s}_k)), \quad (2.3)$$

for $\tau = 0, 1, \ldots, T - 1$, which is called the empirical lag-τ spatial covariance. Note that this is the average (over time) of the cross products of the centered observations at the two

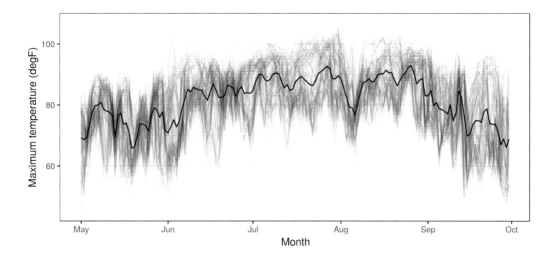

Figure 2.15: `Tmax` data (in °F), from the NOAA data set (blue lines, where each blue line corresponds to a station) and the empirical temporal mean $\widehat{\mu}_{z,t}(\cdot)$ (black line) computed from (2.2), and t is in units of days, ranging from 01 May 1993 to 30 September 1993.

locations ($\mathbf{s}_i$ and $\mathbf{s}_k$); that is, (2.3) is a summary of the covariation of these data. It is often useful to consider the $m \times m$ lag-τ empirical spatial covariance matrix, $\widehat{\mathbf{C}}_z^{(\tau)}$, in which the (i, k)th element is given by (2.3). Alternatively, this can be calculated directly by

$$\widehat{\mathbf{C}}_z^{(\tau)} \equiv \frac{1}{T - \tau} \sum_{j=\tau+1}^{T} (\mathbf{Z}_{t_j} - \widehat{\boldsymbol{\mu}}_{z,s})(\mathbf{Z}_{t_j - \tau} - \widehat{\boldsymbol{\mu}}_{z,s})'; \quad \tau = 0, 1, \ldots, T - 1. \qquad (2.4)$$

Thus, in order to find the lag-τ covariance matrices, we consider the cross products of the residual vectors for each spatial location and each time point relative to its corresponding time-averaged empirical spatial mean.

In general, it can be difficult to obtain any intuition from these matrices, since locations in a two-dimensional space do not have a natural ordering. However, one can sometimes gain insight by splitting the domain into "strips" corresponding to one of the spatial dimensions (e.g., longitudinal strips) and then plotting the associated covariance matrices for those strips. For example, Figure 2.16 shows empirical covariance matrices for the maximum temperature in the NOAA data set (after, as shown in Lab 2.3, a quadratic trend in time has been removed), split into four longitudinal strips. Not surprisingly, these empirical spatial covariance matrices reveal the presence of spatial dependence in the residuals. The lag-0 plots seem to be qualitatively similar, suggesting that there is no strong correlational dependence on longitude but that there is a correlational dependence on latitude, with the spatial covariance decreasing with decreasing latitude.

We can also calculate the empirical lag-τ cross-covariance matrix between two spatio-temporal data sets, $\{\mathbf{Z}_{t_j}\}$ and $\{\mathbf{X}_{t_j}\}$, where $\{\mathbf{X}_{t_j}\}$ corresponds to data vectors at n different locations (but it is assumed for meaningful comparisons that they correspond to the same time points). In particular, we define this $m \times n$ matrix by

$$\widehat{\mathbf{C}}_{z,x}^{(\tau)} \equiv \frac{1}{T-\tau} \sum_{j=\tau+1}^{T} (\mathbf{Z}_{t_j} - \widehat{\boldsymbol{\mu}}_{z,s})(\mathbf{X}_{t_j-\tau} - \widehat{\boldsymbol{\mu}}_{x,s})', \tag{2.5}$$

for $\tau = 0, 1, \ldots, T-1$, where $\widehat{\boldsymbol{\mu}}_{x,s}$ is the empirical spatial mean vector for the data $\{\mathbf{X}_{t_j}\}$. Cross-covariances may be useful in characterizing the spatio-temporal dependence relationship between two different variables, for example maximum temperature and minimum temperature.

Although not as common in spatio-temporal applications, one can also calculate empirical temporal covariance matrices averaging across space (after *removing temporal means* averaged across space). In this case, the time index is unidimensional and ordered, so one does not have to work as hard on the interpretation as we did with empirical spatial covariance matrices.

2.4.2   Spatio-Temporal Covariograms and Semivariograms

In Chapter 4 we shall see that it is necessary to characterize the joint spatio-temporal dependence structure of a spatio-temporal process in order to perform optimal prediction (i.e., kriging). Thus, for measures of the joint spatio-temporal dependence, we consider empirical spatio-temporal *covariograms* (and their close cousins, *semivariograms*). The biggest difference between what we are doing here and the covariance estimates in the previous section is that we are interested in characterizing the covariability in the spatio-temporal data as a function of specific lags in time *and* in space. Note that the lag in time is a scalar, but the lag in space is a vector (corresponding to the displacement between locations in d-dimensional space).

Consider the empirical spatio-temporal covariance function for various space and time lags. Here, we make an assumption that the first moment (mean) depends on space but not on time and that the second moment (covariance) depends only on the lag differences in space and time. Then the empirical spatio-temporal covariogram for spatial lag $\mathbf{h}$ and time lag τ is given by

$$\widehat{C}_z(\mathbf{h};\tau) = \frac{1}{|N_{\mathbf{s}}(\mathbf{h})|} \frac{1}{|N_t(\tau)|} \sum_{\mathbf{s}_i,\mathbf{s}_k \in N_{\mathbf{s}}(\mathbf{h})} \sum_{t_j,t_\ell \in N_t(\tau)} (Z(\mathbf{s}_i;t_j) - \widehat{\mu}_{z,s}(\mathbf{s}_i))(Z(\mathbf{s}_k;t_\ell) - \widehat{\mu}_{z,s}(\mathbf{s}_k)),$$

$$\tag{2.6}$$

where you will recall that $\widehat{\mu}_{z,s}(\mathbf{s}_i) = (1/T) \sum_{j=1}^{T} Z(\mathbf{s}_i;t_j)$, $N_{\mathbf{s}}(\mathbf{h})$ refers to the pairs of spatial locations with spatial lag within some tolerance of $\mathbf{h}$, $N_t(\tau)$ refers to the pairs of time points with time lag within some tolerance of τ, and $|N(\cdot)|$ refers to the number of elements in $N(\cdot)$. Under *isotropy*, one often considers the lag only as a function of distance, $h = ||\mathbf{h}||$, where $|| \cdot ||$ is the Euclidean norm (see Appendix A).

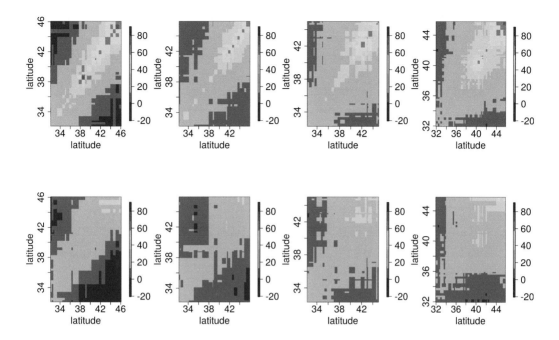

Figure 2.16: Maximum temperature lag-0 (top) and lag-1 (bottom) empirical spatial covariance plots for four longitudinal strips (from left to right, $[-100, -95), [-95, -90), [-90, -85), [-85, -80)$ degrees) in which the domain of interest is subdivided.

Technical Note 2.1: Semivariogram

The semivariogram is defined as

$$\gamma_z(\mathbf{s}_i, \mathbf{s}_k; t_j, t_\ell) \equiv \frac{1}{2}\mathrm{var}(Z(\mathbf{s}_i; t_j) - Z(\mathbf{s}_k; t_\ell)).$$

In the case where the covariance depends only on displacements in space and differences in time, this can be written as

$$
\begin{aligned}
\gamma_z(\mathbf{h}; \tau) &= \frac{1}{2}\mathrm{var}(Z(\mathbf{s} + \mathbf{h}; t + \tau) - Z(\mathbf{s}; t)) \\
&= C_z(\mathbf{0}; 0) - \mathrm{cov}(Z(\mathbf{s} + \mathbf{h}; t + \tau), Z(\mathbf{s}; t)) \\
&= C_z(\mathbf{0}; 0) - C_z(\mathbf{h}; \tau),
\end{aligned}
\tag{2.7}
$$

where $\mathbf{h} = \mathbf{s}_k - \mathbf{s}_i$ is a spatial lag and $\tau = t_\ell - t_j$ is a temporal lag.

Now, (2.7) does not always hold. It is possible that γ_z is a function of spatial lag $\mathbf{h}$ and temporal lag τ, but there is no stationary covariance function $C_z(\mathbf{h}; \tau)$. We generally try to avoid these models of covariability by fitting trend terms that are linear and/or quadratic in spatio-temporal coordinates.

If the covariance function of the process is well defined, then the semivariogram is generally characterized by the nugget effect, the sill, and the partial sill. The nugget effect is given by $\gamma_z(\mathbf{h}; \tau)$ when $\mathbf{h} \to \mathbf{0}$ and $\tau \to 0$, while the sill is $\gamma_z(\mathbf{h}; \tau)$ when $\mathbf{h} \to \infty$ and $\tau \to \infty$. The partial sill is the difference between the sill and the nugget effect. The diagram below shows these components of a semivariogram as a function of spatial distance $\|\mathbf{h}\|$.

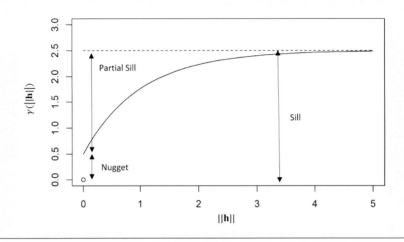

In some kriging applications, one might be interested in looking at the empirical spatio-temporal semivariogram (see Technical Note 2.1). The empirical semivariogram, for the case where the covariance only depends on the displacements in space and the time lags, is obtained from (2.6) as $\hat{\gamma}_z(\mathbf{h}; \tau) = \hat{C}_z(\mathbf{0}; 0) - \hat{C}_z(\mathbf{h}; \tau)$, and so it is easy to go back and forth between the empirical semivariogram and the covariogram in this case (see the caveat in Technical Note 2.1). Assuming a constant spatial mean $\mu_{z,s}$, then (2.7) can be equivalently written as

$$\gamma_z(\mathbf{h}; \tau) = \frac{1}{2} E \left(Z(\mathbf{s} + \mathbf{h}; t + \tau) - Z(\mathbf{s}; t) \right)^2,$$

and hence an alternative estimate is

$$\hat{\gamma}_z(\mathbf{h}; \tau) = \frac{1}{|N_{\mathbf{s}}(\mathbf{h})|} \frac{1}{|N_t(\tau)|} \sum_{\mathbf{s}_i, \mathbf{s}_k \in N_{\mathbf{s}}(\mathbf{h})} \sum_{t_j, t_\ell \in N_t(\tau)} (Z(\mathbf{s}_i; t_j) - Z(\mathbf{s}_k; t_\ell))^2, \qquad (2.8)$$

where the notation in (2.8) is the same as used above in (2.6). Note that this calculation does not need any information about the spatial means. Figure 2.17 shows a semivariogram obtained from the NOAA data set for the maximum temperature data in July 1993.

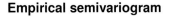

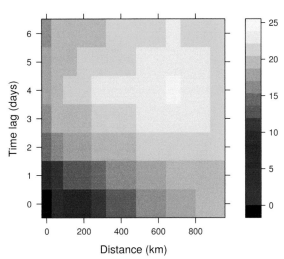

Figure 2.17: Empirical spatio-temporal semivariogram of daily `Tmax` from the NOAA data set during July 2003, computed using the function **variogram** in **gstat**.

2.4.3 Empirical Orthogonal Functions (EOFs)

Empirical orthogonal functions (EOFs) can reveal spatial structure in spatio-temporal data and can also be used for subsequent dimensionality reduction. EOFs came out of the meteorology/climatology literature, and in the context of discrete space and time, EOF analysis is the spatio-temporal manifestation of principal component analysis (PCA) in statistics (see Chapter 5 in Cressie and Wikle, 2011, for an extensive overview). In the terminology of this chapter, one should probably modify "EOFs" to empirical *spatial* orthogonal functions, since they are obtained from an empirical *spatial* covariance matrix, but for legacy reasons we stick with "EOFs." Before we discuss EOFs, we give a brief review of PCA.

Brief Review of Principal Component Analysis

Assume we have two measured traits on a subject of interest (e.g., measurements of $x_1 =$ height (in cm) and $x_2 =$ weight (in kg) in a sample of women in the USA). Figure 2.18 (left panel) shows a (simulated) plot of what such data might look like for $m = 500$ individuals. We note that these data are quite correlated, as expected. Now, we wish to construct new variables that are linear combinations of the measured traits, say $a_1 = w_{11}x_1 + w_{12}x_2$ and $a_2 = w_{21}x_1 + w_{22}x_2$. One way to think of this is that we are "projecting" the original data onto new axes given by the variables a_1 and a_2. Figure 2.18 (center and right panels) shows

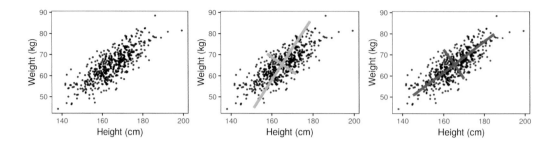

Figure 2.18: Simulated height (in cm) versus weight (in kg) for $m = 500$ females in the USA (left) with two orthogonal projections (center and right). The right panel shows the optimal PCA projection.

two possible projections, which differ according to the values we choose for the weights, $\{w_{11}, w_{12}, w_{21}, w_{22}\}$. Note that in the case of the right-hand panel in Figure 2.18, the new axis a_1 aligns with the axis of largest variation, and the new axis a_2 corresponds to the axis of largest variation perpendicular (orthogonal) to the axis a_1. Maximizing these axes of variation subject to orthogonality helps us think about decomposing the data into lower-dimensional representations in an optimal way. That is, the new variable on the axis a_1 represents the optimal linear combination of the data that accounts for the most variation in the original data. If the variation along the other axis (a_2) is fairly small relative to a_1, then it might be sufficient just to consider a_1 to represent the data.

How does one go about choosing the weights $\{w_{ij}\}$? Let $\mathbf{x}_i = (x_{1i}, \ldots, x_{pi})'$ be a random vector with variance–covariance matrix $\mathbf{C}_x$. Note from Appendix A that by spectral decomposition, a $p \times p$ non-negative-definite, symmetric, real matrix, $\mathbf{\Sigma}$, can be *diagonalized* such that $\mathbf{W}'\mathbf{\Sigma}\mathbf{W} = \mathbf{\Lambda}$ (i.e., $\mathbf{\Sigma} = \mathbf{W}\mathbf{\Lambda}\mathbf{W}'$), where $\mathbf{\Lambda}$ is a diagonal matrix containing the eigenvalues $\{\lambda_i\}$ of $\mathbf{\Sigma}$ (where $\lambda_1 \geq \lambda_2 \geq \ldots \geq \lambda_p \geq 0$) and $\mathbf{W} = [\mathbf{w}_1\ \mathbf{w}_2\ \ldots\ \mathbf{w}_p]$ is the associated matrix of orthogonal eigenvectors, $\{\mathbf{w}_i\}$ (i.e., $\mathbf{W}\mathbf{W}' = \mathbf{W}'\mathbf{W} = \mathbf{I}$); thus, $\mathbf{C}_x = \mathbf{W}\mathbf{\Lambda}_x\mathbf{W}'$. It can be shown that these eigenvectors give the optimal weights, so that $\mathbf{w}_1$ are the weights for a_1 and $\mathbf{w}_2$ are the weights for a_2, and so on.

As an example, consider the variance–covariance matrix associated with the simulated height and weight traits, where $p = 2$:

$$\mathbf{C}_x = \begin{pmatrix} 81 & 50 \\ 50 & 49 \end{pmatrix}.$$

Then $\mathbf{W}$ and $\mathbf{\Lambda}_x$ are given (using the function `eigen` in R) by

$$\mathbf{W} = \begin{pmatrix} -0.8077 & 0.5896 \\ -0.5896 & -0.8077 \end{pmatrix}, \quad \mathbf{\Lambda}_x = \begin{pmatrix} 117.5 & 0 \\ 0 & 12.5 \end{pmatrix}.$$

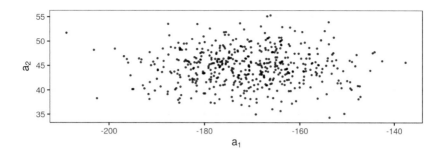

Figure 2.19: Principal components corresponding to the simulated data in Figure 2.18.

So, for each of the observation vectors, $\{\mathbf{x}_i, i = 1, \ldots, 500\}$, we make new variables

$$a_{1i} = -0.8077 x_{1i} - 0.5896 x_{2i}$$

$$a_{2i} = 0.5896 x_{1i} - 0.8077 x_{2i}.$$

These coefficients (which are the data projected onto axes (a_1, a_2)) are plotted in Figure 2.19. Note that these new variables are uncorrelated (no slant to the points in the plot) and the first axis (a_1) corresponds to the one that has the most variability. In PCA, one sometimes attempts to interpret the "loadings" given by $\{\mathbf{w}_i : i = 1, \ldots, p\}$ (or some scaled version of them). That is, one contrasts the signs and magnitudes of the loadings within a given eigenvector (e.g., the first eigenvector, $\mathbf{w}_1 = (-0.8077, -0.5896)'$, suggests that both height and weight are important and vary in the same way, so that the first principal component might represent an overall "size" attribute).

The notions presented in the example above extend to more than just two traits and, in general, the principal-component decomposition has some nice properties. For example, the kth eigenvalue is the variance of the associated linear combination of the elements of $\mathbf{x}$; that is, $\mathrm{var}(a_k) = \mathrm{var}(\mathbf{w}_k' \mathbf{x}) = \lambda_k$. In addition,

$$\mathrm{var}(x_1) + \ldots \mathrm{var}(x_p) = \mathrm{trace}(\mathbf{C}_x) = \lambda_1 + \ldots + \lambda_p = \mathrm{var}(a_1) + \ldots + \mathrm{var}(a_p).$$

Thus, one can consider the proportion of the total variance accounted for by the kth principal component, which is $\lambda_k / \sum_{j=1}^{p} \lambda_j$. In the example above, the first principal component accounts for about 90% of the variance in the original data (i.e., $\lambda_1 / (\lambda_1 + \lambda_2) = 117.5/130 = 0.90$).

Of course, in practice we would not know the covariance matrix, $\mathbf{C}_x$, but we can calculate an empirical covariance matrix using (2.4) with $\tau = 0$, $\{\mathbf{Z}_{t_j}\}$ replaced by $\{\mathbf{x}_i\}$, and $\widehat{\boldsymbol{\mu}}_{z,s}$ replaced by $(1/500) \sum_{i=1}^{500} \mathbf{x}_i$. In that case, the spectral decomposition of $\widehat{\mathbf{C}}_x$ gives empirical estimates of the eigenvectors $\widehat{\mathbf{W}}$ and eigenvalues $\widehat{\boldsymbol{\Lambda}}_x$. The analysis then proceeds with these empirical estimates.

R tip: The PCA routine `prcomp` is included with base R. When the `plot` function is used on an object returned by `prcomp`, the variances of the principal components are displayed. The function `biplot` returns a plot showing how the observations relate to the principal components.

Empirical Orthogonal Functions

The study of EOFs is related to PCA in the sense that the "traits" of the multivariate data vector now are spatially indexed, and the samples are usually taken over time. It is shown in Cressie and Wikle (2011, Chapter 5) that the EOFs can be obtained from the data through either a spectral decomposition of an empirical (spatial or temporal) covariance matrix or a singular value decomposition (SVD) of a centered data matrix (see Technical Note 2.2).

Let $\mathbf{Z}_{t_j} \equiv (Z(\mathbf{s}_1; t_j), \ldots, Z(\mathbf{s}_m; t_j))'$ for $j = 1, \ldots, T$. Using (2.4) to estimate the lag-0 spatial covariance matrix, $\widehat{\mathbf{C}}_z^{(0)}$ (which is symmetric and non-negative-definite), the PCA decomposition is given by the spectral decomposition

$$\widehat{\mathbf{C}}_z^{(0)} = \boldsymbol{\Psi}\boldsymbol{\Lambda}\boldsymbol{\Psi}', \tag{2.9}$$

where $\boldsymbol{\Psi} \equiv (\boldsymbol{\psi}_1, \ldots, \boldsymbol{\psi}_m)$ is a matrix of spatially indexed eigenvectors given by the vectors $\boldsymbol{\psi}_k \equiv (\psi_k(\mathbf{s}_1), \ldots, \psi_k(\mathbf{s}_m))'$ for $k = 1, \ldots, m$, and $\boldsymbol{\Lambda} \equiv \mathrm{diag}(\lambda_1, \ldots, \lambda_m)$ is a diagonal matrix of corresponding non-negative eigenvalues (decreasing down the diagonal). The eigenvectors are called "EOFs" and are often plotted as spatial maps (since they are spatially indexed, which is also why $\boldsymbol{\Psi}$ is used to distinguish them from the more general PCA weights, $\mathbf{W}$, above). For $k = 1, \ldots, m$, the so-called *kth principal component (PC) time series* are given by $a_k(t_j) \equiv \boldsymbol{\psi}_k'\mathbf{Z}_{t_j}$, where $j = 1, \ldots, T$. From PCA considerations, the EOFs have the nice property that $\boldsymbol{\psi}_1$ provides the linear coefficients such that $\mathrm{var}(a_1) = \lambda_1$ is maximized, $\boldsymbol{\psi}_2$ provides the linear coefficients such that $\mathrm{var}(a_2) = \lambda_2$ accounts for the next largest variance such that $\mathrm{cov}(a_1, a_2) = 0$, and so on. As with the principal components in PCA, the EOFs form a discrete orthonormal basis (i.e., $\boldsymbol{\Psi}'\boldsymbol{\Psi} = \boldsymbol{\Psi}\boldsymbol{\Psi}' = \mathbf{I}$).

There are two primary uses for EOFs. First, it is sometimes the case that one can gain some understanding about important spatial patterns of variability in a sequence of spatio-temporal data by examining the EOF coefficient maps (loadings). But care must be taken not to interpret the EOF spatial structures in terms of dynamical or kinematic properties of the underlying process (see, for example, Monahan et al., 2009). Second, these bases can be quite useful for dimension reduction in a random-effects spatial or spatio-temporal representation (see Section 4.4), although again, in general, they are not "optimal" bases in terms of reduced-order dynamical systems.

Technical Note 2.2: Calculating EOFs

As stated above, EOFs can be calculated directly from the spectral decomposition of the empirical lag-0 spatial covariance matrix (2.9). However, they are more often obtained directly through a *singular value decomposition* (SVD, see Appendix A), which provides computational benefits in some situations. To see the equivalence, first we show how to calculate the empirical covariance-based EOFs. Let $\mathbf{Z} \equiv [\mathbf{Z}_1, \ldots, \mathbf{Z}_T]'$ be the $T \times m$ space-wide data matrix and then let $\widetilde{\mathbf{Z}}$ be the "detrended" and scaled data matrix,

$$\widetilde{\mathbf{Z}} \equiv \frac{1}{\sqrt{T-1}}(\mathbf{Z} - \mathbf{1}_T \widehat{\boldsymbol{\mu}}'_{z,s}), \tag{2.10}$$

where $\mathbf{1}_T$ is a T-dimensional vector of ones and $\widehat{\boldsymbol{\mu}}_{z,s}$ is the spatial mean vector given by (2.1). Then it is easy to show that

$$\mathbf{C}_z^{(0)} = \widetilde{\mathbf{Z}}'\widetilde{\mathbf{Z}} = \boldsymbol{\Psi}\boldsymbol{\Lambda}\boldsymbol{\Psi}', \tag{2.11}$$

and the principal component (PC) time series are given by the columns of $\mathbf{A} = (\sqrt{T-1})\widetilde{\mathbf{Z}}\boldsymbol{\Psi}$; that is, they are projections of the detrended data matrix onto the EOF basis functions, $\boldsymbol{\Psi}$. The *normalized PC time series* are then given by $\mathbf{A}_{\mathrm{norm}} \equiv \mathbf{A}\boldsymbol{\Lambda}^{-1/2}$; these are just the PC time series divided by their standard deviation (i.e., the square root of the associated eigenvalue), so that the temporal variance of the normalized time series is equal to one. This normalization allows the m time series to be plotted on the same scale, leaving their relative importance to be captured by their corresponding eigenvalues.

Now, consider the SVD of the detrended and scaled data matrix,

$$\widetilde{\mathbf{Z}} = \mathbf{U}\mathbf{D}\mathbf{V}', \tag{2.12}$$

where $\mathbf{U}$ is the $T \times T$ matrix of left singular vectors, $\mathbf{D}$ is a $T \times m$ matrix containing singular values on the main diagonal, and $\mathbf{V}$ is an $m \times m$ matrix containing the right singular vectors, where both $\mathbf{U}$ and $\mathbf{V}$ are orthonormal matrices. Upon substituting (2.12) into (2.11), it is easy to see that the EOFs are given by $\boldsymbol{\Psi} = \mathbf{V}$, and $\boldsymbol{\Lambda} = \mathbf{D}'\mathbf{D}$. In addition, it is straightforward to show that $\mathbf{A} = (\sqrt{T-1})\mathbf{U}\mathbf{D}$ and that the first m columns of $(\sqrt{T-1})\mathbf{U}$ correspond to the normalized PC time series, $\mathbf{A}_{\mathrm{norm}}$. Thus, the advantages of the SVD calculation approach are: (1) we do not need to calculate the empirical spatial covariance matrix; (2) we get the normalized PC time series and EOFs simultaneously; and (3) the procedure still works when $T < m$. The case of $T < m$ can be problematic in the covariance context since then $\mathbf{C}_z^{(0)}$ is not positive-definite, although, as shown in Cressie and Wikle (2011, Section 5.3.4), in this case one can still calculate the EOFs and PC time series.

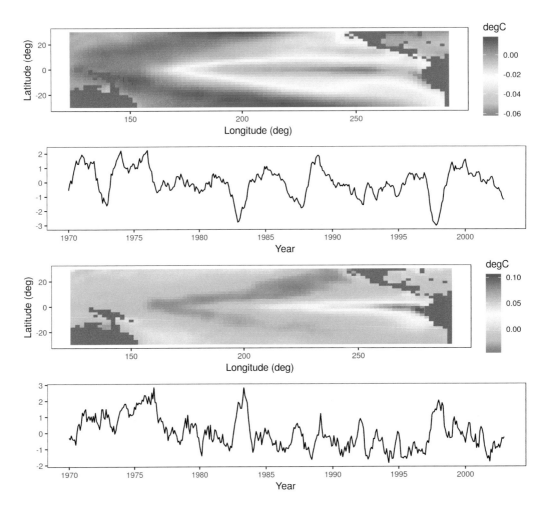

Figure 2.20: The first two empirical orthogonal functions and normalized principal-component time series for the SST data set obtained using an SVD of a space-wide matrix.

Figures 2.20 and 2.21 show the first four EOFs and PC time series for the SST data set. In this case, the number of spatial locations $m = 2261$, and the number of time points $T = 399$. The first four EOFs account for slightly more than 60% of the variation in the data. The EOF spatial patterns show strong variability in the eastern and central tropical Pacific, and they are known to be related to the El Niño and La Niña climate patterns that dominate the tropical Pacific SST variability. The corresponding PC time series (particularly for the first EOF) show time periods at which the data project very strongly on this spatial pattern (both in terms of large positive and large negative values), and it can be shown that these times correspond to strong El Niño and La Niña events, respectively.

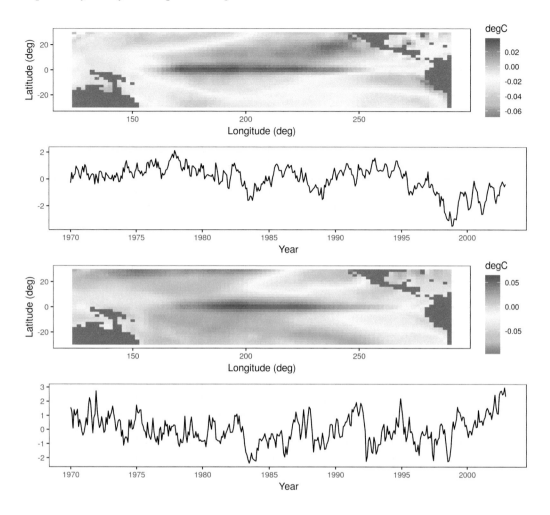

Figure 2.21: The third and fourth empirical orthogonal functions and normalized principal-component time series for the SST data set obtained using an SVD of a space-wide matrix.

How many EOFs should one consider? This is a long-standing question in PCA, and there are numerous suggestions. Perhaps the simplest is just to consider the number of EOFs that account for some desired proportion of overall variance. Alternatively, one can produce a *scree plot*, which is a plot of the relative variance associated with each eigenvalue of the EOF as a function of the index of that EOF (see Figure 2.22), and where the sum of all relative variances is 1. One typically sees a fairly quick drop in relative variance with increasing order of the eigenvalue, and then the variance reduction flattens out. It is sometimes recommended that one only focus on those EOFs before the index that begins the flat part of the curve; this choice of index can be a bit subjective. One can also get a sense as to the "significance" of each component by comparing the relative variances to those in an

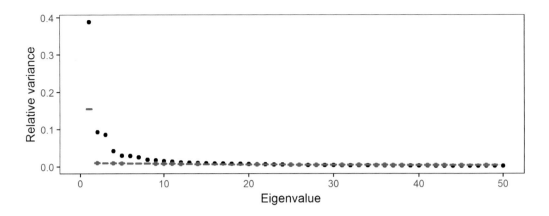

Figure 2.22: Scree plot for the EOF analysis of the SST data. The black symbols correspond to the relative variance associated with the ordered eigenvalues. The red symbols correspond to (very tight) boxplots of the relative variance associated with the eigenvalues from 100 EOF analyses in which the SST values at the spatial locations were randomly permuted for each time point.

EOF analysis in which the values for each spatial location are randomly permuted at each time (see, for example, Hastie et al., 2009, Chapter 14). Then, one plots the scree plot with the actual data superimposed on the permuted data. We recommend that the EOFs retained are around the index at which the two "curves" intersect. For example, the black symbols in Figure 2.22 correspond to the relative variance associated with the first 50 EOFs for the SST data, and the red symbols are the very tight boxplots of relative variances obtained from EOF analyses of 100 random permutations of the data. One can see that by about index 12, the scree plot of the actual data and the boxplots are starting to intersect, suggesting that there is very little "real" variability being accounted for by the EOFs with indices greater than about 12.

Some Technical Comments on Empirical Orthogonal Functions

The EOF decomposition is sometimes derived in a continuous-space context through a Karhunen–Loève expansion, with eigenvalues and eigenfunctions obtained through a solution of a Fredholm integral equation (see the overview in Cressie and Wikle, 2011, Section 5.3). This is relevant, as it shows why one should account for the area/support associated with each spatial observation when working in a discrete-space EOF environment. In particular, one should multiply the elements of the eigenvectors by the square root of the length, area, or volume of the spatial support associated with that spatial observation (e.g., Cohen and Jones, 1969). For example, consider spatial location s_i; for each of the k eigenvectors, one should multiply $\psi_k(s_i)$ by $\sqrt{e_i}$, where e_i is the length, area, or volume associated with location $\mathbf{s}_i$ (and we assume that not all of the $\{e_i\}$ are identical). This modification to the eigenvectors $\boldsymbol{\psi}_1, \ldots, \boldsymbol{\psi}_k$ must be done before calculating the PC time series.

Although most EOF analyses in the spatio-temporal context consider spatial EOFs and PC time series, one can certainly consider the analogous decomposition in which the EOFs are time-series bases and the projection of the data onto these bases is given by PC spatial fields. Implementation is straightforward – one either works with the temporal covariance matrix (averaging over spatial location) or considers the SVD of an $m \times T$ (temporally detrended) data matrix. EOF time series are used as temporal basis functions in a spatio-temporal model in Lab 4.3.

It is also important to note that in cases where EOF analysis is used for dimension reduction (see Section 4.3), it is often necessary to either interpolate the EOFs in a sensible manner (e.g., Obled and Creutin, 1986) or "pre-interpolate" the data onto a finely gridded spatial domain.

Finally, there are many extensions to the basic EOF analysis presented here, including so-called complex EOFs, cyclostationary EOFs, multivariate EOFs, and extended EOFs. These all have particular utility depending on the type of data and the goal of the analysis. For example, complex EOFs are used for trying to identify propagating features that account for a significant amount of variation in the data. Cyclostationary EOFs are appropriate when there are strong periodicities in the data and spatial variation is expected to shift dramatically within this periodicity. Multivariate EOFs are considered when multivariate spatial data are observed at the same time points. Extended EOFs are useful for understanding spatial patterns associated with temporal lags. These methods are described in more detail in Cressie and Wikle (2011, Section 5.3) and the references therein. In Lab 2.3 we will demonstrate the "classic" EOF analysis in R.

2.4.4 Spatio-Temporal Canonical Correlation Analysis

In multivariate statistics, canonical correlation analysis (CCA) seeks to create new variables that are linear combinations of two multivariate data sets (separately) such that the correlations between these new variables are maximized (e.g., Hotelling, 1936). Such methods can be extended to the case where the two data sets are indexed in space and time, typically where a spatial location corresponds to a "trait" in a multivariate set of "traits" (this terminology is borrowed from psychometrics). Time corresponds to the samples. (Note that just as with EOFs, one can reverse the roles of space and time in this setting as well.) A spatio-temporal CCA (ST-CCA) is given below where spatial location corresponds to the multivariate trait.

Assume that we have two data sets that have the same temporal domain of interest but potentially different spatial domains. In particular, consider the data sets given by the collection of spatial vectors $\{\mathbf{Z}_{t_j} \equiv (Z(\mathbf{s}_1; t_j), \ldots, Z(\mathbf{s}_m; t_j))' : j = 1, \ldots, T\}$, and $\{\mathbf{X}_{t_j} \equiv (X(\mathbf{r}_1; t_j), \ldots, X(\mathbf{r}_n; t_j))' : j = 1, \ldots, T\}$. Now, consider the two new variables that are linear combinations of $\mathbf{Z}_{t_j}$ and $\mathbf{X}_{t_j}$, respectively:

$$a_k(t_j) = \sum_{i=1}^{m} \xi_{ik}\, Z(\mathbf{s}_i; t_j) = \boldsymbol{\xi}_k' \mathbf{Z}_{t_j}, \qquad (2.13)$$

$$b_k(t_j) = \sum_{\ell=1}^{n} \psi_{\ell k}\, X(\mathbf{r}_\ell; t_j) = \boldsymbol{\psi}_k' \mathbf{X}_{t_j}. \qquad (2.14)$$

For suitable choices of weights (see below), the kth *canonical correlation*, for $k = 1, 2, \ldots, \min\{n, m\}$, is then simply the correlation between a_k and b_k,

$$r_k \equiv \mathrm{corr}(a_k, b_k) = \frac{\mathrm{cov}(a_k, b_k)}{\sqrt{\mathrm{var}(a_k)}\sqrt{\mathrm{var}(b_k)}},$$

which can also be written as

$$r_k = \frac{\boldsymbol{\xi}_k' \mathbf{C}_{z,x}^{(0)} \boldsymbol{\psi}_k}{(\boldsymbol{\xi}_k' \mathbf{C}_z^{(0)} \boldsymbol{\xi}_k)^{1/2} (\boldsymbol{\psi}_k' \mathbf{C}_x^{(0)} \boldsymbol{\psi}_k)^{1/2}}, \qquad (2.15)$$

where the variance–covariance matrices $\mathbf{C}_z^{(0)}$ and $\mathbf{C}_x^{(0)}$ are of dimension $m \times m$ and $n \times n$, respectively, and the cross-covariance matrix $\mathbf{C}_{z,x}^{(0)} \equiv \mathrm{cov}(\mathbf{Z}, \mathbf{X})$ has dimension $m \times n$. So the first pair of canonical variables corresponds to the weights $\boldsymbol{\xi}_1$ and $\boldsymbol{\psi}_1$ that maximize r_1 in (2.15). In addition, we standardize these weights such that the new canonical variables have unit variance. Given this first pair of canonical variables, we can then find a second pair, $\boldsymbol{\xi}_2$ and $\boldsymbol{\psi}_2$, associated with $\{a_2, b_2\}$ that are uncorrelated with $\{a_1, b_1\}$, have unit variance, and maximize r_2 in (2.15). This procedure continues so that the kth set of canonical variables are the linear combinations, $\{a_k, b_k\}$, that have unit variance, are uncorrelated with the previous $k - 1$ canonical variable pairs, and maximize r_k in (2.15). A specific procedure for calculating ST-CCA is given in Technical Note 2.3.

Because the weights given by $\boldsymbol{\xi}_k$ and $\boldsymbol{\psi}_k$ are indexed in space, they can be plotted as spatial maps, and the associated canonical variables can be plotted as time series. From an interpretation perspective, the time series of the first few canonical variables typically match up fairly closely (given they are optimized to maximize correlation), and the spatial patterns in the weights show the areas in space that are most responsible for the high correlations. Like EOFs, principal components, and other such approaches, one has to be careful with the interpretation of canonical variables beyond the first pair, given the restriction that CCA time series are uncorrelated. In addition, given that high canonical correlations within a canonical pair naturally result from this procedure, one has to be careful in evaluating the importance of that correlation. One way to do this is to randomly permute the spatial locations in the $\mathbf{Z}_{t_j}$ and $\mathbf{X}_{t_j}$ data vectors (separately) and recalculate the ST-CCA many times, thereby giving a permutation-based range of canonical correlations when there is no real structural relationship between the variables.

In addition to the consideration of two separate data sets, one can perform an ST-CCA between $\mathbf{Z}_{t_j}$ and, say, $\mathbf{X}_{t_j} \equiv \mathbf{Z}_{t_j - \tau}$, a τ-lagged version of the $\mathbf{Z}_{t_j}$ data. This "one-field ST-CCA" is often useful for exploratory data analysis or for generating a forecast of a spatial field. Some binning of the spatio-temporal data into temporal bins lagged by τ may be needed in practice.

Finally, in practice, because the covariance matrices required to implement ST-CCA are often fairly noisy (and even singular), depending on the sample size, we typically first project the data into a lower dimension using EOFs for computational stability (see Cressie and Wikle, 2011, Section 5.6.1). This is the approach we take in Lab 2.3.

As an example of ST-CCA, we consider a one-field ST-CCA on the SST data set. In particular, we are interested in forecasting SST seven months in the future, so we let the data $\mathbf{X}$ be the lag $\tau = 7$ month SST data and the data $\mathbf{Z}$ be the same SSTs with no lag. However, because $T < \max\{m, n\}$ for these data, we first project the data onto the first 10 EOFs (which account for about 74% of the variance in the data). For the projected data, Figure 2.23 shows the first canonical variables (i.e., $\{a_1(t_j), b_1(t_j) : j = 1, \ldots, T\}$), plotted as individual time series and which correspond to a canonical correlation of $r_1 = 0.843$. Figure 2.24 shows the corresponding spatial-weights maps for $\boldsymbol{\xi}_1$ and $\boldsymbol{\psi}_1$, respectively. In this example, it can be seen from the time-series plots that the series are quite highly correlated, and it can be shown that the large peaks correspond to known El Niño Southern Oscillation (ENSO) events. Similarly, the left panel of Figure 2.24 suggests a precursor pattern to the SST field in the right panel.

Technical Note 2.3: Calculating ST-CCA

First, let $k = 1$ and, because $\mathbf{C}_z^{(0)}$ and $\mathbf{C}_x^{(0)}$ are positive-definite, note that we can write $\mathbf{C}_z^{(0)} = (\mathbf{C}_z^{(0)})^{1/2}(\mathbf{C}_z^{(0)})^{1/2}$ and $\mathbf{C}_x^{(0)} = (\mathbf{C}_x^{(0)})^{1/2}(\mathbf{C}_x^{(0)})^{1/2}$ (see Appendix A). Thus, from (2.15), the square of the canonical correlation can be written as

$$r_1^2 = \frac{[\tilde{\boldsymbol{\xi}}_1'(\mathbf{C}_z^{(0)})^{-1/2}\mathbf{C}_{z,x}^{(0)}(\mathbf{C}_x^{(0)})^{-1/2}\tilde{\boldsymbol{\psi}}_1]^2}{(\tilde{\boldsymbol{\xi}}_1'\tilde{\boldsymbol{\xi}}_1)(\tilde{\boldsymbol{\psi}}_1'\tilde{\boldsymbol{\psi}}_1)}, \tag{2.16}$$

with $\tilde{\boldsymbol{\xi}}_1 \equiv (\mathbf{C}_z^{(0)})^{1/2}\boldsymbol{\xi}_1$ and $\tilde{\boldsymbol{\psi}}_1 \equiv (\mathbf{C}_x^{(0)})^{1/2}\boldsymbol{\psi}_1$, the so-called normalized weights. The CCA problem is now solved if we can find the $\tilde{\boldsymbol{\xi}}_1$ and $\tilde{\boldsymbol{\psi}}_1$ that maximize (2.16). In the multivariate statistics literature (e.g., Johnson and Wichern, 1992, p. 463) it is well known that r_1^2 corresponds to the largest singular value of the singular value decomposition (SVD; see Appendix A) of

$$(\mathbf{C}_z^{(0)})^{-1/2}\mathbf{C}_{z,x}^{(0)}(\mathbf{C}_x^{(0)})^{-1/2}, \tag{2.17}$$

where the normalized weight vectors $\tilde{\boldsymbol{\xi}}_1$ and $\tilde{\boldsymbol{\psi}}_1$ are the left and right singular vectors, respectively. Then we can obtain the unnormalized weights, $\boldsymbol{\xi}_1$ and $\boldsymbol{\psi}_1$, through $\boldsymbol{\xi}_1 \equiv$

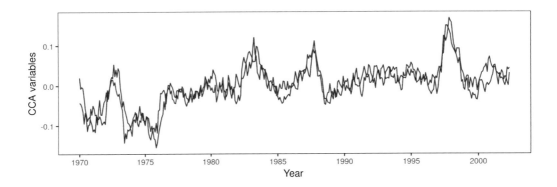

Figure 2.23: Time series of the first canonical variables, $\{a_1, b_1\}$, for $\tau = 7$ month lagged monthly SST anomalies at time $t_j - \tau$ (blue) and those at time t_j (red).

$(\mathbf{C}_z^{(0)})^{-1/2}\widetilde{\boldsymbol{\xi}}_1$ and $\boldsymbol{\psi}_1 \equiv (\mathbf{C}_x^{(0)})^{-1/2}\widetilde{\boldsymbol{\psi}}_1$, respectively. As mentioned above, these are the first ST-CCA pattern maps. The corresponding time series of ST-CCA canonical variables are then calculated directly from $a_1(t_j) = \boldsymbol{\xi}_1'\mathbf{Z}_{t_j}$ and $b_1(t_j) = \boldsymbol{\psi}_1'\mathbf{X}_{t_j}$, for $j = 1, \ldots, T$. More generally, $\widetilde{\boldsymbol{\xi}}_k$ and $\widetilde{\boldsymbol{\psi}}_k$ correspond to the left and right singular vectors associated with the kth singular value (r_k^2) in the SVD of (2.17). Then the unnormalized spatial-weights maps and the canonical time series are obtained analogously to the $k = 1$ case.

In practice, to evaluate the SVD in (2.17), we must first calculate the empirical covariance matrices $\widehat{\mathbf{C}}_z^{(0)}$, $\widehat{\mathbf{C}}_x^{(0)}$ using (2.4), as well as the empirical cross-covariance matrix $\widehat{\mathbf{C}}_{z,x}^{(0)}$ given by (2.5). Finally, we consider the SVD of $(\widehat{\mathbf{C}}_z^{(0)})^{-1/2}\widehat{\mathbf{C}}_{z,x}^{(0)}(\widehat{\mathbf{C}}_x^{(0)})^{-1/2}$. As mentioned in the text, the empirical covariance matrices can be unstable (or singular) unless $T \gg \max(n, m)$, and so it is customary to work in EOF space; that is, project the data for one or both variables onto a lower-dimensional space given by a relatively few EOFs before carrying out ST-CCA.

2.5 Chapter 2 Wrap-Up

There were three main goals in this chapter. First, we wanted to expose the reader to some basic ideas about data structures in R that are useful for working with spatio-temporal data. Next, we wanted to illustrate some useful ways to visualize spatio-temporal data, noting that it can be particularly challenging to visualize dynamical evolution of spatial fields either without collapsing the spatial component onto one spatial dimension (e.g., as with the Hovmöller plots) or through animation. Finally, we wanted to describe some standard

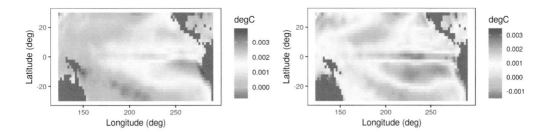

Figure 2.24: Spatial-weights maps corresponding to the linear combination of EOFs used to construct the canonical variables for SST data lagged $\tau = 7$ months (left) and the unlagged SST data (right).

ways to explore spatio-temporal data in preparation for developing models in Chapter 3. In particular, we discussed the exploration of the first moments (means) in space or time, and the second-order structures (covariances) either jointly in space and time, or averaged over one of the dimensions (usually the time dimension) to give covariance and cross-covariance matrices. Stepping up the technical level, we considered eigenvector approaches to explore the structure and potentially reduce the dimensionality of the spatio-temporal data. Specifically, we considered EOFs and ST-CCA. Of these, the EOFs are the most ubiquitous in the literature. Even if the technical details were a bit elaborate, the end result is a powerful and interpretable visualization and exploration of spatio-temporal variability.

You now have the survival skills to start building statistical models for spatio-temporal data, with the goal of spatial prediction, parameter inference, or temporal forecasting. In subsequent chapters, spatio-temporal statistical models will be discussed from an introductory perspective in Chapter 3, from a descriptive perspective in Chapter 4, and from a dynamic perspective in Chapter 5.

Lab 2.1: Data Wrangling

Spatio-temporal modeling and prediction generally involve substantial amounts of data that are available to the user in a variety of forms, but more often than not as tables in CSV files or text files. A considerable amount of time is usually spent in loading the data and pre-processing them in order to put them into a form that is suitable for analysis. Fortunately, there are several packages in R that help the user achieve these goals quickly; here we focus on the packages **dplyr** and **tidyr**, which contain functions particularly suited for the data manipulation techniques that are required. We first load the required packages, as well as **STRbook** (visit `https://spacetimewithr.org` for instructions on how to install **STRbook**).

```
library("dplyr")
library("tidyr")
library("STRbook")
```

As running example we will consider the NOAA data set, which was provided to us as text in tables and is available with the package **STRbook**. There are six data tables:

- `Stationinfo.dat`. This table contains 328 rows (one for each station) and three columns (station ID, latitude coordinate, and longitude coordinate) containing information on the stations' locations.

- `Times_1990.dat`. This table contains 1461 rows (one for each day between 01 January 1990 and 30 December 1993) and four columns (Julian date, year, month, day) containing the data time stamps.

- `Tmax_1990.dat`. This table contains 1461 rows (one for each time point) and 328 columns (one for each station location) containing all maximum temperature data with missing values coded as -9999.

- `Tmin_1990.dat`. Same as `Tmax_1990.dat` but containing minimum temperature data.

- `TDP_1990.dat`. Same as `Tmax_1990.dat` but containing temperature dew point data with missing values coded as -999.90001.

- `Precip_1990.dat`. Same as `Tmax_1990.dat` but containing precipitation data with missing values coded as -99.989998.

The first task is to reconcile all these data into one object. Before seeing how to use the spatio-temporal data classes to do this, we first consider the rather simpler task of reconciling them into a standard R data frame in long format.

Working with Spatio-Temporal Data in Long Format

The station locations, time stamps and maximum temperature data can be loaded into R from **STRbook** as follows.

```
locs <- read.table(system.file("extdata", "Stationinfo.dat",
                               package = "STRbook"),
                col.names = c("id", "lat", "lon"))
Times <- read.table(system.file("extdata", "Times_1990.dat",
                                package = "STRbook"),
                col.names = c("julian", "year", "month", "day"))
Tmax <- read.table(system.file("extdata", "Tmax_1990.dat",
                               package = "STRbook"))
```

In this case, `system.file` and its arguments are used to locate the data within the package **STRbook**, while `read.table` is the most important function used in R for reading data input from text files. By default, `read.table` assumes that data items are separated by a blank space, but this can be changed using the argument `sep`. Other important data input functions worth looking up include `read.csv` for comma-separated value files, and `read.delim`.

Above we have added the column names to the data `locs` and `Times` since these were not available with the original text tables. Since we did not assign column names to `Tmax`, the column names are the default ones assigned by `read.table`, that is, `V1`, `V2`, `...`, `V328`. As these do not relate to the station ID in any way, we rename these columns as appropriate using the data in `locs`.

```
names(Tmax) <- locs$id
```

The other data can be loaded in a similar way to `Tmax`; we denote the resulting variables as `Tmin`, `TDP`, and `Precip`, respectively. One can, and should, use the functions **head** and **tail** to check that the loaded data are sensible.

Consider now the maximum-temperature data in the NOAA data set. Since each row in `Tmax` is associated with a time point, we can attach it columnwise to the data frame `Times` using **cbind**.

```
Tmax <- cbind(Times, Tmax)
head(names(Tmax), 10)
```

```
## [1] "julian" "year"   "month" "day"    "3804"  "3809"
## [7] "3810"   "3811"   "3812"  "3813"
```

Now `Tmax` contains the time information in the first four columns and temperature data in the other columns. To put `Tmax` into long format we need to identify a *key–value* pair. In our case, the data are in space-wide format where the *keys* are the station IDs and the *values* are the maximum temperatures (which we store in a field named `z`). The function we use to put the data frame into long format is **gather**. This function takes the data as first argument, the key–value pair, and then the next arguments are the names of any columns to exclude as values (in this case those relating to the time stamp).

```
Tmax_long <- gather(Tmax, id, z, -julian, -year, -month, -day)
head(Tmax_long)
```

```
##    julian year month day   id  z
## 1 726834 1990     1   1 3804 35
## 2 726835 1990     1   2 3804 42
## 3 726836 1990     1   3 3804 49
## 4 726837 1990     1   4 3804 59
```

```
## 5 726838 1990      1     5 3804 41
## 6 726839 1990      1     6 3804 45
```

Note how **gather** has helped us achieve our goal: we now have a single row per measurement and multiple rows may be associated with the same time point. As is, the column `id` is of class `character` since it was extracted from the column names. Since the station ID is an integer it is more natural to ensure the field is of class `integer`.

```
Tmax_long$id <- as.integer(Tmax_long$id)
```

There is little use to keep missing data (coded as -9999 in our case) when the data are in long format. To filter out these data we can use the function **filter**. Frequently it is better to use an *inequality* criterion (e.g., less than) when filtering in this way rather than an *equivalence* criterion (is equal to) due to truncation error when storing data. This is what we do below, and filter out data with values less than -9998 rather than data with values equal to -9999. This is particularly important when processing the other variables, such as precipitation, where the missing value is -99.989998.

```
nrow(Tmax_long)
```

```
## [1] 479208
```

```
Tmax_long <- filter(Tmax_long, !(z <= -9998))
nrow(Tmax_long)
```

```
## [1] 196253
```

Note how the number of rows in our data set (returned from the function **nrow**) has now decreased by more than half. One may also use the R function **subset**; however, **filter** tends to be faster for large data sets. Both **subset** and **filter** take a logical expression as instruction on how to filter out unwanted rows. As with **gather**, the column names in the logical expression do not appear as strings. In R this method of providing arguments is known as *non-standard evaluation*, and we shall see several instances of it in the course of the Labs.

Now assume we wish to include minimum temperature and the other variables inside this data frame too. The first thing we need to do is first make sure every measurement `z` is attributed to a process. In our case, we need to add a column, say `proc`, indicating what process the measurement relates to. There are a few ways in which to add a column to a data frame; here we shall introduce the function **mutate**, which will facilitate operations in the following Labs.

```
Tmax_long <- mutate(Tmax_long, proc = "Tmax")
head(Tmax_long)

##   julian year month day   id   z proc
## 1 726834 1990     1   1 3804 35 Tmax
## 2 726835 1990     1   2 3804 42 Tmax
## 3 726836 1990     1   3 3804 49 Tmax
## 4 726837 1990     1   4 3804 59 Tmax
## 5 726838 1990     1   5 3804 41 Tmax
## 6 726839 1990     1   6 3804 45 Tmax
```

Now repeat the same procedure with the other variables to obtain data frames `Tmin_long`, `TDP_long`, and `Precip_long` (remember the different codings for the missing values!). To save time, the resulting data frames can also be loaded directly from **STRbook** as follows.

```
data(Tmin_long, package = "STRbook")
data(TDP_long, package = "STRbook")
data(Precip_long, package = "STRbook")
```

We can now construct our final data frame in long format by simply concatenating all these (rowwise) together using the function **rbind**.

```
NOAA_df_1990 <- rbind(Tmax_long, Tmin_long, TDP_long, Precip_long)
```

There are many advantages of having data in long form. For example, it makes grouping and summarizing particularly easy. Let us say we want to find the mean value for each variable in each year. We do this using the functions `group_by` and `summarise`. The function `group_by` creates a *grouped data frame*, while `summarise` does an operation *on each group within the grouped data frame*.

```
summ <- group_by(NOAA_df_1990, year, proc) %>%   # groupings
        summarise(mean_proc = mean(z))            # operation
```

Alternatively, we may wish to find out the number of days on which it did not rain at each station in June of every year. We can first filter out the other variables and then use **summarise**.

```
NOAA_precip <- filter(NOAA_df_1990, proc == "Precip" & month == 6)
summ <- group_by(NOAA_precip, year, id) %>%
        summarise(days_no_precip = sum(z == 0))
head(summ)
```

```
## # A tibble: 6 x 3
## # Groups:    year [1]
##    year    id days_no_precip
##    <int> <int>        <int>
## 1  1990  3804           19
## 2  1990  3810           26
## 3  1990  3811           21
## 4  1990  3812           24
## 5  1990  3813           25
## 6  1990  3816           23
```

The median number of days with no recorded precipitation was

```
median(summ$days_no_precip)
```

```
## [1] 20
```

In the R code above, we have used the operator %>%, known as the *pipe* operator. This operator has its own nuances and should be used with care, but we find it provides a clear desciption of the processing pipeline a data set is passed through. We shall always use this operator as x %>% f(y), which is shorthand for f(x,y). For example, the June summaries above can be found equivalently using the commands

```
grps <- group_by(NOAA_precip, year, id)
summ <- summarise(grps, days_no_precip = sum(z == 0))
```

There are other useful commands in **dplyr** that we use in other Labs. First, the function **arrange** sorts by a column. For example, NOAA_df_1990 is sorted first by station ID, and then by time (Julian date). The following code sorts the data first by time and then by station ID.

```
NOAA_df_sorted <- arrange(NOAA_df_1990, julian, id)
```

Calling **head**(NOAA_df_sorted) reveals that no measurements on temperature dew point are available for the first few days of the data set.

Another useful function is **select**, which can be used to select or discard columns. For example, in the following, df1 selects only the Julian date and the measurement while df2 contains all columns except the Julian date.

```
df1 <- select(NOAA_df_1990, julian, z)
df2 <- select(NOAA_df_1990, -julian)
```

At present, our long data frame contains no spatial information attached to it. However, for each station ID we have an associated coordinate in the data frame locs. We can merge locs to NOAA_df_1990 using the function `left_join`; this is considerably faster than the function `merge`. With `left_join` we need to supply the column field name by which we are merging. In our case, the field common to both data sets is `"id"`.

```
NOAA_df_1990 <- left_join(NOAA_df_1990, locs, by = "id")
```

Finally, it may be the case that one wishes to revert from long format to either space-wide or time-wide format. The reverse function of `gather` is `spread`. This also works by identifying the key–value pair in the data frame; the values are then "widened" into a table while the keys are used to label the columns. For example, the code below constructs a space-wide data frame of maximum temperatures, with each row denoting a different date and each column containing data `z` from a specific station `id`.

```
Tmax_long_sel <- select(Tmax_long, julian, id, z)
Tmax_wide <- spread(Tmax_long_sel, id, z)
dim(Tmax_wide)
```

```
## [1] 1461  138
```

The first column is the Julian date. Should one wish to construct a standard matrix containing these data, then one can simply drop this column and convert as follows.

```
M <- select(Tmax_wide, -julian) %>% as.matrix()
```

Working with Spatio-Temporal Data Classes

Next, we convert the data into objects of class `STIDF` and `STFDF`; in these class names "DF" is short for "data frame," which indicates that in addition to the spatio-temporal locations (which only need `STI` or `STF` objects), the objects will also contain data. These classes are defined in the package **spacetime**. Since sometimes we construct spatio-temporal objects using spatial objects we also need to load the package **sp**. For details on these classes see Pebesma (2012).

```
library("sp")
library("spacetime")
```

Constructing an `STIDF` Object

The spatio-temporal object for irregular data, `STIDF`, can be constructed using two functions: **stConstruct** and **STIDF**. Let us focus on the maximum temperature in `Tmax_long`. The only thing we need to do before we call **stConstruct** is to define a formal time stamp from the `year`, `month`, `day` fields. First, we construct a field with the date in `year–month–day` format using the function **paste**, which concatenates strings together. Instead of typing `NOAA_df_1990$year`, `NOAA_df_1990$month` and `NOAA_df_1990$day` we embed the **paste** function within the function **with** to reduce code length.

```
NOAA_df_1990$date <- with(NOAA_df_1990,
                     paste(year, month, day, sep = "-"))
head(NOAA_df_1990$date, 4)    # show first four elements

## [1] "1990-1-1" "1990-1-2" "1990-1-3" "1990-1-4"
```

The field `date` is of type `character`. This field can now be converted into a `Date` object using `as.Date`.

```
NOAA_df_1990$date <- as.Date(NOAA_df_1990$date)
class(NOAA_df_1990$date)

## [1] "Date"
```

Now we have everything in place to construct the spatio-temporal object of class `STIDF` for maximum temperature. The easiest way to do this is using **stConstruct**, in which we provide the data frame in long format and indicate which are the spatial and temporal coordinates. This is the bare minimum required for constructing a spatio-temporal data set.

```
Tmax_long2 <- filter(NOAA_df_1990, proc == "Tmax")
STObj <- stConstruct(x = Tmax_long2,         # data set
                space = c("lon", "lat"),  # spatial fields
                time = "date")            # time field
class(STObj)

## [1] "STIDF"
## attr(,"package")
## [1] "spacetime"
```

The function **class** can be used to confirm we have successfully generated an object of class `STIDF`. There are several other options that can be used with **stConstruct**. For example, one can set the coordinate reference system or specify whether the time field indicates an instance or an interval. Type **help**(stConstruct) into the R console for more details.

The function **STIDF** is slightly different from **stConstruct** as it requires one to also specify the spatial part as an object of class `Spatial` from the package **sp**. In our case, the spatial component is simply an object containing irregularly spaced data, which in the package **sp** is a `SpatialPoints` object. A `SpatialPoints` object may be constructed using the function **SpatialPoints** and by supplying the coordinates as arguments. As with `stConstruct`, several other arguments can also be supplied to `SpatialPoints`; see the help file of `SpatialPoints` for more details.

```
spat_part <- SpatialPoints(coords = Tmax_long2[, c("lon", "lat")])
temp_part <- Tmax_long2$date
STObj2 <- STIDF(sp = spat_part,
                time = temp_part,
                data = select(Tmax_long2, -date, -lon, -lat))
class(STObj2)

## [1] "STIDF"
## attr(,"package")
## [1] "spacetime"
```

Constructing an `STFDF` Object

A similar approach can be used to construct an `STFDF` object instead of an `STIDF` object. When the spatial points are fixed in time, we only need to provide as many spatial coordinates as there are spatial points, in this case those of the station locations. We also need to provide the regular time stamps, that is, one for each day between 01 January 1990 and 30 December 1993. Finally, the data can be provided both in space-wide or time-wide format with `stConstruct`, and in long format with `STFDF`. Here we show how to use `STFDF`.

The spatial and temporal parts can be obtained from the original data as follows.

```
spat_part <- SpatialPoints(coords = locs[, c("lon", "lat")])
temp_part <- with(Times,
                  paste(year, month, day, sep = "-"))
temp_part <- as.Date(temp_part)
```

The data need to be provided in long format, but now they must contain all the missing values too since a data point must be provided for every spatial and temporal combination. To get the data into long format we use `gather`.

```
Tmax_long3 <- gather(Tmax, id, z, -julian, -year, -month, -day)
```

It is very important that the data frame in long format supplied to `STFDF` has the spatial index moving faster than the temporal index, and that the order of the spatial index is the same as that of the spatial component supplied.

```
Tmax_long3$id <- as.integer(Tmax_long3$id)
Tmax_long3 <- arrange(Tmax_long3, julian, id)
```

Confirming that the spatial ordering in `Tmax_long3` is the correct one can be done as follows.

```
all(unique(Tmax_long3$id) == locs$id)

## [1] TRUE
```

We are now ready to construct the `STFDF`.

```
STObj3 <- STFDF(sp = spat_part,
                time = temp_part,
                data = Tmax_long3)
class(STObj3)

## [1] "STFDF"
## attr(,"package")
## [1] "spacetime"
```

Since we will be using `STObj3` often in the Labs we further equip it with a coordinate reference system (see Bivand et al., 2013, for details on these reference systems),

```
proj4string(STObj3) <- CRS("+proj=longlat +ellps=WGS84")
```

and replace the missing values (currently coded as -9999) with `NA`s.

```
STObj3$z[STObj3$z == -9999] <- NA
```

For ease of access, this object is saved as a data file in **STRbook** and can be loaded using the command **data**(`"STObj3", package = "STRbook"`).

Lab 2.2: Visualization

In this Lab we shall visualize maximum temperature data in the NOAA data set. Specifically, we consider the maximum recorded temperature between May 1993 and September 1993 (inclusive). The packages we need are **animation**, **dplyr**, **ggplot2**, **gstat**, **maps**, and **STRbook**.

```
library("animation")
library("dplyr")
library("ggplot2")
library("gstat")
library("maps")
library("STRbook")
```

In order to ensure consistency of results and visualizations we fix the seed to 1.

```
set.seed(1)
```

We now load the data set and take a subset of it using the function `filter`.

```
data("NOAA_df_1990", package = "STRbook")
Tmax <- filter(NOAA_df_1990,        # subset the data
               proc == "Tmax" &     # only max temperature
               month %in% 5:9 &     # May to September
               year == 1993)        # year of 1993
```

The data frame we shall work with is hence denoted by `Tmax`. The first six records in `Tmax` are:

```
Tmax %>% select(lon, lat, date, julian, z) %>% head()
```

```
##          lon    lat        date julian   z
## 1 -81.43333 39.35 1993-05-01 728050 82
## 2 -81.43333 39.35 1993-05-02 728051 84
## 3 -81.43333 39.35 1993-05-03 728052 79
## 4 -81.43333 39.35 1993-05-04 728053 72
## 5 -81.43333 39.35 1993-05-05 728054 73
## 6 -81.43333 39.35 1993-05-06 728055 78
```

The first record has a Julian date of 728050, corresponding to 01 May 1993. To ease the following operations, we create a new variable t that is equal to 1 when `julian == 728050` and increases by 1 for each day in the record.

```
Tmax$t <- Tmax$julian - 728049      # create a new time variable
```

The first task faced by the spatio-temporal modeler is data visualization. This is an important preliminary task that needs to be carried out prior to the exploratory-data-analysis stage and the modeling stages. Throughout, we shall make extensive use of the *grammar of graphics* package **ggplot2**, which is a convenient way to plot and visualize data and results in R. The book by Wickham (2016) provides a comprehensive introduction to **ggplot2**.

Spatial Plots

Visualization techniques vary with the data being analyzed. The NOAA data are collected at stations that are fixed in space; therefore, initial plots should give the modeler an idea of the overall spatial variation of the observed data. If there are many time points, usually only a selection of time points are chosen for visualization. In this case we choose three time points.

```
Tmax_1 <- subset(Tmax, t %in% c(1, 15, 30))  # extract data
```

The variable `Tmax_1` contains the data associated with the first, fifteenth, and thirtieth day in `Tmax`. We now plot this data subset using **ggplot2**. Note that the function `col_scale`, below, is simply a wrapper for the **ggplot2** function `scale_colour_distiller`, and is provided with **STRbook**.

```
NOAA_plot <- ggplot(Tmax_1) +                 # plot points
    geom_point(aes(x = lon, y = lat,          # lon and lat
                   colour = z),               # attribute color
               size = 2) +                    # make all points larger
    col_scale(name = "degF") +                # attach color scale
    xlab("Longitude (deg)") +                 # x-axis label
    ylab("Latitude (deg)") +                  # y-axis label
    geom_path(data = map_data("state"),       # add US states map
          aes(x = long, y = lat, group = group)) +
    facet_grid(~date) +                       # facet by time
    coord_fixed(xlim = c(-105, -75),
                ylim = c(25, 50))  +          # zoom in
    theme_bw()                                # B&W theme
```

`NOAA_plot` is a plot of the spatial locations of the stations. The function `aes` (short for aesthetics) for `geom_point` identifies which field in the data frame `Tmax_1` is the x-coordinate and which is the y-coordinate. **ggplot2** also allows one to attribute color (and size, if desired) to other fields in a similar fashion. The command `print`(NOAA_plot) generates the figure shown in Figure 2.1. As can be seen, the stations are approximately regularly spaced within the domain.

When working with geographic data, it is also good practice to put the spatial locations of the data into perspective, by plotting country or state boundaries together with the data locations. Above, the US state boundaries are obtained from the **maps** package through the command `map_data("state")`. The boundaries are then overlaid on the plot using `geom_path`, which simply joins the points and draws the resulting path with x against y. Projections can be applied by adding another layer to the **ggplot2** object using `coord_map`. For example adding + `coord_map`(projection = "sinusoidal") will plot using a sinusoidal projection. One can also plot in three dimensions by using projection = "ortho".

In this example we have used **ggplot2** to plot *point-referenced data*. Plots of regular lattice data, such as those shown in Figure 2.2, are generated similarly by using `geom_tile` instead. Plots of irregular lattice data are generated using `geom_polygon`. As an example of the latter, consider the BEA income data set. These data can be loaded from **STRbook** as follows.

```
data("BEA", package = "STRbook")
head(BEA %>% select(-Description), 3)
```

```
##            NAME10 X1970 X1980 X1990
## 6      Adair, MO  2723  7399 12755
## 9     Andrew, MO  3577  7937 15059
## 12 Atchison, MO  3770  5743 14748
```

From the first three records, we can see that the data set contains the personal income, in dollars, by county and by year for the years 1970, 1980 and 1990. These data need to be merged with Missouri county data which contain geospatial information. These county data, which are also available in **STRbook**, were originally processed from a shapefile that was freely available online.[6]

```
data("MOcounties", package = "STRbook")
head(MOcounties %>% select(long, lat, NAME10), 3)
```

```
##        long      lat    NAME10
## 1 627911.9 4473554 Clark, MO
## 2 627921.4 4473559 Clark, MO
## 3 627923.0 4473560 Clark, MO
```

The data set contains the boundary points for the counties, amongst several other variables which we do not explore here. For example, to plot the boundary of the first county one can simply type

```
County1 <- filter(MOcounties, NAME10 == "Clark, MO")
plot(County1$long, County1$lat)
```

To add the BEA income data to the county data containing geospatial information we use `left_join`.

```
MOcounties <- left_join(MOcounties, BEA, by = "NAME10")
```

Now it is just a matter of calling `ggplot` with `geom_polygon` to display the BEA income data as spatial polygons. We also use `geom_path` to draw the county boundaries. Below we show the code for 1970; similar code would be needed for 1980 and 1990. Note the use of the group argument to identify which points correspond to which county. The resulting plots are shown in Figure 2.4.

---

[6]http://msdis-archive.missouri.edu/archive/metadata_gos/MO_2010_TIGER_Census_County_Boundaries.xml

```
g1 <- ggplot(MOcounties) +
   geom_polygon(aes(x = long, y = lat,        # county boundary
                    group = NAME10,            # county group
                    fill = log(X1970))) +      # log of income
   geom_path(aes(x = long, y = lat,           # county boundary
                 group = NAME10)) +            # county group
   fill_scale(limits = c(7.5,10.2),
              name = "log($)")   +
   coord_fixed() + ggtitle("1970") +           # annotations
   xlab("x (m)") + ylab("y (m)") + theme_bw()
```

Type `print(g1)` in the R console to display the plot.

Time-Series Plots

Next, we look at the time series associated with the maximum temperature data in the NOAA data set. One can plot the time series at all 139 weather stations (and this is recommended); here we look at the time series at a set of stations selected at random. We first obtain the set of unique station identifiers, choose 10 at random from these, and extract the data associated with these 10 stations from the data set.

```
UIDs <- unique(Tmax$id)                        # extract IDs
UIDs_sub <- sample(UIDs, 10)                    # sample 10 IDs
Tmax_sub <- filter(Tmax, id %in% UIDs_sub)      # subset data
```

To visualize the time series at these stations, we use *facets*. When given a long data frame, one can first subdivide the data frame into groups and generate a plot for each group. The following code displays the time series at each station. The command we use is `facet_wrap`, which automatically adjusts the number of rows and columns in which to display the facets. The command `facet_grid` instead uses columns for one grouping variable and rows for a second grouping variable, if specified.

```
TmaxTS <- ggplot(Tmax_sub) +
   geom_line(aes(x = t, y = z)) + # line plot of z against t
   facet_wrap(~id, ncol = 5) +     # facet by station
   xlab("Day number (days)") +     # x label
   ylab("Tmax (degF)") +           # y label
   theme_bw() +                    # BW theme
   theme(panel.spacing = unit(1, "lines")) # facet spacing
```

The argument `~id` supplied to `facet_wrap` is a `formula` in R. In this case, the formula is used to denote the groups by which we are faceting. The syntax x~y can be used to facet by two variables. The command `print(TmaxTS)` produces Figure 2.9.

Hovmöller Plots

A Hovmöller plot is a two-dimensional space-time visualization, where space is collapsed (projected or averaged) onto one dimension; the second dimension then denotes time. A Hovmöller plot can be generated relatively easily if the data are on a space-time grid, but unfortunately this is rarely the case! This is where data-wrangling techniques such as those explored in Lab 2.1 come in handy.

Consider the latitudinal Hovmöller plot. The first step is to generate a regular grid of, say, 25 spatial points and 100 temporal points using the function `expand.grid`, with limits set to the latitudinal and temporal limits available in the data set.

```
lim_lat <- range(Tmax$lat)          # latitude range
lim_t <- range(Tmax$t)              # time range
lat_axis <- seq(lim_lat[1],         # latitude axis
                lim_lat[2],
                length=25)
t_axis <- seq(lim_t[1],             # time axis
              lim_t[2],
              length=100)
lat_t_grid <- expand.grid(lat = lat_axis,
                          t = t_axis)
```

We next need to associate each station's latitudinal coordinate with the closest one on the grid. This can be done by finding the distance from the station's latitudinal coordinate to each point of the grid, finding which gridpoint is the closest, and allocating that to it. We store the gridded data in `Tmax_grid`.

```
Tmax_grid <- Tmax
dists <- abs(outer(Tmax$lat, lat_axis, "-"))
Tmax_grid$lat <- lat_axis[apply(dists, 1, which.min)]
```

Now that we have associated each station with a latitudinal coordinate, all that is left is to group by latitude and time, and then we average all station values falling in the latitude–time bands.

```
Tmax_lat_Hov <- group_by(Tmax_grid, lat, t) %>%
                summarise(z = mean(z))
```

In this case, every latitude–time band contains at least one data point, so that the Hovmöller plot contains no missing points on the established grid. This may not always be the case, and simple interpolation methods, such as `interp` from the **akima** package, can be used to fill out grid cells with no data.

Plotting gridded data is facilitated using the **ggplot2** function `geom_tile`. The function `geom_tile` is similar to `geom_point`, except that it assumes regularly spaced data and automatically uses rectangular patches in the plot. Since rectangular patches are "filled," we use the **STRbook** function `fill_scale` instead of `col_scale`, which takes the legend title in the argument `name`.

```
Hovmoller_lat <- ggplot(Tmax_lat_Hov) +              # take data
        geom_tile(aes(x = lat, y = t, fill = z)) + # plot
        fill_scale(name = "degF") +     # add color scale
        scale_y_reverse() +             # rev y scale
        ylab("Day number (days)") +     # add y label
        xlab("Latitude (degrees)") +    # add x label
        theme_bw()                      # change theme
```

The function `scale_y_reverse` ensures that time increases from top to bottom, as is typical in Hovmöller plots. We can generate a longitude-based Hovmöller plot in the same way. The resulting Hovmöller plots are shown in Figure 2.11.

Animations

To generate an animation in R, one can use the package **animation**. First, we define a function that plots a spatial map of the maximum temperature as a function of time:

```
Tmax_t <- function(tau) {
    Tmax_sub <- filter(Tmax, t == tau)          # subset data
    ggplot(Tmax_sub) +
        geom_point(aes(x = lon,y = lat, colour = z),    # plot
                   size = 4) +                          # pt. size
        col_scale(name = "z", limits = c(40, 110)) +
        theme_bw() # B&W theme
}
```

The function above takes a day number `tau`, filters the data frame according to the day number, and then plots the maximum temperature at the stations as a spatial map.

Next, we construct a function that plots the data for every day in the data set. The function that generates the animation within an HTML webpage is **saveHTML**. This takes the function that plots the sequence of images and embeds them in a webpage (by default named `index.html`) using JavaScript. The function **saveHTML** takes many arguments; type the command

```
help(saveHTML)
```

in the R console for more details.

```
gen_anim <- function() {
    for(t in lim_t[1]:lim_t[2]){   # for each time point
        plot(Tmax_t(t))            # plot data at this time point
    }
}

ani.options(interval = 0.2)        # 0.2s interval between frames
saveHTML(gen_anim(),               # run the main function
        autoplay = FALSE,          # do not play on load
        loop = FALSE,              # do not loop
        verbose = FALSE,           # no verbose
        outdir = ".",              # save to current dir
        single.opts = "'controls': ['first', 'previous',
                                    'play', 'next', 'last',
                                    'loop', 'speed'],
                                    'delayMin': 0",
        htmlfile = "NOAA_anim.html")  # save filename
```

To view the animation, load `NOAA_anim.html` from your working directory. The animation reveals dynamics within the spatio-temporal data that are not apparent using other visualization methods. For example, the maximum temperature clearly drifts from west to east at several points during the animation. This suggests that a dynamic spatio-temporal model that can capture this drift could provide a good fit to these data.

Lab 2.3: Exploratory Data Analysis

In this Lab we carry out exploratory data analysis (EDA), which typically requires visualization techniques similar to those utilized in Lab 2.2. There are several ways in which to carry out EDA with spatio-temporal data; in this Lab we consider the construction and visualization of the empirical means and covariances, the use of empirical orthogonal functions and their associated principal component time series, semivariogram analysis, and spatio-temporal canonical correlation analysis.

For the first part of the Lab, as in Lab 2.2, we shall consider the daily maximum temperatures in the NOAA data set between May 1993 and September 1993 (inclusive). The packages we need are **CCA, dplyr, ggplot2, gstat, sp, spacetime, STRbook** and **tidyr**.

```
library("CCA")
library("dplyr")
library("ggplot2")
library("gstat")
library("sp")
library("spacetime")
library("STRbook")
```

```
library("tidyr")
```

In order to ensure consistency of results and visualizations, we fix the seed to 1.

```
set.seed(1)
```

We now load the NOAA data set using the **data** command. To keep the data size manageable, we take a subset of it corresponding to the maximum daily temperatures in the months May–September 1993. As in Lab 2.2 we also add a new variable `t` which starts at 1 at the beginning of the data set and increases by 1 each day.

```
data("NOAA_df_1990", package = "STRbook")
Tmax <- filter(NOAA_df_1990,       # subset the data
               proc == "Tmax" &    # only max temperature
               month %in% 5:9 &    # May to September
               year == 1993)       # year of 1993
Tmax$t <- Tmax$julian - 728049     # create a new time variable
```

Empirical Spatial Means

The empirical spatial mean of our data is given by (2.1). The empirical spatial mean is a spatial quantity that can be stored in a new data frame that contains the spatial locations and the respective average maximum temperature at each location. These, and other data manipulations to follow, can be carried out easily using the tools we learned in Lab 2.1. We group by longitude and latitude, and then we compute the average maximum temperature at each of the separate longitude–latitude coordinates.

```
spat_av <- group_by(Tmax, lat, lon) %>%   # group by lon-lat
           summarise(mu_emp = mean(z))     # mean for each lon-lat
```

We can now plot the average maximum temperature per station and see how this varies according to longitude and latitude. The following plots are shown in Figure 2.14.

```
lat_means <- ggplot(spat_av) +
             geom_point(aes(lat, mu_emp)) +
             xlab("Latitude (deg)") +
             ylab("Maximum temperature (degF)") + theme_bw()

lon_means <- ggplot(spat_av) +
             geom_point(aes(lon, mu_emp)) +
             xlab("Longitude (deg)") +
             ylab("Maximum temperature (degF)") + theme_bw()
```

Empirical Temporal Means

We now generate the plot of Figure 2.15. The empirical temporal mean can be computed easily using the tools we learned in Lab 2.1: first, group the data by time; and second, summarize using the **summarise** function.

```
Tmax_av <- group_by(Tmax, date) %>%
          summarise(meanTmax = mean(z))
```

The variable `Tmax_av` is a data frame containing the average maximum temperature on each day (averaged across all the stations). This can be visualized easily, together with the original raw data, using **ggplot2**.

```
gTmaxav <-
    ggplot() +
    geom_line(data = Tmax,aes(x = date, y = z, group = id),
              colour = "blue", alpha = 0.04) +
    geom_line(data = Tmax_av, aes(x = date, y = meanTmax)) +
    xlab("Month") + ylab("Maximum temperature (degF)") +
    theme_bw()
```

Empirical Covariances

Before obtaining the empirical covariances, it is important that all trends are removed (not just the intercept). One simple way to do this is to first fit a linear model (that has spatial and/or temporal covariates) to the data. Then plot the empirical covariances of the detrended data (i.e., the residuals). Linear-model fitting proceeds with use of the `lm` function in R. The residuals from `lm` can then be incorporated into the original data frame `Tmax`.

In the plots of Figure 2.9 we observed a quadratic tendency of temperature over the chosen time span. Therefore, in what follows, we consider time and time squared as covariates. Note the use of the function `I`. This is required for R to interpret the power sign "^" as an arithmetic operator instead of a formula operator.

```
lm1 <- lm(z ~ lat + t + I(t^2), data = Tmax) # fit a linear model
Tmax$residuals <- residuals(lm1)               # store the residuals
```

We also need to consider the spatial locations of the stations, which we extract from `Tmax` used above.

```
spat_df <- filter(Tmax, t == 1) %>% # lon/lat coords of stations
           select(lon, lat)  %>%   # select lon/lat only
           arrange(lon, lat)        # sort ascending by lon/lat
m <- nrow(spat_av)                  # number of stations
```

The most straightforward way to compute the empirical covariance matrix (2.4) is using the `cov` function in R. When there are missing data, the usual way forward is to drop all records that are not complete (provided there are not too many of these). Specifically, if any of the elements in $\mathbf{Z}_{t_j}$ or $\mathbf{Z}_{t_j - \tau}$ are missing, the associated term in the summation of (2.4) is ignored altogether. The function `cov` implements this when the argument `use = 'complete.obs'` is supplied. If there are too many records that are incomplete, imputation, or the consideration of only subsets of stations, might be required.

In order to compute the empirical covariance matrices, we first need to put the data into space-wide format using `spread`.

```
X <- select(Tmax, lon, lat, residuals, t) %>%  # select columns
        spread(t, residuals) %>%                # make time-wide
        select(-lon, -lat) %>%                  # drop coord info
        t()                                     # make space-wide
```

Now it is simply a matter of calling `cov(X, use = 'complete.obs')` for computing the lag-0 empirical covariance matrix. For the lag-1 empirical covariance matrix we compute the covariance between the residuals from X excluding the first time point and X excluding the last time point.

```
Lag0_cov <- cov(X, use = 'complete.obs')
Lag1_cov <- cov(X[-1, ], X[-nrow(X),], use = 'complete.obs')
```

In practice, it is very hard to gain any intuition from these matrices, since points in a two-dimensional space do not have any specific ordering. One can, for example, order the stations by longitude and then plot the permuted spatial covariance matrix, but this works best when the domain of interest is rectangular with a longitude span that is much larger than the latitude span. In our case, with a roughly square domain, a workaround is to split the domain into either latitudinal or longitudinal strips, and then plot the spatial covariance matrix associated with each strip. In the following, we split the domain into four longitudinal strips (similar code can be used to generate latitudinal strips).

```
spat_df$n <- 1:nrow(spat_df)        # assign an index to each station
lim_lon <- range(spat_df$lon)       # range of lon coordinates
lon_strips <- seq(lim_lon[1],       # create 4 long. strip boundaries
                  lim_lon[2],
                  length = 5)
spat_df$lon_strip <- cut(spat_df$lon,      # bin the lon into
                         lon_strips,       # their respective bins
                         labels = FALSE,   # don't assign labels
                         include.lowest = TRUE) # include edges
```

The first six records of `spat_df` are:

```
head(spat_df)    # print the first 6 records of spat_df
```

```
##          lon      lat n lon_strip
## 1 -99.96667 37.76667 1         1
## 2 -99.76667 36.30000 2         1
## 3 -99.68333 32.43333 3         1
## 4 -99.05000 35.00000 4         1
## 5 -98.81667 38.86666 5         1
## 6 -98.51667 33.98333 6         1
```

Now that we know in which strip each station falls, we can subset the station data frame by strip and then sort the subsetted data frame by latitude. In **STRbook** we provide a function **plot_cov_strips** that takes an empirical covariance matrix C and a data frame in the same format as spat_df, and then plots the covariance matrix associated with each longitudinal strip. Plotting requires the package **fields**. We can plot the resulting lag-0 and lag-1 covariance matrices using the following code.

```
plot_cov_strips(Lag0_cov, spat_df)    # plot the lag-0 matrices
plot_cov_strips(Lag1_cov, spat_df)    # plot the lag-1 matrices
```

As expected (see Figure 2.16), the empirical spatial covariance matrices reveal the presence of spatial correlation in the residuals. The four lag-0 plots seem to be qualitatively similar, suggesting that there is no strong dependence on longitude. However, there is a dependence on latitude, and the spatial covariance appears to decrease with decreasing latitude. This dependence is a type of spatial *non-stationarity*, and such plots can be used to assess whether non-stationary spatio-temporal models are required or not.

Similar code can be used to generate spatial correlation (instead of covariance) image plots.

Semivariogram Analysis

From now on, in order to simplify computations, we will use a subset of the data containing only observations in July. Computing the empirical semivariogram is much faster when using objects of class STFDF rather than STIDF since the regular space-time structure can be exploited. We hence take STObj3 computed in Lab 2.1 (load using **data**(STObj3)) and subset the month of July 1993 as follows.

```
data("STObj3", package = "STRbook")
STObj4 <- STObj3[, "1993-07-01::1993-07-31"]
```

For computing the sample semivariogram we use the function **variogram**.[7] We bin the distances between measurement locations into bins of size 60 km, and consider at most six time lags.

---

[7]Although the function is named "variogram," it is in fact the sample semivariogram that is computed.

```
vv <- variogram(object = z~1 + lat,  # fixed effect component
                data = STObj4,       # July data
                width = 80,          # spatial bin (80 km)
                cutoff = 1000,       # consider pts < 1000 km apart
                tlags = 0.01:6.01)   # 0 days to 6 days
```

The command **plot**(vv) produces Figure 2.17. The plot suggests that there are considerable spatio-temporal correlations in the data; spatio-temporal modeling of the residuals is thus warranted.

Empirical Orthogonal Functions

Empirical orthogonal functions (EOFs) can reveal spatial structure in the data and can also be used for subsequent dimensionality reduction. EOFs can be obtained from the data through either a spectral decomposition of the covariance matrix or a singular value decomposition (SVD) of the detrended space-time data matrix. The data matrix has to be in space-wide format (i.e., where space varies along the columns and time varies along the rows).

For this part of the Lab we use the SST data set. The SST data set does not contain any missing values, which renders our task slightly easier than when data are missing. When data are missing, one typically needs to consider interpolation, median polishing, or other imputation methods to fill in the missing values prior to computing the EOFs.

First we load the sea-land mask, the lon-lat coordinates of the SST grid, and the SST data set itself which is in time-wide format.

```
data("SSTlandmask", package = "STRbook")
data("SSTlonlat", package = "STRbook")
data("SSTdata", package = "STRbook")
```

Since SSTdata contains readings over land,[8] we delete these using SSTlandmask. Further, in order to consider whole years only, we take the first 396 months (33 years) of the data, containing SST values spanning 1970–2002.

```
delete_rows <- which(SSTlandmask == 1)
SSTdata <- SSTdata[-delete_rows, 1:396]
```

From (2.10) recall that prior to carrying out an SVD, we need to put the data set into space-wide format, mean-correct it, and then standardize it. Since SSTdata is in time-wide format, we first transpose it to make it space-wide.

---

[8] The land SST data are "pseudo-data," and just there to help analysts re-grid the SST data to different resolutions.

```
## Put data into space-wide form
Z <- t(SSTdata)
dim(Z)
```

```
## [1]  396 2261
```

Note that Z is of size 396×2261, and it is hence in space-wide format as required. Equation (2.10) is implemented as follows.

```
## First find the matrix we need to subtract:
spat_mean <- apply(SSTdata, 1, mean)
nT <- ncol(SSTdata)
```

```
## Then subtract and standardize:
Zspat_detrend <- Z - outer(rep(1, nT), spat_mean)
Zt <- 1/sqrt(nT - 1)*Zspat_detrend
```

Finally, to carry out the SVD we run

```
E <- svd(Zt)
```

The SVD returns a list E containing the matrices **V**, **U**, and the singular values diag(**D**). The matrix **V** contains the EOFs in space-wide format. We change the column names of this matrix, and append the lon-lat coordinates to it as follows.

```
V <- E$v
colnames(E$v) <- paste0("EOF", 1:ncol(SSTdata)) # label columns
EOFs <- cbind(SSTlonlat[-delete_rows, ], E$v)
head(EOFs[, 1:6])
```

```
##      lon lat          EOF1          EOF2         EOF3          EOF4
## 16  154 -29 -0.004915064 -0.012129566 -0.02882162  8.540892e-05
## 17  156 -29 -0.001412275 -0.002276177 -0.02552841  6.726077e-03
## 18  158 -29  0.000245909  0.002298082 -0.01933020  8.591251e-03
## 19  160 -29  0.001454972  0.002303585 -0.01905901  1.025538e-02
## 20  162 -29  0.002265778  0.001643138 -0.02251571  1.125295e-02
## 21  164 -29  0.003598762  0.003910823 -0.02311128  1.002285e-02
```

The matrix **U** returned from **svd** contains the principal component time series in wide-table format (i.e., each column corresponds to a time series associated with an EOF). Here we use the function **gather** in the package **tidyr** that reverses the operation **spread**. That is, the function takes a spatio-temporal data set in wide-table format and puts it into long-table format. We instruct the function to gather every column except the column denoting time, and we assign the key–value pair EOF-PC:

```
TS <- data.frame(E$u) %>%              # convert U to data frame
    mutate(t = 1:nrow(E$u)) %>%        # add a time field
    gather(EOF, PC, -t)                # put columns (except time)
                                       # into long-table format with
                                       # EOF-PC as key-value pair
```

Finally, the normalized time series are given by:

```
TS$nPC <- TS$PC * sqrt(nT-1)
```

We now can use the visualization tools discussed earlier to visualize the EOFs and the (normalized) principal component time series during July 2003. In Figures 2.20 and 2.21, we show the first three EOFs and the first three principal component time series. We can use the following code to illustrate the first EOF:

```
ggplot(EOFs) + geom_tile(aes(x = lon, y = lat, fill = EOF1)) +
    fill_scale(name = "degC") + theme_bw() +
    xlab("Longitude (deg)") + ylab("Latitude (deg)")
```

Plotting of other EOFs and principal component time series is left as an excercise to the reader. The EOFs reveal interesting spatial structure in the residuals. The second EOF is a west–east gradient, while the third EOF again reveals a temporally dependent north–south gradient. This north–south gradient has a lower effect in the initial part of the time series, and a higher effect towards the end.

EOFs can also be constructed by using `eof` in the package **spacetime**. With the latter, one must cast the data into an `STFDF` object using the function `stConstruct` before calling the function `eof`. The last example in the help file of `stConstruct` shows how one can do this from a space-wide matrix. The function `eof` uses `prcomp` (short for principal component analysis) to find the EOFs, which in turn uses `svd`.

Spatio-Temporal Canonical Correlation Analysis

We can carry out a canonical correlation analysis (CCA) using the package **CCA** in R. One cannot implement CCA on the raw data since $T < n$. Instead we carry out CCA on the SST projected onto EOF space, specifically the first 10 EOFs which explain just over 74% of the variance of the signal (you can show this from the singular values in the object E). In this example we consider the problem of long-lead prediction, and we check whether SST is a useful predictor for SST in 7 months' time. To this end, we split the data set into two parts, one containing SST and another containing SST lagged by 7 months.

```
nEOF <- 10
EOFset1 <- E$u[1:(nT-7), 1:nEOF] * sqrt(nT - 1)
EOFset2 <- E$u[8:nT, 1:nEOF] * sqrt(nT - 1)
```

The CCA is carried out by running the function `cancor`.

```
cc <- cancor(EOFset1, EOFset2)    # compute CCA
options(digits = 3)               # print to three d.p.
print(cc$cor[1:5])                # print
```

```
## [1] 0.843 0.758 0.649 0.584 0.463
```

```
print(cc$cor[6:10])
```

```
## [1] 0.4137 0.3067 0.2058 0.0700 0.0273
```

The returned quantity `cc$cor` provides the correlations between the canonical variates of the unshifted and shifted SSTs in EOF space. The correlations decrease, as expected, but the first two canonical variates are highly correlated. The time series of the first canonical variables can be found by multiplying the EOF weights with the computed coefficients as follows (see (2.13) and (2.14)).

```
CCA_df <- data.frame(t = 1:(nT - 7),
                     CCAvar1 = (EOFset1 %*% cc$xcoef[,1])[,1],
                     CCAvar2 = (EOFset2 %*% cc$ycoef[,1])[,1])
```

A plot can be made using standard **ggplot2** commands.

```
t_breaks <- seq(1, nT, by = 60)      # breaks for x-labels
year_breaks <- seq(1970,2002,by=5)   # labels for x-axis
g <- ggplot(CCA_df) +
    geom_line(aes(t, CCAvar1), col = "dark blue") +
    geom_line(aes(t, CCAvar2), col = "dark red") +
    scale_x_continuous(breaks = t_breaks, labels = year_breaks) +
    ylab("CCA variables") + xlab("Year") + theme_bw()
```

The plot of the time series of the first canonical variables is shown in Figure 2.23. The plot shows a high correlation between the first pair of canonical variables. What are these canonical variables? They are simply a linear combination of the EOFs, where the linear weights are given in `cc$xcoef[,1]` and `cc$ycoef[,1]`, respectively.

```
EOFs_CCA <- EOFs[,1:4] # first two columns are lon-lat
EOFs_CCA[,3] <- c(as.matrix(EOFs[,3:12]) %*% cc$xcoef[,1])
EOFs_CCA[,4] <- c(as.matrix(EOFs[,3:12]) %*% cc$ycoef[,1])
```

Plotting of the weights as spatial maps is straightforward and left as an exercise. We plot weights (recall these are just linear combination of EOFs) for the lagged SSTs and the unlagged SSTs in Figure 2.24.

Chapter 3

Spatio-Temporal Statistical Models

As you read this chapter and the next two, remind yourself that what you see in data may be different than what you might expect to see. Your view might be obstructed and/or not in sharp focus. Spatial predictive models can fill in the gaps and clear up your vision, but what you see in the data is still a "guess" at what is really there. We use statistical prediction methods that quantify these guesses with their associated prediction variances. Now, up the ante – include time as well and try to forecast the future ... even in places where there are no current or past data! We show how this is possible in the pages that follow.

Spatio-temporal prediction based on spatio-temporal statistical modeling is a central theme of this book. Importantly, our type of prediction comes with prediction variances that quantify the uncertainty in the prediction. Predicting the future is notoriously hard, but at least the spatio-temporal prediction variances can quantify how hard it is – if you use the "right" model! In this spatio-temporal setting, what if your goal is not to predict new values but to study the impact of covariates on a response? As we shall see, the same statistical models that are useful for prediction also allow us to infer important relationships between covariates and responses.

We see *three principal goals for spatio-temporal statistical modeling*:

1. predicting a plausible value of a response variable at some location in space within the time span of the observations and reporting the uncertainty of that prediction;

2. performing scientific inference about the importance of covariates on the response variable in the presence of spatio-temporal dependence; and

3. forecasting the future value of the response variable at some location, along with the uncertainty of that forecast.

It is important to note that our observations associated with each of these goals will always include measurement error and will often be incomplete, in the sense that there are some locations in space and time that have missing observations. When modeling to accomplish any of the goals above, we have to be able to take into account these data issues, and also that our model is almost surely "wrong." As the famous aphorism by George Box goes, "all models are wrong but some are useful" (Box, 1976, 1979). Our task is to maximize the "usefulness" and to minimize the "wrongness."

The primary purpose of this chapter is to present an example illustrating each of the three goals given above, along with a potential modeling solution that initially does *not* account for a spatio-temporal error process. This will allow us to illustrate some of the benefits and shortcomings of standard approaches and show why it is often better to consider statistical models that *do* account for spatio-temporal dependent errors (see Chapters 4 and 5). This will also give you a chance to use some of the visualization and exploratory techniques you learned in Chapter 2, and the R Labs at the end of this chapter will further develop your R programming and analysis skills for spatio-temporal data, in preparation for later chapters.

3.1   Spatio-Temporal Prediction

To start with, consider the prediction (i.e., "interpolation") of maximum daily temperatures on 15 July 1993 at the location denoted by the triangle in the top panel of Figure 3.1, given observations on the same variable on the same date at 138 measurement locations in the central USA (NOAA data set). We seek a predictor, and it is easy to imagine visually how we might construct one – we somehow just combine the nearest observations. Indeed, as mentioned in Section 1.2.2, Tobler's "law" suggests that we should give more weight to nearby observations when we interpolate. But, why stop with just space? We also have other observations at different time points, so we should consider nearby observations in both space *and* time, as shown in the bottom panel of Figure 3.1. We have already shown in Chapter 2 (e.g., Figure 2.9) that there is strong spatio-temporal dependence in these data. Since we have observations at times before and after 15 July 1993, this application is an example of *smoothing* – that is, we seek a smoothing predictor. If we only had observations up to 15 July, then we would seek a *filtering* predictor for the entire temperature field on 15 July 1993, and a *forecasting* predictor for the entire field at any time after 15 July 1993. Discussion of the distinction between the three types of spatio-temporal predictor is given in Section 1.3.

Deterministic Prediction

Perhaps the simplest way to perform spatio-temporal prediction would be to follow Tobler's law and simply average the data in such a way as to give more weight to the nearest observations in space and time. The most obvious way to do this is through *inverse distance weighting* (IDW). Suppose we have spatio-temporal data given by

$$\{Z(\mathbf{s}_{11}; t_1), Z(\mathbf{s}_{21}; t_1), \ldots, Z(\mathbf{s}_{m_1 1}; t_1), \ldots, Z(\mathbf{s}_{1T}; t_T), Z(\mathbf{s}_{2T}; t_T), \ldots, Z(\mathbf{s}_{m_T T}; t_T)\},$$

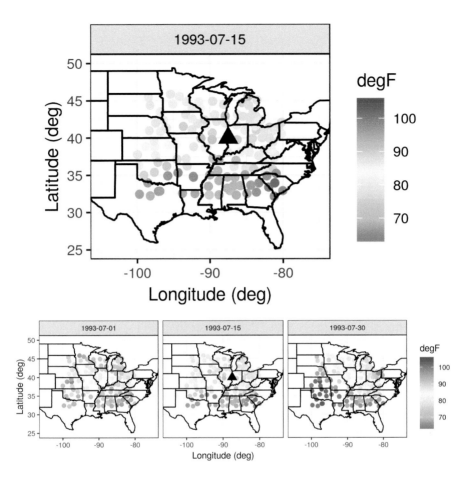

Figure 3.1: Top: NOAA maximum daily temperature observations for 15 July 1993 (degrees Fahrenheit). Bottom: NOAA maximum daily temperature observations for 01, 15, and 30 July 1993 (degrees Fahrenheit). The triangle corresponds to a spatial location and time point $\{s_0; t_0\}$ for which we would like to obtain a prediction of the maximum daily temperatures.

where for each time t_j we have m_j observations. Then the IDW predictor at some location s_0 and time t_0 (where, in this smoothing-predictor case, we assume that $t_1 \leq t_0 \leq t_T$) is given by

$$\widehat{Z}(s_0; t_0) = \sum_{j=1}^{T} \sum_{i=1}^{m_j} w_{ij}(s_0; t_0) Z(s_{ij}; t_j), \qquad (3.1)$$

where

$$w_{ij}(\mathbf{s}_0; t_0) \equiv \frac{\widetilde{w}_{ij}(\mathbf{s}_0; t_0)}{\sum_{k=1}^{T} \sum_{\ell=1}^{m_k} \widetilde{w}_{\ell k}(\mathbf{s}_0; t_0)}, \tag{3.2}$$

$$\widetilde{w}_{ij}(\mathbf{s}_0; t_0) \equiv \frac{1}{d((\mathbf{s}_{ij}; t_j), (\mathbf{s}_0; t_0))^{\alpha}}, \tag{3.3}$$

$d((\mathbf{s}_{ij}; t_i), (\mathbf{s}_0; t_0))$ is the "distance" between the spatio-temporal location $(\mathbf{s}_{ij}; t_j)$ and the prediction location $(\mathbf{s}_0; t_0)$, and the power coefficient α is a positive real number that controls the amount of smoothing (e.g., often $\alpha = 2$, but it does not have to be). The notation makes this look more complicated than it actually is: IDW is simply a weighted average of the data points, giving the closest locations more weight (while requiring that the weights sum to 1). You are free to choose your preferred distance $d(\cdot, \cdot)$; a simple one is the Euclidean distance (although this implicitly treats space and time in the same way, which may not be appropriate; see Section 4.2.3). Note that if we were interested in predicting at a different spatio-temporal location, we would necessarily get different weights, but in a way that respects Tobler's first law of geography. Also note that some practitioners require an "exact interpolator" in the sense that if the prediction location $(\mathbf{s}_0; t_0)$ corresponds to a data location, they want the prediction to be exactly the same as the data value (so, not a smoothed estimate there). The formula in (3.1) gives an exact interpolator. Thus, $\widehat{Z}(\mathbf{s}_0; t_0) = Z(\mathbf{s}_{k\ell}; t_\ell)$ if a data location $(\mathbf{s}_{k\ell}; t_\ell)$ corresponds to the prediction location $(\mathbf{s}_0; t_0)$ (since $\alpha > 0$, $(\mathbf{s}_0; t_0)$ being a data location implies that the right-hand side of (3.3) is infinite, so it gets a weight of 1 in (3.2)). As discussed in Cressie (1993, p. 379), exact interpolators can be problematic when one has measurement uncertainty, and one way to obtain a smoothing predictor is to use weights in (3.3) proportional to $1/(d(\cdot, \cdot) + c)^{\alpha}$, where $c > 0$. (Setting $c = 0$ reverts to the exact interpolator.)

> **R tip:** Computing distances between a single set of coordinates can be done in base R using the function **dist**. To compute distances between two sets of coordinates, it is more convenient to use the function **rdist** in the package **fields**, or the function **spDists** in the package **sp**, both of which take two sets of coordinates as arguments. The latter also works with Spatial objects defined in the package **sp**.

The left panel in Figure 3.2 shows predictions of maximum temperature for six days within the month of July 1993 using 30 days of July 1993 data, where data from 14 July 1993 was omitted. These predictions were obtained using IDW with $\alpha = 5$. In this example, setting α to a smaller value (such as 2) gives a smoother surface since more weight is given to observations that are "far" from the prediction locations. In deterministic interpolators, smoothing parameters such as α are usually chosen using a procedure known as cross-validation (see Technical Note 3.1 and the left panel in Figure 3.3). From the

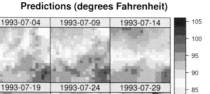

Figure 3.2: Predictions of `Tmax` in degrees Fahrenheit for the maximum temperature in the NOAA data set within a square box enclosing the domain of interest for six days (each five days apart) spanning the temporal window of the data, 01 July 1993 to 30 July 1993, using (left) inverse distance weighting functionality from the R package **gstat** with inverse distance power $\alpha = 5$ and (right) a Gaussian radial basis kernel with bandwidth $\theta = 0.5$. Data for 14 July 1993 were omitted from the original data set.

IDW prediction in Figure 3.2, we observe that our predictions on the day with no data look smoother than those on days for which we have data. We shall see in Chapter 4 that this is typical of most predictors, including stochastic ones that are optimal in the sense of minimizing the mean squared prediction error (MSPE).

In general, IDW is a type of spatio-temporal *kernel predictor*. That is, in (3.3) we can let

$$\widetilde{w}_{ij}(\mathbf{s}_0; t_0) = k((\mathbf{s}_{ij}; t_j), (\mathbf{s}_0; t_0); \theta),$$

where $k((\mathbf{s}_{ij}; t_j), (\mathbf{s}_0; t_0); \theta)$ is a *kernel function* (i.e., a function that quantifies the similarity between two locations) that depends on the distance between $(\mathbf{s}_{ij}; t_j)$ and $(\mathbf{s}_0; t_0)$ and some *bandwidth* parameter, θ. Specifically, the bandwidth controls the "width" of the kernel, so a larger bandwidth averages more observations (and produces smoother prediction fields) than a narrow bandwidth. A classic example of a kernel function is the *Gaussian radial basis kernel*

$$k((\mathbf{s}_{ij}; t_j), (\mathbf{s}_0; t_0); \theta) \equiv \exp\left(-\frac{1}{\theta} d((\mathbf{s}_{ij}; t_j), (\mathbf{s}_0; t_0))^2\right), \tag{3.4}$$

where the bandwidth parameter θ is proportional to the variance parameter in a normal (Gaussian) distribution. Many other kernels exist (e.g., tricube, bisquare, Epanechnikov), some of which have *compact support* (i.e., provide zero weight beyond a certain distance threshold). If we write $d(\cdot, \cdot)^\alpha = \exp(\alpha \log d(\cdot, \cdot))$, it is clear that α in IDW plays the role of the bandwidth parameter and IDW has non-compact support. The right panel of Figure 3.2 shows an interpolation of the NOAA temperature data using a Gaussian radial

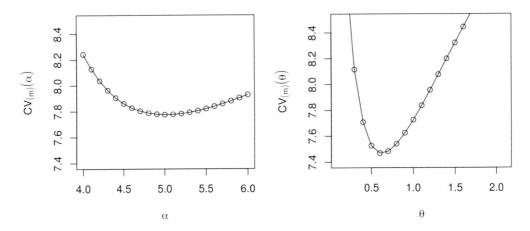

Figure 3.3: The leave-one-out cross-validation score $CV_{(m)}$ (see Technical Note 3.1) for different values of α and θ when doing IDW prediction (left) and Gaussian kernel prediction (right) of maximum temperature in the NOAA data set in July 1993.

basis kernel with $\theta = 0.5$. As in IDW, θ is usually chosen by cross-validation (see the right panel in Figure 3.3).

Traditional implementations of deterministic methods do not explicitly account for measurement uncertainty in the data nor do they provide model-based estimates of the prediction uncertainty. One might argue that, for non-exact interpolators, one is implicitly removing (filtering or smoothing) the observation error with the averaging that takes place as part of the interpolation. However, there is no mechanism to incorporate explicit knowledge of the magnitude of the measurement error. Regarding prediction uncertainty of deterministic predictors, we can get estimates of the overall quality of predictions by doing cross-validation (see Technical Note 3.1). Recall that we have also suggested using cross-validation to select the degree of smoothing (e.g., the α parameter in IDW and, more generally, the θ parameter in the kernel-based prediction). As an example, in Figure 3.3 we show the leave-one-out cross-validation (LOOCV) MSPE score for different values of α and θ (lower cross-validation scores are better) when doing IDW and Gaussian kernel smoothing for the NOAA maximum temperature data set in July 1993. These cross-validation analyses suggest that $\alpha = 5$ and $\theta = 0.6$ are likely to give the best out-of-sample predictions for this specific example. In addition, note that the lowest cross-validation score for the Gaussian kernel smoother is lower (i.e., better) than the lowest cross-validation score for IDW. This suggests that the Gaussian kernel smoother is likely to be a better predictor than the IDW smoother for these data.

Cross-validation can also be used to compare models through their predictions, as the following Technical Note 3.1 explains.

Technical Note 3.1: Cross-Validation

Cross-validation seeks to evaluate model predictions by splitting up the data into a training sample and a validation sample, then fitting the model with the training sample and evaluating it with the validation sample. In *K-fold cross-validation* we randomly split the available data into K roughly equal-size components (or "folds"). Each fold is held out, the model is trained on the remaining $K - 1$ folds, and then the model is evaluated on the fold that was held out. Specifically, for $k = 1, \ldots, K$ folds, fit the model with the kth fold removed, and obtain predictions $\widehat{Z}_i^{(-k)}$ for $i = 1, \ldots, m_k$, where m_k is the number of data in the kth fold. We then select a metric by which we evaluate the predictions relative to the held-out samples. For example, if we were interested in the mean squared prediction error (MSPE), we would compute $MSPE_k = (1/m_k) \sum_{i=1}^{m_k} (Z_i - \widehat{Z}_i^{(-k)})^2$ for the m_k observations in the kth fold, $k = 1, \ldots, K$. The K-fold cross-validation score is then

$$CV_{(K)} = \frac{1}{K} \sum_{k=1}^{K} MSPE_k.$$

It has been shown empirically that good choices for the number of folds are $K = 5$ and $K = 10$.

A special case of K-fold cross-validation occurs when $K = m$. This is called *leave-one-out cross-validation* (LOOCV). In this case, only a single observation is used for validation and the remaining observations are used to make up the training set. This is repeated for all m observations. The LOOCV score is then

$$CV_{(m)} = \frac{1}{m} \sum_{i=1}^{m} MSPE_i.$$

LOOCV typically has low bias as an estimate of the expected squared error of a test sample, but it can also have high variance. This is why the choice of $K = 5$ or $K = 10$ often provides a better compromise between bias and variance. It is also the case that LOOCV can be computationally expensive to implement in general, since it requires the model to be fitted m times (although there are notable exceptions such as with the predicted residual error sum of squares (PRESS) statistic in multiple linear regression models; see Appendix B). For more details on cross-validation, see Hastie et al. (2009, Section 7.10).

R tip: K-fold cross-validation is an "embarrassingly parallel" problem since all the K validations can be done simultaneously. There are several packages in R that enable this, with **parallel** and **foreach** among the most popular. The vignettes in these packages contain more information on how to use them for multicore computing.

3.2 Regression (Trend-Surface) Estimation

In Section 3.1 we presented some simple deterministic predictors to obtain predictions at spatio-temporal locations given a spatio-temporal data set. We can also use a basic statistical regression model to obtain predictions for such data, assuming that all of the spatio-temporal dependence can be accounted for by "trend" (i.e., covariate) terms. Such a model has the advantage of being exceptionally simple to implement in almost any software package. In addition, a regression model explicitly accounts for model error (usually assumed independent), and it also allows us to obtain a model-based prediction-error variance, although cross-validation scores still provide useful insight into model performance.

Consider a regression model that attempts to account for spatial and temporal trends. To make the notation a bit simpler, we consider the case where we have observations at discrete times $\{t_j : j = 1, \ldots, T\}$ for all spatial data locations $\{\mathbf{s}_i : i = 1, \ldots, m\}$. For example,

$$Z(\mathbf{s}_i; t_j) = \beta_0 + \beta_1 X_1(\mathbf{s}_i; t_j) + \ldots + \beta_p X_p(\mathbf{s}_i; t_j) + e(\mathbf{s}_i; t_j), \qquad (3.5)$$

where β_0 is the intercept and β_k $(k > 0)$ is a regression coefficient associated with $X_k(\mathbf{s}_i; t_j)$, the kth covariate at spatial location $\mathbf{s}_i$ and time t_j. We also assume for the moment *iid* errors such that $e(\mathbf{s}_i; t_j) \sim \text{ indep. } N(0, \sigma_e^2)$ for all $\{\mathbf{s}_i; t_j\}$ where there are data, and note that $N(\mu, \sigma^2)$ corresponds to a normal distribution with mean μ and variance σ^2. The covariates $X_k(\mathbf{s}_i; t_j)$ may describe explanatory features, such as elevation, that vary spatially but are temporally invariant (on the scales of interest here), time trends (such as an overall seasonal effect) that are spatially invariant but temporally varying, or other variables such as humidity, that are both spatially and temporally varying. We might also consider spatio-temporal "basis functions" that can be used to reconstruct the observed data.

We take a little space here to discuss *basis functions* beyond the brief explanation given in Chapter 1. What are basis functions? Imagine that we have a complex curve or surface in space. We are often able to decompose this curve or surface as a linear combination of some "elemental" basis functions. For example,

$$Y(\mathbf{s}) = \alpha_1 \phi_1(\mathbf{s}) + \alpha_2 \phi_2(\mathbf{s}) + \ldots + \alpha_r \phi_r(\mathbf{s}), \qquad (3.6)$$

where $\{\alpha_i\}$ are constants and $\{\phi_i(\mathbf{s})\}$ are known *basis functions*. We can think of the coefficients $\{\alpha_i\}$ as weights that describe how important each basis function is in representing the function $Y(\mathbf{s})$. The basis functions can be *local* with compact support, or can

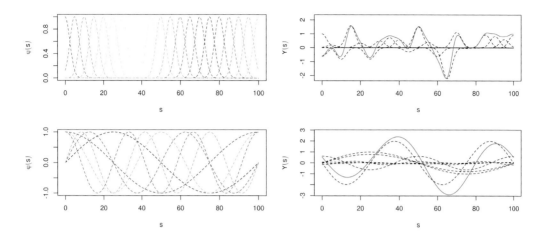

Figure 3.4: Left: Local (top) and global (bottom) basis functions over a one-dimensional spatial domain. Different colors are used to denote different basis functions. Right: Linear combination (red curve) of the individual basis functions (dashed lines depicted in the left panels). In this case, the coefficients $\{\alpha_i\}$ give the relative importance of the basis functions (curves).

be *global*, taking values across the whole domain (see Figure 3.4). In statistics, when $Y(\mathbf{s})$ is a random process, we typically assume the basis functions are known and the coefficients (weights) are random. The expression in (3.6) could be written as a function of time t, or most generally as a function of $\mathbf{s}$ and t. In time series, the domain over which the basis functions take their values is the one-dimensional real line, whereas in spatial statistics, the domain is typically one-dimensional space (see Figure 3.4) or two-dimensional space (see Figure 3.5); in spatio-temporal statistics, the domain is over both space and time. Examples of basis functions include polynomials, splines, wavelets, sines and cosines, among many others. We often construct spatio-temporal basis functions via a tensor product of spatial basis functions and temporal basis functions (see Technical Note 4.1).

Now consider the maximum daily temperature `Tmax` in the NOAA data set for the month of July 1993, where we have observations at $m = 138$ common spatial locations $\{\mathbf{s}_i : i = 1, \ldots, m\}$ for $\{t_j : j = 1, \ldots, T = 31\}$ days. In this case, we could account for spatial trends by allowing the covariates $\{X_k\}$ to correspond to the spatio-temporal coordinate, and/or their transformations and interactions. For example, let $\mathbf{s}_i \equiv (s_{1,i}, s_{2,i})'$, and consider a linear model with the following basis functions:

- overall mean: $X_0(\mathbf{s}_i; t_j) = 1$, for all $\mathbf{s}_i$ and t_j;
- linear in *lon*-coordinate: $X_1(\mathbf{s}_i; t_j) = s_{1,i}$, for all t_j,
- linear in *lat*-coordinate: $X_2(\mathbf{s}_i; t_j) = s_{2,i}$, for all t_j;
- linear time (day) trend: $X_3(\mathbf{s}_i; t_j) = t_j$, for all $\mathbf{s}_i$;

Spatial Basis Functions   **Coefficients**   **Process**

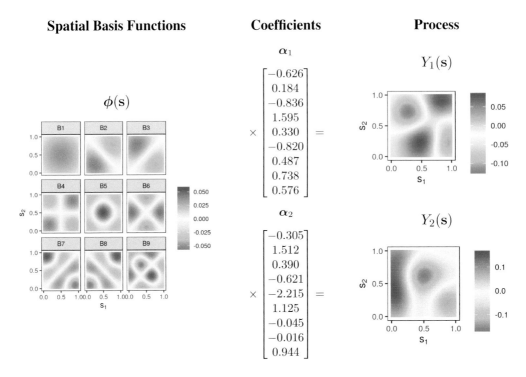

Figure 3.5: Two-dimensional spatial basis functions and associated coefficients $\boldsymbol{\alpha}_1$ and $\boldsymbol{\alpha}_2$ that lead to two different spatial-process realizations, $Y_1(\mathbf{s})$ and $Y_2(\mathbf{s})$, respectively.

- *lon–lat* interaction: $X_4(\mathbf{s}_i; t_j) = s_{1,i}\, s_{2,i}$, for all t_j;
- *lon–t* interaction: $X_5(\mathbf{s}_i; t_j) = s_{1,i}\, t_j$, for all $s_{2,i}$;
- *lat–t* interaction: $X_6(\mathbf{s}_i; t_j) = s_{2,i}\, t_j$, for all $s_{1,i}$;
- additional spatial-only basis functions: $X_k(\mathbf{s}_i; t_j) = \phi_{k-6}(\mathbf{s}_i), k = 7, \ldots, 18$, for all t_j (see Figure 3.6).

Note that the space and time coordinates used in $X_0, \ldots, X_6$ can be thought of as basis functions; we choose the separate notation between these latitude, longitude, and time trend covariates and the spatial-only basis functions (denoted $\{\phi_k : k = 1, \ldots, 12\}$) given in Figure 3.6 for the sake of interpretability. In this example, there is an intercept and $p = 18$ regression coefficients.

The regression model given in (3.5) can be fitted via *ordinary least squares* (OLS), in which case we find estimates of the parameters $\beta_0, \beta_1, \ldots, \beta_p$ that minimize the residual sum of squares,

$$RSS = \sum_{j=1}^{T} \sum_{i=1}^{m} (Z(\mathbf{s}_i; t_j) - \widehat{Z}(\mathbf{s}_i; t_j))^2. \tag{3.7}$$

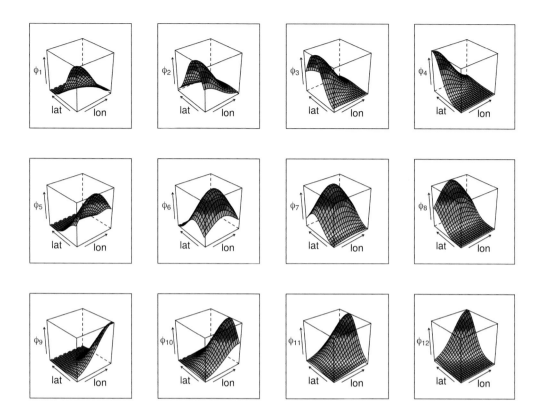

Figure 3.6: The time-invariant basis functions, $\phi_1(\mathbf{s}), \ldots, \phi_{12}(\mathbf{s})$, used for regression prediction of maximum temperature data from the NOAA data set for July 1993.

We denote these estimates by $\{\widehat{\beta}_0, \widehat{\beta}_1, \ldots, \widehat{\beta}_p\}$ and we write $\widehat{Z}(\mathbf{s}; t) = \widehat{\beta}_0 + \widehat{\beta}_1 X_1(\mathbf{s}; t) + \ldots + \widehat{\beta}_p X_p(\mathbf{s}; t)$. (We also obtain an estimate of the variance parameter, namely $\widehat{\sigma}_e^2 = RSS/(mT - p - 1)$.) This then allows us to get predictions for a mean response, or a new response, $Z(\mathbf{s}_0; t_0)$, at any location $\{\mathbf{s}_0; t_0\}$ for which we have covariates. We can also obtain uncertainty estimates for these predictions. The formulas for these estimates and predictors are most easily seen from a matrix representation, as shown in Technical Note 3.3. Figure 3.7 shows the predictions and the prediction standard errors (assuming the regression model with an intercept and $p = 18$) for the maximum temperature data in the NOAA data set in July 1993, with 14 July 1993 omitted when fitting the model. The predictions are much smoother than those found using kernel smoothing (Figure 3.2), a direct result of using basis functions that are spatio-temporally smooth. This is not always the case, and using covariates that are highly spatially varying (e.g., from topography) will yield predictions that also vary substantially with space. Note also from Figure 3.7 that the

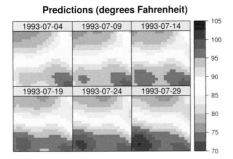
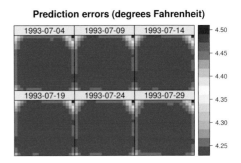

Figure 3.7: Regression predictions (left) and associated prediction standard errors (right) of maximum temperature (in degrees Fahrenheit) within a square box enclosing the domain of interest for six individual days (each 5 days apart) in July 1993 using the R function `lm`. Data for 14 July 1993 were purposely omitted from the original data set during fitting.

prediction standard errors do not show much structure because the Xs are accounting for most of the spatio-temporal variation in the data. Uncertainty increases at the domain edges where prediction becomes extrapolation.

It is important to mention here that the regression model given in equation (3.5) does not explicitly account for measurement errors in the responses, and thus that variation due to measurement error is confounded with the variation due to lack of fit in the residual variance σ_e^2. We account explicitly for this measurement-error variation (and small-scale spatio-temporal variation) in Chapters 4 and 5. In addition, note that the regression predictor can be considered a type of kernel predictor (see Appendix B).

> **R tip:** Basis functions such as those depicted in Figure 3.6 can be easily constructed using the package **FRK**, which we explore further in Chapter 4. Basis functions can be constructed at multiple resolutions, can be spatial-only (as used here) or also spatio-temporal. See Lab 3.2 for more details.

3.2.1 Model Diagnostics: Dependent Errors

When we first learn how to do regression modeling in statistics, we gain an appreciation for the importance of model diagnostics to verify the assumptions of the model. For example, we look for the presence of outliers, influential observations, non-constant error variance, non-normality, dependence in the errors, and so forth (see, for example, Kutner et al., 2004). It is particularly important to consider the possibility of dependent errors in the case where

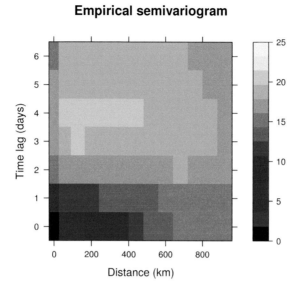

Figure 3.8: Empirical spatio-temporal semivariogram of the residuals after fitting a linear model to daily maximum temperatures in the NOAA data set during July 2003, computed using the function **variogram** in **gstat**.

the data are indexed in space or time (see Chapter 6 for more detailed discussion about model evaluation). From an exploratory perspective, one can calculate the spatio-temporal covariogram (or semivariogram), discussed in Chapter 2, from the residuals, $\widehat{e}(\mathbf{s}_i; t_j) \equiv Z(\mathbf{s}_i; t_j) - \widehat{Z}(\mathbf{s}_i; t_j)$, and look for dependence structure as a function of spatial and temporal lags. As seen in Figure 3.8, there is ample spatial and temporal structure in the residuals. It is instructive to compare Figure 3.8 with the empirical semivariogram calculated from the original data set and given in Figure 2.17. The former has a lower sill, and therefore the basis functions and the other covariates have been able to explain some of the spatio-temporal variability in the data, but clearly not all of it.

More formally, one can apply a statistical test for *temporal* dependence such as the Durbin–Watson test (see Technical Note 3.2), and if the data correspond to areal regions in two-dimensional space, one can use a test for *spatial* dependence such as Moran's *I* test (see Technical Note 3.2). In looking at *spatio-temporal* dependence, we can consider the "space-time index" (STI) approach of Henebry (1995), which is a type of Moran's *I* statistic for spatio-temporal data (see Cressie and Wikle, 2011, p. 303). This approach was developed for areal regions that have a known adjacency structure. In principle, this can be extended to the case of spatio-temporal data with continuous spatial support; see Lab 3.2.

Alternatively, we can consider a spatio-temporal analog to the Durbin–Watson test. Cressie and Wikle (2011, p.131) give a statistic based on the empirical (spatial) semivariogram that can be extended to the spatio-temporal setting. In particular, let

$$F \equiv \left| \frac{\widehat{\gamma}_e(||\mathbf{h}_1||; \tau_1)}{\widehat{\sigma}_e^2} - 1 \right|,$$

where $\widehat{\gamma}_e(||\mathbf{h}_1||; \tau_1)$ is the empirical semivariogram estimate at the smallest possible spatial ($||\mathbf{h}_1||$) and temporal (τ_1) lags (see Technical Note 2.1), and $\widehat{\sigma}_e^2$ is the regression-error-variance estimate (see Technical Note 3.3). If this value of F is "large," we reject the null hypothesis of spatio-temporal independence. We can evaluate what is "large" in this case by doing a permutation test of the null hypothesis of independence, which does not depend on any distributional assumptions on the test statistic, F. In this case, the data locations (in space and time) are randomly permuted and F is calculated for many such permutation samples. If the statistic F calculated with the observed data is below the 2.5th percentile or above the 97.5th percentile of these permutation samples, then we reject the null hypothesis of spatio-temporal independence (at the 5% level of significance), which suggests that the data are dependent.

Technical Note 3.2: Durbin–Watson and Moran's I Tests

One of the most used tests for serial dependence in time-series residuals is the *Durbin–Watson* test (e.g., Kutner et al., 2004, p. 487). Let $\widehat{e}_t = Z_t - \widehat{Z}_t$ be the residual from some fitted time-series model for which we have T observations $\{Z_t\}$. The Durbin–Watson test statistic is given by

$$d = \frac{\sum_{t=2}^{T}(\widehat{e}_t - \widehat{e}_{t-1})^2}{\sum_{t=1}^{T} \widehat{e}_t^2}.$$

The intuition for this test is that if residuals are highly (positively) correlated, then $\widehat{e}_t - \widehat{e}_{t-1}$ is small relative to $\widehat{e}_t$ and so, as d gets closer to 0, there is more evidence of positive serial dependence (e.g., a "rule of thumb" suggests that values less than 1 indicate strong positive serial dependence). In contrast, as the value of d gets larger (it is bounded above by 4), it is indicative of no positive serial dependence. This test can be formalized with appropriate upper and lower critical values for d, and statistical software packages can easily calculate these, as well as the analogous test for negative serial dependence.

One of the most commonly used tests for spatial dependence for spatial lattice data is *Moran's I* test (e.g., Waller and Gotway, 2004, Section 7.4). This test can be applied to the data directly, or to the residuals from some spatial regression model. Let $\{Z_i : i = 1, \ldots, m\}$ represent spatially referenced data (or residuals) for m spatial locations. Then, Moran's I statistic is calculated as

$$I = \frac{m \sum_{i=1}^{m} \sum_{j=1}^{m} w_{ij}(Z_i - \bar{Z})(Z_j - \bar{Z})}{(\sum_{i=1}^{m} \sum_{j=1}^{m} w_{ij})(\sum_{i=1}^{m}(Z_i - \bar{Z})^2)}, \tag{3.8}$$

where $\bar{Z} = (1/m) \sum_{i=1}^{m} Z_i$ is the spatial mean and w_{ij} are spatial adjacency "weights" between locations i and j (where we require $w_{ii} = 0$, for all $i = 1, \ldots, m$). Thus, Moran's I statistic is simply a weighted form of the usual Pearson correlation coefficient, where the weights are the spatial proximity weights, and it takes values between -1 and 1. If (3.8) is positive, then neighboring regions tend to have similar values, and if it is negative, then neighboring regions tend to have different values. Appropriate critical values or p-values are easily obtained in many software packages.

Note that there are additional measures of temporal dependence (e.g., the Ljung–Box test; see Shumway and Stoffer, 2006, p.129) and spatial dependence (e.g., the Geary C test; see Waller and Gotway, 2004, Section 7.4).

It is very common, when studying environmental phenomena, that a linear model of some covariates will not explain all the observed spatio-temporal variability. Thus, fitting such a model will frequently result in residuals that are spatially and temporally correlated. This is not surprising, since several environmental processes are certainly more complex than could be described by simple geographical and temporal trend terms. In Figure 3.9 we show the time series of the residuals at two observation locations $(81.38°\text{W}, 35.73°\text{N})$ and $(83.32°\text{W}, 37.60°\text{N})$, respectively, and the spatial residuals between 24 July and 31 July 1993. The residual time series exhibit considerable *temporal* correlation (i.e., residuals close together in time tend to be more similar than residuals far apart in time), and the spatial residuals exhibit clear *spatial* correlation (i.e., residuals close together in space tend to be more similar than residuals far apart in space). In Lab 3.2 we go further and use the Durbin–Watson and Moran's I tests to reject the null hypotheses of no temporal or spatial correlation in these residuals.

Given that our diagnostics have suggested there is spatio-temporal dependence in the errors after fitting the trend surface, what can we do? Readers who are familiar with more complicated regression procedures might suggest that we could use a generalized least squares (GLS) procedure that explicitly accounts for the dependence in the errors. That is, GLS relaxes the assumption of independence in the errors, so that $e(\mathbf{s}_i; t_j)$ and $e(\mathbf{s}_\ell; t_k)$ could be correlated. In this case, the vector of errors, $\mathbf{e} \equiv (e(\mathbf{s}_1; t_1), \ldots, e(\mathbf{s}_m; t_T))'$, has the multivariate normal distribution $\mathbf{e} \sim N(\mathbf{0}, \mathbf{C}_e)$, where $\mathbf{C}_e$ is a spatio-temporal covariance matrix. But do we know in advance what this covariance matrix is? Typically, no – and it is further complicated by the fact that to predict at spatio-temporal locations for which we do not have data, we need to know what the error dependence is between *any* two locations in time and space within our prediction domain, not just those for which we have observations! This aspect of spatio-temporal prediction will be a primary focus of Chapter 4.

One might ask, what is the problem with ignoring the dependence in the errors when doing OLS regression? The answer depends somewhat on the goal. It is fairly easy to show

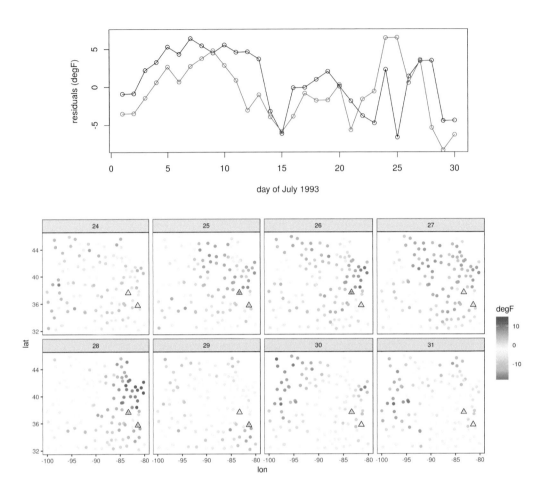

Figure 3.9: (Top) Temporal residuals at station 3810 (black line) and station 3889 (red line), and (bottom) spatial residuals between 24 and 31 July 1993, inclusive, when fitting the regression (trend) model described in Section 3.2 to the maximum temperature data in the NOAA data set in July 1993 (but recall that in the fitting we excluded data from 14 July 1993). The triangles denote the two station locations.

that the OLS parameter estimates and predictions are still unbiased even if one has ignored the dependence in the errors. But ignoring the dependence tends to give inappropriate standard errors and prediction standard errors. In the case of positive dependence (which is the most common case in spatio-temporal data – recall Tobler's law), the standard errors and prediction standard errors are underestimated if one ignores dependence, giving a false sense of how good the estimates and predictions really are. This issue comes up again in Section 3.2.2.

Technical Note 3.3: Ordinary Least Squares Regression: Matrix Representation

Consider an m-dimensional response vector, $\mathbf{Z} = (Z_1, \ldots, Z_m)'$, and an $m \times (p+1)$ matrix of predictors, $\mathbf{X}$, where we assume that the first column of this matrix contains a vector of 1s for the model intercept. That is,

$$\mathbf{X} = \begin{bmatrix} 1 & x_{11} & \cdots & x_{1p} \\ 1 & x_{21} & \cdots & x_{2p} \\ \vdots & \vdots & \ddots & \vdots \\ 1 & x_{m1} & \cdots & x_{mp} \end{bmatrix}.$$

Then the regression equation is given by

$$\mathbf{Z} = \mathbf{X}\boldsymbol{\beta} + \mathbf{e},$$

where $\boldsymbol{\beta}$ is a $(p+1)$-dimensional parameter vector, and the error vector, $\mathbf{e} = (e_1, \ldots, e_m)'$, has the multivariate normal distribution $\mathbf{e} \sim N(\mathbf{0}, \sigma_e^2 \mathbf{I})$, where $\mathbf{I}$ is an $m \times m$ identity matrix. The ordinary least squares parameter estimates are given by $\widehat{\boldsymbol{\beta}} = (\mathbf{X}'\mathbf{X})^{-1}\mathbf{X}'\mathbf{Z}$, and the variance–covariance matrix for these estimates is given by $\widehat{\sigma}_e^2(\mathbf{X}'\mathbf{X})^{-1}$, with $\widehat{\sigma}_e^2 = (1/(m-p-1)) \sum_i (Z_i - \widehat{Z}_i)^2$. The estimated mean response and prediction, $\widehat{Z}_i$, is given by $\widehat{Z}_i = \mathbf{x}_i'\widehat{\boldsymbol{\beta}}$, where $\mathbf{x}_i'$ is the ith row of $\mathbf{X}$. Further, the variance of the jth regression-coefficient estimator, $\widehat{\beta}_j$, is given by the jth diagonal element of $\widehat{\sigma}_e^2(\mathbf{X}'\mathbf{X})^{-1}$. If $\widehat{Z}_i$ is an estimate of the mean response, then an estimate of its variance is given by $\widehat{\sigma}_e^2(\mathbf{x}_i'(\mathbf{X}'\mathbf{X})^{-1}\mathbf{x}_i)$. If one is predicting a new observation, say Z_h, the prediction is $\widehat{Z}_h = \mathbf{x}_h'\widehat{\boldsymbol{\beta}}$, and the *prediction variance* is estimated by $\widehat{\sigma}_e^2(1 + \mathbf{x}_h'(\mathbf{X}'\mathbf{X})^{-1}\mathbf{x}_h)$. Derivations and details can be found in textbooks on multiple regression (see, for example, Kutner et al., 2004).

3.2.2 Parameter Inference for Spatio-Temporal Data

In many scientific applications of spatio-temporal modeling, one may only be interested in whether the covariates (the Xs) are important in the model for explanation rather than for prediction. Such examples typically include scientifically meaningful covariates, such as a habitat covariate (X) related to the relative abundance (Z) of an animal in some area, or whether some demographic variable (X) is associated with household income (Z). In this section, for illustration we again consider the maximum temperature data in the NOAA data set – specifically, we consider the regression model given in Section 3.2, but our focus here is on the regression parameters. For example, do we need the longitude-by-latitude spatial interaction term (X_4) or the latitude-by-day term (X_6) in the regression?

Table 3.1: Estimated regression coefficients and the standard errors (within parentheses) for the linear regression model of Section 3.2 when using ordinary least squares (OLS) and generalized least squares (GLS). One, two, and three asterisks are used to denote significance at the 10%, 5%, and 1% levels of significance, respectively.

	Dependent variable:	
	Max. Temperature (°F)	
	$\hat{\beta}_{\mathrm{ols}}\ (SE(\hat{\beta}_{\mathrm{ols}}))$	$\hat{\beta}_{\mathrm{gls}}\ (SE(\hat{\beta}_{\mathrm{gls}}))$
Intercept	192.240** (97.854)	195.320** (98.845)
Longitude	1.757 (1.088)	1.780 (1.097)
Latitude	−1.317 (2.556)	−0.974 (2.597)
Day	−1.216*** (0.134)	−1.237*** (0.136)
Longitude × Latitude	−0.026 (0.028)	−0.022 (0.029)
Longitude × Day	−0.023*** (0.001)	−0.023*** (0.001)
Latitude × Day	−0.019*** (0.002)	−0.019*** (0.002)
α_1	16.647*** (4.832)	19.174*** (4.849)
α_2	18.528*** (3.056)	16.224*** (3.125)
α_3	−6.607** (3.172)	−4.204 (3.199)
α_4	30.545*** (4.370)	27.500*** (4.493)
α_5	14.739*** (2.747)	13.957*** (2.759)
α_6	−17.541*** (3.423)	−15.779*** (3.461)
α_7	28.472*** (3.552)	25.985*** (3.613)
α_8	−27.348*** (3.164)	−25.230*** (3.202)
α_9	−10.235** (4.457)	−7.401 (4.556)
α_{10}	10.558*** (3.327)	8.561** (3.396)
α_{11}	−22.758*** (3.533)	−19.834*** (3.569)
α_{12}	21.864*** (4.813)	17.771*** (5.041)
Observations	3,989	3,989

Note: $^{*}p < 0.1;\ ^{**}p < 0.05;\ ^{***}p < 0.01$

The middle column of Table 3.1 shows the OLS parameter estimates and their standard errors (i.e., square root of their variances) from the OLS fit of this regression model, assuming independent errors. The standard errors suggest that longitude, latitude, and the longitude–latitude interaction, are not important in the model given all of the other variables included in the model, based on the observation that their confidence intervals cover zero. It might be surprising to think that latitude is not important here, since we saw in Chapter 2 that there is a clear latitudinal dependence in temperature for these data (it is typically cooler the further north you go in the central USA). But recall that when interpreting parameters in multiple regression we are considering their importance *in the presence of* all of the other variables in the model. Thus, this result may be due to the fact that there are interactions of the latitude effect with longitude and/or time, or it could be due to other factors. We discuss some of these below.

As discussed in Section 3.2, the residuals from this regression fit exhibit spatio-temporal dependence, and thus the OLS assumption of independent errors is violated, which calls into question the validity of the standard errors given in the middle column of Table 3.1. As already mentioned, in the case of positive dependence (present in the residuals here) the standard errors are underestimated, potentially implying that a covariate is important in the model when it really is not. In the right-hand column we show the estimates and standard errors after fitting using GLS, where the covariance of the errors is assumed, *a priori*, to be a function of distance in space and time, specifically constructed from a Gaussian kernel with bandwidth 0.5 (see Lab 3.2 for details). Note that all the standard errors are larger, and some of our conclusions have changed regarding which effects are significant, and which are not.

Readers who are familiar with regression analysis may also recall that there are other factors that might affect the standard errors given in Table 3.1. For example, the presence of moderate to serious *multicollinearity* in the covariates (e.g., when some linear combination of Xs is approximately equal to one or more of the other X variables) can inflate the standard errors. In Lab 3.2, we see the effect of adding another basis function, $\phi_{13}(\mathbf{s})$, that is a slightly noisy version of $\phi_5(\mathbf{s})$. Without $\phi_{13}(\mathbf{s})$, the effect of $\phi_5(\mathbf{s})$ is considered significant in the model (see Table 3.1). However, the estimate of α_5 is *not* significant at the 1% level when both $\phi_5(\mathbf{s})$ and $\phi_{13}(\mathbf{s})$ are included in the model.

Inference can also be affected by *confounding*, in which interpretation or significance is substantially altered when an important variable is ignored, or perhaps when an extraneous variable is included in the model. Since we typically do not know or have access to all of the important variables in a regression, this is often a problem. Indeed, one of the interpretations of dependent errors in spatial, time-series, and spatio-temporal models is that they probably represent the effects of covariates that were left out of the model. As we describe in Chapter 4, this implies that there can be confounding between the spatio-temporally dependent random errors and covariates of primary interest, which can affect parameter inference and accompanying interpretations. But, if our goal is spatio-temporal prediction, this confounding is not necessarily a bad thing, since building dependence into the model is somewhat of an "insurance policy" against our model missing important covariates.

R tip: Several excellent packages can be used to neatly display results from models in R, such as **xtable** and **stargazer**. All tables in this chapter were produced using **stargazer**.

3.2.3 Variable Selection

As mentioned in the previous section, it can be the case that when p (the number of co-variates) is fairly large, we do not believe that all of them are truly related to the response, and we are interested in choosing which are the most important. This is generally called *variable selection*. Outside the context of regression, Chapter 6 considers the more general problem of *model selection*.

It would be ideal if we could test all possible combinations of all p covariates and determine which one gives the best predictive ability. This can be done if p is small, but it quickly becomes problematic for large p as there are 2^p possible models that would have to be considered, assuming all of them have an intercept parameter. Alternatively, we can consider a *best subsets* procedure that uses a special algorithm (such as the "leaps and bounds algorithm") to efficiently find a few of the best models for a given number of covariates (see, for example, Kutner et al., 2004).

Another option is to use an automated selection algorithm such as *forward selection*. In this case, we start with a model that includes just the intercept, and then we find which covariate reduces the error sums of squares (or some other chosen model-selection criterion) the most. That covariate is added to the model, and we then consider which of the remaining $(p - 1)$ gives the best two-variable model. We continue this until some pre-specified stopping rule is reached. In the context of the regression with the NOAA data set, Table 3.2 shows the best candidate models for one to four variables (in addition to the intercept), as obtained by the forward-selection algorithm using the function `step` in R; here the Akaike information criterion (AIC, see Section 6.4.4) was adopted as the model-selection criterion. Note how the residual standard error decreases sharply with the inclusion of one covariate (in this case, latitude) and slowly thereafter. We have already seen that there is considerable correlation between maximum temperature and latitude, so this is not surprising. As further evidence of this, note that latitude, which was not significant in the full model, is the single most important variable according to forward stepwise selection. But when the latitude-by-day interaction term enters the model, the parameter estimate for latitude decreases noticeably. For comparison, Table 3.3 shows the same forward-selection analysis but now using the residual sum of squares (RSS) as the model-selection criterion. Note that this still has latitude as the most important single variable, but the longitude-by-day interaction is the second variable entered into the model (followed by the latitude-by-day variable), and the day variable is not included. This shows that the choice of criterion can make a substantial difference when doing stepwise selection: the AIC criterion penalizes for model complexity (i.e., the number of variables in the model), whereas the RSS criterion does not.

Alternative stepwise methods include *backward-selection* and *mixed-selection* algorithms (see James et al., 2013, Chapter 6). Note that no stepwise procedure is guaranteed to give the best model other than for the single-covariate case, but these methods can provide potential candidate models that are reasonable. It is also important to realize that the

Table 3.2: Estimated regression coefficients for the linear regression model of Section 3.2 when using ordinary least squares to estimate the parameters and forward selection based on the AIC, starting from the intercept-only model. One, two, and three asterisks are used to denote significance at the 10%, 5%, and 1% levels of significance, respectively. Note that the residual standard error when fitting the full model ($p = 18$) was 4.22.

		Dependent variable:			
		Max. Temperature (°F)			
		$\hat{\beta}_{\text{ols}}$			
	(1)	(2)	(3)	(4)	(5)
Intercept	88.673***	148.940***	147.840***	136.810***	138.420***
Latitude		−1.559***	−1.559***	−1.274***	−1.273***
Day			0.069***	0.755***	0.755***
Latitude × Day				−0.018***	−0.018***
Longitude					0.019
Observations	3,989	3,989	3,989	3,989	3,989
Residual Std. Error	7.726	4.710	4.669	4.626	4.625

Note: $^{*}p < 0.1$; $^{**}p < 0.05$; $^{***}p < 0.01$

forward-selection procedure can be used in the "large p, small n" case where one has more covariates p than observations n, at least up to models of size $n − 1$, which is increasingly common in "big data" statistical-learning applications (James et al., 2013). (Note that in this book we prefer to use m instead of n to represent sample size for spatio-temporal data.)

The subset-selection methods discussed above penalize model complexity at the expense of model fit by removing variables. This is a manifestation of a common problem in statistics, balancing the trade-off between variance and bias. That is, these methods trade some bias for variance reduction by removing variables. Another approach to this problem in regression is to constrain the least squares estimates in such a way that the regression coefficients are *regularized* (or shrunk) towards zero, hence adding bias. The two most-used approaches for regularization in regression are *ridge regression* and the *lasso*. These are briefly described in Technical Note 3.4.

Technical Note 3.4: Ridge and Lasso Regression
Recall that the OLS spatio-temporal regression estimates are found by minimizing the RSS given in (3.7). One can consider a *regularization* in which a penalty term is added to the RSS that effectively *shrinks* the regression parameter estimates towards zero. Specif-

Table 3.3: Same as Table 3.2 but using a forward-selection criterion given by the total residual sum of squares.

		Dependent variable:			
		Max. Temperature (°F)			
		$\hat{\beta}_{\text{ols}}$			
	(1)	(2)	(3)	(4)	(5)
Intercept	88.673***	148.940***	147.780***	140.420***	122.020***
Latitude		−1.559***	−1.560***	−1.366***	−0.838***
Longitude × Day			−0.001***	−0.006***	−0.011***
Latitude × Day				−0.012***	−0.023***
α_{10}					−6.927***
Observations	3,989	3,989	3,989	3,989	3,989
Residual Std. Error	7.726	4.710	4.661	4.607	4.470

Note: $^{*}p < 0.1;\ ^{**}p < 0.05;\ ^{***}p < 0.01$

ically, consider estimates of β that come from a penalized (regularization) form of the RSS given by

$$\sum_{j=1}^{T}\sum_{i=1}^{m} [Z(\mathbf{s}_i; t_j) - (\beta_0 + \beta_1 X_1(\mathbf{s}_i; t_j) + \ldots + \beta_p X_p(\mathbf{s}_i; t_j))]^2 + \lambda \sum_{\ell=1}^{p} |\beta_\ell|^q, \quad (3.9)$$

where λ is a tuning parameter and $\sum_{\ell=1}^{p} |\beta_\ell|^q$ is the penalty term. Note that the penalty term does not include the intercept parameter β_0. When $q = 2$, the estimates, say $\widehat{\boldsymbol{\beta}}_R$, are said to be *ridge regression* estimates, and when $q = 1$ the estimates, say $\widehat{\boldsymbol{\beta}}_L$, are *lasso* estimates. Clearly, $q = 2$ corresponds to the square of an L_2-norm penalty and $q = 1$ corresponds to an L_1-norm penalty; recall that the L_2-norm of a vector $\mathbf{a} = (a_1, \ldots, a_q)'$ is given by $\sqrt{\sum_{k=1}^{q} a_k^2}$, and the L_1-norm is given by $\sum_{k=1}^{q} |a_k|$.

Thus, minimizing (3.9) with respect to the regression coefficients subject to these penalty constraints attempts to balance the model fit (variance) given by the first term and shrinking the parameters towards zero (adding bias) via the penalty term. It is clear that both the ridge-regression estimates, $\widehat{\boldsymbol{\beta}}_R$, and the lasso estimates, $\widehat{\boldsymbol{\beta}}_L$, should be closer to zero than the equivalent OLS estimates. (When $\lambda = 0$, the ridge or lasso estimates are just the OLS estimates.) A potential advantage of the lasso is that it can shrink parameters *exactly* to zero (unlike ridge regression, which only shrinks towards zero). This provides a more explicit form of variable selection. More general regularization in regression can be achieved by assigning prior distributions to the parameters β and considering the analysis from a Bayesian perspective. Indeed, both ridge and lasso regression have equivalent Bayesian formulations. In practice, one selects the tuning parameter λ by cross-validation. Note also that these penalized regression estimates are not scale-invariant, so one typically scales (and centers) the Xs when implementing ridge or lasso regression. See James et al. (2013) for more information about these procedures.

3.3 Spatio-Temporal Forecasting

As an example of the third goal of spatio-temporal modeling, suppose we want to forecast the sea surface temperature (SST) in the tropical Pacific Ocean six months from now. For example, the top left panel of Figure 3.10 shows SST anomalies, which are just deviations from long-term monthly averages, for April 1997, and the bottom right panel shows the SST anomalies for October 1997. You might ask why we would be interested in predicting SST six months ahead. As it turns out, the so-called El Niño Southern Oscillation (ENSO) phenomenon is in this region, which is characterized by frequent (but not regular) periods of warmer-than-normal and cooler-than-normal ocean temperatures, and ENSO has a dramatic effect on worldwide weather patterns and associated impacts (e.g., droughts, floods, tropical storms, tornadoes). Thus, being able to predict these warmer (El Niño) or cooler (La Niña) periods can help with resource and disaster planning. The series of plots shown in Figure 3.10 corresponds to a major El Niño event.

One way we might try to forecast the SST anomalies into the future is to use regression. For example, the Southern Oscillation Index (SOI) is a well-known indicator of ENSO that is regularly recorded; here we consider it at monthly time steps. In what follows, we use the SOI index at time t (e.g., April 1997) to forecast the SST at time $t + \tau$ (e.g., October 1997, where $\tau = 6$ months). We do this for each spatial location separately, so that each oceanic pixel in the domain shown in Figure 3.10 gets its own simple linear regression (including an intercept coefficient and a coefficient corresponding to the lagged SOI value). The top panels in Figure 3.11 show the intercept (left) and SOI regression coefficient (right) for the regression fit at each location. Note the fairly distinct pattern in these coefficients that corresponds to the El Niño warm region in Figure 3.10 – clearly, these estimated regression coefficients exhibit quite strong spatial dependencies. The middle panels in Figure 3.11 show contour plots of the the actual anomalies for October 1997 (left), as well as the pixelwise simple-linear-regression forecast based on SOI from April 1997 (right; note the different color scale). The associated regression-forecast prediction standard error (see Technical Note 3.3) is given in the bottom panel.

It is clear that although the forecast in the middle-right panel of Figure 3.10 captures the broad El Niño feature, it is very biased towards a cooler anomaly than that observed.

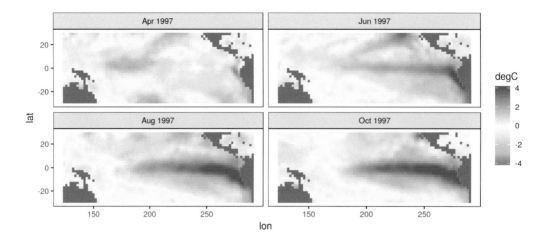

Figure 3.10: Tropical Pacific Ocean SST anomalies for April, June, August, and October 1997. This series of plots shows the onset of the extreme El Niño event that happened in the Northern Hemisphere in the fall of 1997.

This illustrates that we likely need additional information to perform a long-lead forecast of SST, something we discuss in more detail using dynamic models in Chapter 5. This example also shows that it might be helpful to account for the fact that these regression-coefficient estimates show such strong spatial dependence. This is often the case in time-series regressions at nearby spatial locations, and we shall see another example of this in Section 4.4.

> **R tip:** Fitting multiple models to groups of data in a single data frame in long format has been made easy and computationally efficient using functions in the packages **tidyr**, **purrr**, and **broom**. Take a look at Labs 3.2 and 3.3 to see how multiple models, predictions, and tests can be easily carried out using these packages.

3.4   Non-Gaussian Errors

You have probably already heard about the normal distribution that was used to describe the regression errors in the previous sections. The name "normal" seems to imply that any other distribution is abnormal – not so! Data that are binary or counts or skewed are common and of great interest to scientists and statisticians. Consequently, in spatial and spatio-temporal statistics we use the terminology *Gaussian distribution* and "Gau" instead of "*N*," which

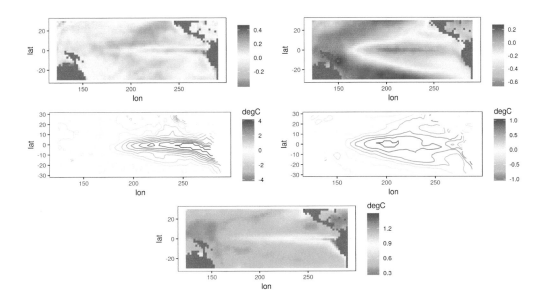

Figure 3.11: Top: Spatially varying estimated intercept (left) and spatially varying estimated regression coefficient of lagged SOI (right) in a simple linear regression of lagged SOI on the SST anomalies at each oceanic spatial location (fitted individually). Middle: Contour plots of SST anomalies for October 1997 (left) and six-month-ahead forecast (right) based on a simple linear regression model regressed on the SOI value in April 1997. Note the different color scales. Bottom: Prediction standard error for the forecast (see Technical Note 3.3).

falls into line with the well-known Gaussian processes defined in time or in Euclidean space (see Section 4.2). There are many off-the-shelf methods that can be used for non-Gaussian modeling – both from the statistics perspective and from the machine-learning perspective. By "machine learning" we are referring to methods that do not explicitly account for the *random* spatio-temporal nature of the data. From the statistical perspective, we could simply use a *generalized linear model* (GLM) or a *generalized additive model* (GAM) to analyze spatio-temporal data.

3.4.1 Generalized Linear Models and Generalized Additive Models

The basic GLM has two components, a random component and a systematic component. The random component assumes that observations, conditioned on their respective means and (in some cases) scaling parameters, are independent and come from the *exponential family* of distributions. That is,

$$Z(\mathbf{s}_i; t_j)|Y(\mathbf{s}_i; t_j), \gamma \ \sim \ indep. \ EF(Y(\mathbf{s}_i; t_j); \gamma), \tag{3.10}$$

where $EF(\cdot)$ refers to the exponential family, $Y(\mathbf{s}_i; t_j)$ is the mean, and γ is a scale parameter (see, for example, McCulloch and Searle, 2001, for details). Members of the exponential family include common distributions such as the normal (Gaussian), Poisson, binomial, and gamma distributions.

The systematic component of the GLM then specifies a relationship between the mean response and the covariates. In particular, the systematic component consists of a *link function* that transforms the mean response and then expresses this transformed mean in terms of a linear function of the covariates. In our notation, this is given by

$$g(Y(\mathbf{s}; t)) = \beta_0 + \beta_1 X_1(\mathbf{s}; t) + \beta_2 X_2(\mathbf{s}; t) + \ldots + \beta_p X_p(\mathbf{s}; t), \qquad (3.11)$$

where $g(\cdot)$ is some specified monotonic link function. Note that in a classic GLM there is no additive random effect term in (3.11), but this can be added to make the model a *generalized linear mixed model* (GLMM), where "mixed" refers to having both fixed and random effects in the model for $g(Y(\mathbf{s}; t))$.

The GAM is also composed of a random component and a systematic component. The random component is the same as for the GLM, namely (3.10). In addition, like the GLM, the systematic component of the GAM also considers a transformation of the mean response related to the covariates, but it assumes a more flexible function of the covariates. That is,

$$g(Y(\mathbf{s}; t)) = \beta_0 + f_1(X_1(\mathbf{s}; t)) + f_2(X_2(\mathbf{s}; t)) + \ldots + f_p(X_p(\mathbf{s}; t)), \qquad (3.12)$$

where the functions $\{f_k(\cdot)\}$ can have a specified parametric form (such as polynomials in the covariate), or, more generally, they can be some smooth function specified semi-parametrically or nonparametrically. Often, $f_k(\cdot)$ is written as a basis expansion (see Wood, 2017, for more details). Thus, the GLM is a special parametric case of the GAM. These models can be quite flexible. Again, note that a random effect can be added to (3.12), as with the GLM, in which case the model becomes a *generalized additive mixed model* (GAMM).

As with normal (Gaussian) error regression, so long as covariates (or functions of these in the case of GAMs) are available at any location in the space-time domain, GLMs or GAMs can be used for spatio-temporal prediction. Whether or not this accommodates sufficiently the dependence in the observations depends on the specific data set and the covariates that are available. A straightforward way to fit a GLM in R is to use the function `glm`. In Lab 3.4 we fit a GLM to the Carolina wren counts in the BBS data set, where we assume a Poisson response and a log link. We consider the same classes of covariates used in the regression example in Section 3.2, where the response was `Tmax` in the NOAA data set. The latent mean surface is given by (3.11) (with estimated regression parameters $\boldsymbol{\beta}$) and is illustrated in Figure 3.12. This latent spatial surface captures the large-scale trends, but it is unable to reproduce the small-scale spatial and temporal fluctuations in the Carolina wren intensity, and the residuals show both temporal and spatial correlation. We could accommodate this additional dependence structure by adding more basis functions and treating their

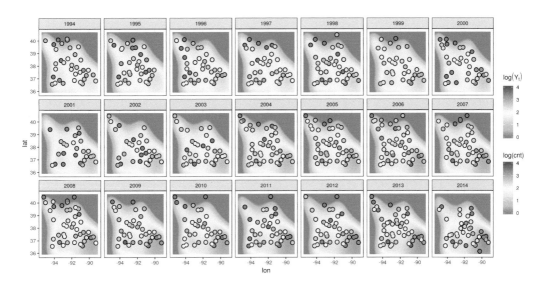

Figure 3.12: Prediction of $\log Y(\cdot)$ for the Carolina wren sighting data set on a grid between $t = 1$ (the year 1994) and $t = 21$ (2014) based on a Poisson response model, implemented with the function `glm`. The log of the observed count is shown in circles using the same color scale.

regression coefficients as fixed effects, but this will likely result in overfitting. In Chapter 4 we explore the use of random effects to circumvent this problem.

Recall that it is useful to consider residuals in the linear-regression context to evaluate the model fit and potential violations of model assumptions. In the context of GLMs, we typically consider a special type of residual when the data are not assumed to come from a Gaussian distribution. Technical Note 3.5 defines so-called *deviance residuals* and *Pearson (chi-squared) residuals*, which are often used for GLM model evaluation (see, for example, McCullagh and Nelder, 1989). Heuristically, examining these residuals for spatio-temporal structure can often suggest that additional spatial, temporal, or spatio-temporal random effects are needed in the model, or that a different response model is warranted (e.g., to account for over-dispersion; see Lab 3.4).

Technical Note 3.5: Deviance and Pearson Residuals

One way to consider the agreement between a model and data is to compare the predictions of the model to a "saturated" model that fits the data exactly. In GLMs, this corresponds to the notion of *deviance*. Specifically, suppose we have a model for an m-vector of data $\mathbf{Z}$ that depends on parameters $\boldsymbol{\theta}_{\text{model}}$ and has a log-likelihood given by

$\ell(\mathbf{Z}; \boldsymbol{\theta}_{\mathrm{model}})$. We then define the deviance as

$$D(\mathbf{Z}; \widehat{\boldsymbol{\theta}}_{\mathrm{model}}) = 2\{\ell(\mathbf{Z}; \widehat{\boldsymbol{\theta}}_{\mathrm{sat}}) - \ell(\mathbf{Z}; \widehat{\boldsymbol{\theta}}_{\mathrm{model}})\} = \sum_{i=1}^{m} D(Z_i; \widehat{\boldsymbol{\theta}}_{\mathrm{model}}),$$

where $\ell(\mathbf{Z}; \widehat{\boldsymbol{\theta}}_{\mathrm{sat}})$ is the log-likelihood for the so-called *saturated model*, which is the model that has one parameter per observation (i.e., that fits the data exactly). Note that $D(Z_i; \widehat{\boldsymbol{\theta}}_{\mathrm{model}})$ corresponds to the contribution of observation Z_i to the deviance given the parameter estimates $\boldsymbol{\theta}_{\mathrm{model}}$. The deviance is just 2 times the log-likelihood *ratio* of the full (saturated) model relative to the reduced model of interest. We then define the *deviance residual* as

$$r_{d,i} \equiv \mathrm{sign}(Z_i - \widehat{\mu}_i)\sqrt{D(Z_i; \widehat{\boldsymbol{\theta}}_{\mathrm{model}})}, \qquad (3.13)$$

where $\widehat{\mu}_i$ corresponds to $E(Z_i | \widehat{\boldsymbol{\theta}}_{\mathrm{model}})$, the estimate of the mean response from the model given parameter estimates, $\widehat{\boldsymbol{\theta}}_{\mathrm{model}}$. The $\mathrm{sign}(\cdot)$ function in (3.13) assigns the sign of the residual to indicate whether the mean response is less than or greater than the observation. In practice, we often consider standardized deviance residuals (see, for example, McCullagh and Nelder, 1989).

Alternatively, we can define a standardized residual that more directly considers the difference between the data and the estimated mean response. That is,

$$r_{p,i} \equiv \frac{(Z_i - \widehat{\mu}_i)^2}{V(\widehat{\mu}_i)},$$

where $V(\widehat{\mu}_i)$ is called the *variance function*, and it is generally a function of the mean response (except when the likelihood is Gaussian). The specific form of the variance function depends on the form of the data likelihood. The unsigned residual, $r_{p,i}$, is known as a Pearson residual (or Pearson chi-squared residual) because the sum of these residuals for all $i = 1, \ldots, m$ gives a Pearson chi-squared statistic, which can be used for formal hypothesis tests of model adequacy (see, for example, McCullagh and Nelder, 1989).

3.5   Hierarchical Spatio-Temporal Statistical Models

The previous sections showed that it may be possible to accomplish the goals of spatio-temporal modeling without using specialized methodology. However, it was also clear from those examples that there are some serious limitations with the standard methodology. In particular, our methods should be able to include measurement uncertainty explicitly, they should have the ability to predict at locations in time or space, and they should allow us to perform parameter inference when there are dependent errors. In the remainder of this book we shall describe models that can deal with these problems.

To put our spatio-temporal statistical models into perspective, we consider a hierarchical spatio-temporal model that includes at least two stages. Specifically,

$$observations \quad = \quad true\ process \quad + \quad observation\ error, \tag{3.14}$$
$$true\ process \quad = \quad regression\ component \quad + \quad dependent\ random\ process, \tag{3.15}$$

where (3.14) and (3.15) are the first two stages of the hierarchical-statistical-model paradigm presented in Chapter 1. There are two general approaches to modeling the last term in (3.15): the *descriptive* approach and the *dynamic* approach; see Section 1.2.1. The descriptive approach is considered in Chapter 4 and offers a more traditional perspective. In that case, the dependent random process in (3.15) is defined in terms of the first-order and second-order moments (means, variances, and covariances) of its marginal distribution. This framework is not particularly concerned with the underlying causal structure that leads to dependence in the random process. Rather, it is most useful for the first two goals presented in Section 1.2: spatio-temporal prediction and parameter inference.

In contrast, we consider the dynamic approach in Chapter 5. In that case, the modeling effort is focused on conditional distributions that describe the evolution of the dependent random process in time; it is most useful for the third goal – forecasting (but also can be used for the other two goals). We note that the conditional perspective can also be considered in the context of mixed-effects descriptive models, with or without a dynamic specification, as we discuss in Section 4.4.

3.6 Chapter 3 Wrap-Up

The primary purpose of this chapter was to discuss in detail the three goals of spatio-temporal statistical modeling: predicting at a new location in space given spatio-temporal data; doing parameter inference with spatio-temporal data; and forecasting a new value at a future time. We have also emphasized the importance of quantifying the uncertainty in our predictions, parameter estimates, and forecasts. We showed that deterministic methods for spatio-temporal prediction are sensible in that they typically follow Tobler's law and give more weight to nearby observations in space and time; however, they do not provide direct estimates of the prediction uncertainty. We then showed that one could use a (linear) regression model with spatio-temporal data and that, as long as the residuals do not have spatio-temporal dependence, it is easy to obtain statistically optimal predictions and, potentially, statistically optimal forecasts. With respect to parameter inference, we showed that the linear-regression approach is again relevant but that our inference can be misleading in the presence of unmodeled extra variation, dependent errors, multicollinearity, and confounding. Finally, we showed that standard generalized linear models or generalized additive models can be used for many problems with non-Gaussian data. But again, without random effects to account for extra variation and dependence, these models are likely to give inappropriate prediction uncertainty and inferences.

The methods presented in this chapter are very common throughout the literature, and the references provided in the chapter are excellent places to find additional background material. Of course, topics such as interpolation, regression, and generalized linear models are discussed in a wide variety of textbooks and online resources, and the interested reader should have no trouble finding additional references.

In the next two chapters, we explore what to do when there *is* spatio-temporal dependence beyond what can be explained by covariates. We shall cover descriptive models that focus more on the specification of spatio-temporal covariance functions in Chapter 4, and dynamic models that focus explicitly on the evolution of spatial processes through time in Chapter 5. These two chapters together make up the "protein" in the book, and the material in them will have a decidedly more technical flavor. More powerful, more flexible, but more complex, dependent processes require a higher technical level than is usually found in introductory statistical-modeling courses. That said, we maintain an emphasis on describing the motivations for our methods and on their implementation in the associated R Labs.

Lab 3.1: Deterministic Prediction Methods

Inverse Distance Weighting

Inverse distance weighting (IDW) is one of the simplest deterministic spatio-temporal interpolation methods. It can be implemented easily in R using the function `idw` in the package **gstat**, or from scratch, and in this Lab we shall demonstrate both approaches. We require the following packages.

```
library("dplyr")
library("fields")
library("ggplot2")
library("gstat")
library("RColorBrewer")
library("sp")
library("spacetime")
library("STRbook")
```

We consider the maximum temperature field in the NOAA data set for the month of July 1993. These data can be obtained from the data NOAA_df_1990 using the **filter** function in **dplyr**.

```
data("NOAA_df_1990", package = "STRbook")
Tmax <- filter(NOAA_df_1990,          # subset the data
```

```
       proc == "Tmax" &      # only max temperature
       month == 7 &          # July
       year == 1993)         # year of 1993
```

We next construct the three-dimensional spatio-temporal prediction grid using `expand.grid`. We consider a 20 × 20 grid in longitude and latitude and a sequence of 6 days regularly arranged in the month.

```
pred_grid <- expand.grid(lon = seq(-100, -80, length = 20),
                         lat = seq(32, 46, length = 20),
                         day = seq(4, 29, length = 6))
```

The function in **gstat** that does the inverse distance weighting, `idw`, takes the following arguments: `formula`, which identifies the variable to interpolate; `locations`, which identifies the spatial and temporal variables; `data`, which can take the data in a data frame; `newdata`, which contains the space-time grid locations at which to interpolate; and `idp`, which corresponds to α in (3.3). The larger α (`idp`) is, the less the smoothing. This parameter is typically set using cross-validation, which we explore later in this Lab; here we fix $\alpha = 5$. We run `idw` below with the variable Tmax, omitting data on 14 July 1993.

```
Tmax_no_14 <- filter(Tmax, !(day == 14))             # remove day 14
Tmax_July_idw <- idw(formula = z ~ 1,                # dep. variable
                     locations = ~ lon + lat + day,  # inputs
                     data = Tmax_no_14,              # data set
                     newdata = pred_grid,            # prediction grid
                     idp = 5)                        # inv. dist. pow.
```

The output `Tmax_July_idw` contains the fields `lon`, `lat`, `day`, and `var1.pred` corresponding to the IDW interpolation over the prediction grid. This data frame can be plotted using **ggplot2** commands as follows.

```
ggplot(Tmax_July_idw) +
    geom_tile(aes(x = lon, y = lat,
                  fill = var1.pred)) +
    fill_scale(name = "degF") +      # attach color scale
    xlab("Longitude (deg)") +        # x-axis label
    ylab("Latitude (deg)") +         # y-axis label
    facet_wrap(~ day, ncol = 3) +    # facet by day
    coord_fixed(xlim = c(-100, -80),
                ylim = c(32, 46)) +  # zoom in
    theme_bw()                       # B&W theme
```

A similar plot to the one above, but produced using `stplot` instead, is shown in the left panel of Figure 3.2. Notice how the day with missing data is "smoothed out" when compared to the others. As an exercise, you can redo IDW including the 14 July 1993 in the data set, and observe how the prediction changes for that day.

Implementing IDW from First Principles

It is often preferable to implement simple algorithms, like IDW, from scratch, as doing so increases code versatility (e.g., it facilitates implementation of a cross-validation study). Reducing dependence on other packages will also help the code last the test of time (as it becomes immune to package changes).

We showed that IDW is a kernel predictor and yields the kernel weights given by (3.3). To construct these kernel weights we first need to find the distances between all prediction locations and data locations, take their reciprocals and raise them to the power (`idp`) of α. Pairwise distances between two arbitrary sets of points are most easily computed using the `rdist` function in the package **fields**. Since we wish to generate these kernel weights for different observation and prediction sets and different bandwidth parameters, we create a function `Wt_IDW` that generates the required kernel-weights matrix.

```
pred_obs_dist_mat <- rdist(select(pred_grid, lon, lat, day),
                           select(Tmax_no_14, lon, lat, day))
Wt_IDW <- function(theta, dist_mat) 1/dist_mat^theta
Wtilde <- Wt_IDW(theta = 5, dist_mat = pred_obs_dist_mat)
```

The matrix `Wtilde` now contains all the $\tilde{w}_{ij}$ described in (3.3); that is, the (k,l)th element in `Wtilde` contains the distance between the kth prediction location and the lth observation location, raised to the power of 5, and reciprocated.

Next, we compute the weights in (3.2). These are just the kernel weights normalized by the sum of all kernel weights associated with each prediction location. Normalizing the weights at every location can be done easily using `rowSums` in R.

```
Wtilde_rsums <- rowSums(Wtilde)
W <- Wtilde/Wtilde_rsums
```

The resulting matrix `W` is the weight matrix, sometimes known as the *influence matrix*. The predictions are then given by (3.1), which is just the influence matrix multiplied by the data.

```
z_pred_IDW <- as.numeric(W %*% Tmax_no_14$z)
```

One can informally verify the computed predictions by comparing them to those given by `idw` in **gstat**. We see that the two results are very close; numerical mismatches of this order of magnitude are likely to arise from the slightly different way the IDW weights are computed in **gstat** (and it is possible that you get different, but still small, mismatches on your computer).

```
summary(Tmax_July_idw$var1.pred - z_pred_IDW)
```

```
##      Min.   1st Qu.   Median      Mean   3rd Qu.      Max.
## -1.62e-12 -1.85e-13  0.00e+00 -1.00e-15  1.99e-13  1.16e-12
```

Generic Kernel Smoothing and Cross-Validation

One advantage of implementing IDW from scratch is that now we can change the kernel function to whatever we want and compare predictions from different kernel functions. We implement a kernel smoother below, where the kernel is a Gaussian radial basis function given by (3.4) with $\theta = 0.5$.

```
theta <- 0.5                          # set bandwidth
Wt_Gauss <- function(theta, dist_mat) exp(-dist_mat^2/theta)
Wtilde <- Wt_Gauss(theta = 0.5, dist_mat = pred_obs_dist_mat)
Wtilde_rsums <- rowSums(Wtilde)       # normalizing factors
W <- Wtilde/Wtilde_rsums              # normalized kernel weights
z_pred2 <- W %*% Tmax_no_14$z         # predictions
```

The vector `z_pred2` can be assigned to the prediction grid `pred_grid` and plotted using **ggplot2** as shown above. Note that the the the predictions are similar, but not identical, to those produced by IDW. But which predictions are the best in terms of squared prediction error? A method commonly applied to assess goodness of fit is known as *cross-validation* (CV). CV also allows us to choose bandwidth parameters (i.e., α or θ) that are optimal for a given data set. See Section 6.1.3 for more discussion on CV.

To carry out CV, we need to fit the model using a subset of the data (known as the training set), predict at the data locations that were omitted (known as the validation set), and compute a *discrepancy*, usually the squared error, between the predicted and observed values. If we leave one data point out at a time, the procedure is known as leave-one-out cross-validation (LOOCV). We denote the sum of the discrepancies for a particular bandwidth parameter θ as the LOOCV score, $CV_{(m)}(\theta)$ (note that m, here, is the number of folds used in the cross-validation; in LOOCV, the number of folds is equal the number of data points, m).

The LOOCV for simple predictors, like kernel smoothers, can be computed analytically without having to refit; see Appendix B. Since the data set is reasonably small, it is feasible here to do the refitting with each data point omitted (since each prediction is just an inner product of two vectors). The simplest way to do LOOCV in this context is to compute the pairwise distances between *all* observation locations and the associated kernel-weight matrix, and then to select the appropriate rows and columns from the resulting matrix to do prediction at a left-out observation; this is repeated for every observation.

The distances between all observations are computed as follows.

```
obs_obs_dist_mat <- rdist(select(Tmax, lon, lat, day),
                          select(Tmax, lon, lat, day))
```

A function that computes the LOOCV score is given as follows.

```
LOOCV_score <- function(Wt_fun, theta, dist_mat, Z) {
  Wtilde <- Wt_fun(theta, dist_mat)
  CV <- 0
  for(i in 1:length(Z)) {
    Wtilde2 <- Wtilde[i,-i]
    W2 <- Wtilde2 / sum(Wtilde2)
    z_pred <- W2 %*% Z[-i]
    CV[i] <- (z_pred - Z[i])^2
  }
  mean(CV)
}
```

The function takes as arguments the kernel function that computes the kernel weights `Wt_fun`; the kernel bandwidth parameter `theta`; the full distance matrix `dist_mat`; and the data `Z`. The function first constructs the kernel-weights matrix for the given bandwith. Then, for the ith observation, it selects the ith row and excludes the ith column from the kernel-weights matrix and assigns the resulting vector to `Wtilde2`. This vector contains the kernel weights for the ith observation location (which is now a prediction location) with the weights contributed by this ith observation removed. This vector is normalized and then cross-multiplied with the data to yield the prediction. This is done for all $i = 1, \ldots, n$, and then the mean of the squared errors is returned. To see which of the two predictors is "better," we now simply call `LOOCV_score` with the two different kernel functions and bandwidths.

```
LOOCV_score(Wt_fun = Wt_IDW,
            theta = 5,
            dist_mat = obs_obs_dist_mat,
            Z = Tmax$z)
```

```
## [1] 7.78
```

```
LOOCV_score(Wt_fun = Wt_Gauss,
            theta = 0.5,
            dist_mat = obs_obs_dist_mat,
            Z = Tmax$z)
```

```
## [1] 7.53
```

Clearly the Gaussian kernel smoother has performed marginally better than IDW in this case. But how do we know the chosen kernel bandwidths are suitable? Currently we do not, as these were set by simply "eye-balling" the predictions and assessing visually whether they looked suitable or not. An objective way to set the bandwidth parameters is to put them equal to those values that minimize the LOOCV scores. This can be done by simply computing `LOOCV_score` for a set, say 21, of plausible bandwidths and finding the minimum. We do this below for both IDW and the Gaussian kernel.

```
theta_IDW <- seq(4, 6, length = 21)
theta_Gauss <- seq(0.1, 2.1, length = 21)
CV_IDW <- CV_Gauss <- 0
for(i in seq_along(theta_IDW)) {
  CV_IDW[i] <- LOOCV_score(Wt_fun = Wt_IDW,
                           theta = theta_IDW[i],
                           dist_mat = obs_obs_dist_mat,
                           Z = Tmax$z)

  CV_Gauss[i] <- LOOCV_score(Wt_fun = Wt_Gauss,
                             theta = theta_Gauss[i],
                             dist_mat = obs_obs_dist_mat,
                             Z = Tmax$z)
}
```

The plots showing the LOOCV scores as a function of α and θ for the IDW and Gaussian kernels, respectively (Figure 3.3), exhibit clear minima when plotted, which is very typical of plots of this kind.

```
par(mfrow = c(1,2))
plot(theta_IDW, CV_IDW,
     xlab = expression(alpha),
     ylab = expression(CV[(m)](alpha)),
     ylim = c(7.4, 8.5), type = 'o')
plot(theta_Gauss, CV_Gauss,
     xlab = expression(theta),
     ylab = expression(CV[(m)](theta)),
     ylim = c(7.4, 8.5), type = 'o')
```

The optimal inverse-power and minimum LOOCV score for IDW are

```
theta_IDW[which.min(CV_IDW)]
```

```
## [1] 5
```

```
min(CV_IDW)
```

```
## [1] 7.78
```

The optimal bandwidth and minimum LOOCV score for the Gaussian kernel smoother are

```
theta_Gauss[which.min(CV_Gauss)]
```

```
## [1] 0.6
```

```
min(CV_Gauss)
```

```
## [1] 7.47
```

Our choice of $\alpha = 5$ was therefore (sufficiently close to) optimal when doing IDW, while a bandwidth of $\theta = 0.6$ is better for the Gaussian kernel than our initial choice of $\theta = 0.5$. It is clear from the results that the Gaussian kernel predictor with $\theta = 0.6$ has, in this example, provided superior performance to IDW with $\alpha = 5$, in terms of mean-squared-prediction error.

Lab 3.2: Trend Prediction

There is considerable in-built functionality in R for linear regression and for carrying out hypothesis tests associated with linear models. Several packages have also been written to extend functionality, and in this Lab we shall make use of **leaps**, which contains function-ality for stepwise regression; **lmtest**, which contains a suite of tests to carry out on fitted linear models; and **nlme**, which is a package generally used for fitting nonlinear mixed effects models (but we shall use it to fit linear models in the presence of correlated errors).

```
library("leaps")
library("lmtest")
library("nlme")
```

In addition, we use **ape**, which is one of several packages that contain functionality for testing spatial or spatio-temporal independence with Moran's I statistic; and we use **FRK**, which contains functionality for constructing the basis functions shown in Figure 3.6. We also make use of **broom** and **purrr** to easily carry out multiple tests on groups within our data set.

```
library("ape")
library("broom")
library("FRK")
library("purrr")
```

We need the following for plotting purposes.

```
library("lattice")
library("ggplot2")
library("RColorBrewer")
```

We also need the usual packages for data wrangling and handling of spatial/spatio-temporal objects as in the previous Labs.

```
library("dplyr")
library("gstat")
library("sp")
library("spacetime")
library("STRbook")
library("tidyr")
```

Fitting the Model

For this Lab we again consider the NOAA data set, specifically the maximum temperature data for the month of July 1993. These data can be extracted as follows.

```
data("NOAA_df_1990", package = "STRbook")
Tmax <- filter(NOAA_df_1990,          # subset the data
                proc == "Tmax" &      # only max temperature
                month == 7 &          # July
                year == 1993)         # year of 1993
```

The linear model we fit has the form (3.5), with the list of basis functions given in Section 3.2. The set of basis functions can be constructed using the function `auto_basis` in **FRK**. The function takes as arguments `data`, which is a spatial object; `nres`, which is the number of "resolutions" to construct; and `type`, which indicates the type of basis function to use. Here we consider a single resolution of the Gaussian radial basis function; see Figure 3.6.

```
G <- auto_basis(data = Tmax[,c("lon","lat")] %>%   # Take Tmax
                        SpatialPoints(),            # To sp obj
                nres = 1,                           # One resolution
                type = "Gaussian")                  # Gaussian BFs
```

These basis functions evaluated at data locations are then the covariates we seek for fitting the data. The functions are evaluated at any arbitrary location using the function `eval_basis`. This function requires the locations as a `matrix` object, and it returns the evaluations as an object of class `Matrix`, which can be easily converted to a `matrix` as follows.

```
S <- eval_basis(basis = G,                             # basis functions
                s = Tmax[,c("lon","lat")] %>%          # spat locations
                    as.matrix()) %>%                   # conv. to matrix
    as.matrix()                                        # results as matrix
colnames(S) <- paste0("B", 1:ncol(S)) # assign column names
```

When fitting the linear model we shall use the convenient notation "." to denote "all variables in the data frame" as covariates. This is particularly useful when we have many covariates, such as the 12 basis functions above. Therefore, we first remove all variables (except the field `id` that we shall omit manually later) that we do not wish to include in the model, and we save the resulting data frame as `Tmax2`.

```
Tmax2 <- cbind(Tmax, S) %>%              # append S to Tmax
        select(-year, -month, -proc,     # and remove vars we
                -julian, -date)          # will not be using in
                                         # the model
```

As we did in Lab 3.1, we also remove 14 July 1993 to see how predictions on this day are affected, given that we have no data on that day.

```
Tmax_no_14 <- filter(Tmax2, !(day == 14))   # remove day 14
```

We now fit the linear model using `lm`. The formula we use is `z ~ (lon + lat + day)^2 + .` which indicates that we have as covariates longitude, latitude, day, and all the interactions between them, as well as the other covariates in the data frame (the 12 basis functions) without interactions.

```
Tmax_July_lm <- lm(z ~ (lon + lat + day)^2 + .,     # model
                    data = select(Tmax_no_14, -id))  # omit id
```

The results of this fit can be viewed using **summary**. Note that latitude is no longer considered a significant effect, largely because of the presence of the latitude-by-day interaction in the model, which is considered significant. The output from **summary** corresponds to what is shown in Table 3.1.

```
Tmax_July_lm %>% summary()

##
## Call:
## lm(formula = z ~ (lon + lat + day)^2 + ., data = select(Tmax_no_14,
##      -id))
##
## Residuals:
##     Min     1Q Median     3Q     Max
## -17.51  -2.48   0.11   2.66  14.17
##
## Coefficients:
##              Estimate Std. Error t value Pr(>|t|)
## (Intercept) 192.24324   97.85413   1.96  0.04953 *
## lon           1.75692    1.08817   1.61  0.10649
## lat          -1.31740    2.55563  -0.52  0.60624
```

```
## day            -1.21646    0.13355   -9.11  < 2e-16 ***
## B1             16.64662    4.83240    3.44  0.00058 ***
## B2             18.52816    3.05608    6.06  1.5e-09 ***
## B3             -6.60690    3.17176   -2.08  0.03731 *
## B4             30.54536    4.36959    6.99  3.2e-12 ***
## B5             14.73915    2.74687    5.37  8.5e-08 ***
## B6            -17.54118    3.42308   -5.12  3.1e-07 ***
## B7             28.47220    3.55190    8.02  1.4e-15 ***
## B8            -27.34815    3.16432   -8.64  < 2e-16 ***
## B9            -10.23478    4.45673   -2.30  0.02170 *
## B10            10.55823    3.32737    3.17  0.00152 **
## B11           -22.75766    3.53251   -6.44  1.3e-10 ***
## B12            21.86438    4.81294    4.54  5.7e-06 ***
## lon:lat        -0.02602    0.02823   -0.92  0.35675
## lon:day        -0.02270    0.00129  -17.62  < 2e-16 ***
## lat:day        -0.01903    0.00188  -10.15  < 2e-16 ***
## ---
## Signif. codes:
## 0 '***' 0.001 '**' 0.01 '*' 0.05 '.' 0.1 ' ' 1
##
## Residual standard error: 4.22 on 3970 degrees of freedom
## Multiple R-squared:  0.702, Adjusted R-squared:  0.701
## F-statistic:  520 on 18 and 3970 DF,  p-value: <2e-16
```

Correlated Errors

As we show later in this Lab, there is clearly correlation in the residuals, indicating that the fixed effects are not able to explain the spatio-temporal variability in the data. If we knew the spatio-temporal covariance function of these errors, we could then use generalized least squares to fit the model. For example, if we knew that the covariance function was a Gaussian function, isotropic, and with a range of 0.5 (see Chapter 4 for more details on covariance functions), then we could fit the model as follows.

```
Tmax_July_gls <- gls(z ~ (lon + lat + day)^2 + .,
                     data = select(Tmax_no_14, -id),
                     correlation = corGaus(value = 0.5,
                                           form = ~ lon + lat + day,
                                           fixed = TRUE))
```

Results of the linear fitting can be seen using `summary`. Note that the estimated coefficients are quite similar to those using linear regression, but the standard errors are larger. The output from `summary` should correspond to what is shown in Table 3.1.

Stepwise Selection

Stepwise selection is a procedure used to find a parsimonious model (where parsimony refers to a model with as few parameters as possible for a given criterion) from a large selection of explanatory variables, such that each variable is included or excluded in a *step*. In the simplest of cases, a step is the introduction of a variable (always the case in forward selection) or the removal of a variable (always the case in backward selection).

The function `step` takes as arguments the initial (usually the intercept) model as an `lm` object, the full model as its `scope` and, if `direction = "forward"`, starts from an intercept model and at each step introduces a new variable that minimizes the Akaike information criterion (AIC) (see Section 6.4.4) of the fitted model. The following `for` loop retrieves the fitted model for each step of the stepwise AIC forward-selection method.

```
Tmax_July_lm4 <- list()    # initialize
for(i in 0:4) {                    # for four steps (after intercept model)
   ## Carry out stepwise forward selection for i steps
   Tmax_July_lm4[[i+1]] <- step(lm(z ~ 1,
                          data = select(Tmax_no_14, -id)),
                          scope = z ~(lon + lat + day)^2 + .,
                          direction = 'forward',
                          steps = i)
}
```

Each model in the list can be analyzed using `summary`, as above.

Notice from the output of `summary` that `Tmax_July_lm4[[5]]` contains the covariate `lon` whose effect is not significant. This is fairly common with stepwise AIC procedures. One is more likely to include covariates whose effects are significant when minimizing the residual sum of squares at each step. This can be carried out using the function `regsubsets` from the **leaps** package, which can be called as follows.

```
regfit.full = regsubsets(z ~ 1 + (lon + lat + day)^2 + .,   # model
                    data = select(Tmax_no_14, -id),
                    method = "forward",
                    nvmax = 4)                            # 4 steps
```

All information from the stepwise-selection procedure is available in the object returned by the `summary` function.

```
regfit.summary <- summary(regfit.full)
```

You can type `regfit.summary` to see which covariates were selected in each step of the algorithm. The outputs from `step` and `regsubsets` are shown in Tables 3.2 and 3.3, respectively.

Multicollinearity

It is fairly common in spatio-temporal modeling to have multicollinearity, both in space and in time. For example, in a spatial setting, average salary might be highly correlated with unemployment levels, but both could be included in a model to explain life expectancy. It is beyond the scope of this book to discuss methods to deal with multicollinearity, but it is important to be aware of its implications.

Consider, for example, a setting where we have a 13th basis function that is simply the 5th basis function corrupted by some noise.

```
set.seed(1) # Fix seed for reproducibility
Tmax_no_14_2 <- Tmax_no_14 %>%
                mutate(B13 = B5 + 0.01*rnorm(nrow(Tmax_no_14)))
```

If we fit the same linear model, but this time we include the 13th basis function, then the effects of both the 5th and the 13th basis functions are no longer considered significant at the 1% level, although the effect of the 5th basis function was considered very significant initially (without the 13th basis function being present).

```
Tmax_July_lm3 <- lm(z ~ (lon + lat + day)^2 + .,
                    data = Tmax_no_14_2 %>%
                        select(-id))
```

```
summary(Tmax_July_lm3)
```

```
##
## Call:
## lm(formula = z ~ (lon + lat + day)^2 + ., data = Tmax_no_14_2 %>%
##     select(-id))
##
## Residuals:
##     Min      1Q  Median      3Q     Max
## -17.787  -2.495   0.103   2.674  14.318
##
## Coefficients:
##             Estimate Std. Error t value Pr(>|t|)
## (Intercept) 195.36509   97.81955    2.00  0.04587 *
## lon           1.78547    1.08775    1.64  0.10079
## lat          -1.41413    2.55484   -0.55  0.57995
## day          -1.21585    0.13349   -9.11  < 2e-16 ***
## B1           16.67581    4.83018    3.45  0.00056 ***
## B2           18.37950    3.05544    6.02  2.0e-09 ***
## B3           -6.47093    3.17092   -2.04  0.04135 *
## B4           30.30399    4.36900    6.94  4.7e-12 ***
## B5            0.60329    7.09215    0.09  0.93221
## B6          -17.32202    3.42300   -5.06  4.4e-07 ***
```

```
## B7               28.13555    3.55367     7.92  3.1e-15 ***
## B8              -27.00216    3.16690    -8.53  < 2e-16 ***
## B9              -10.18176    4.45474    -2.29  0.02233 *
## B10              10.37967    3.32686     3.12  0.00182 **
## B11             -22.41943    3.53434    -6.34  2.5e-10 ***
## B12              21.66451    4.81160     4.50  6.9e-06 ***
## B13              13.99856    6.47562     2.16  0.03070 *
## lon:lat          -0.02694    0.02822    -0.95  0.33981
## lon:day          -0.02263    0.00129   -17.56  < 2e-16 ***
## lat:day          -0.01888    0.00188   -10.07  < 2e-16 ***
## ---
## Signif. codes:
## 0 '***' 0.001 '**' 0.01 '*' 0.05 '.' 0.1 ' ' 1
##
## Residual standard error: 4.22 on 3969 degrees of freedom
## Multiple R-squared:  0.703,Adjusted R-squared:  0.701
## F-statistic:  494 on 19 and 3969 DF,  p-value: <2e-16
```

The introduction of the 13th basis function will not adversely affect the predictions and prediction standard errors, but it does compromise our ability to correctly interpret the fixed effects. Multicollinearity will result in a high positive or negative correlation between the estimators of the regression coefficients. For example, the correlation matrix of the estimators of the fixed effects corresponding to these two basis functions is given by

```
vcov(Tmax_July_lm3)[c("B5", "B13"),c("B5", "B13")] %>%
    cov2cor()
```

```
##            B5     B13
## B5    1.000  -0.922
## B13  -0.922   1.000
```

Analyzing the Residuals

Having fitted a spatio-temporal model, it is good practice to check the residuals. If they are still spatio-temporally correlated, then our model will not have captured adequately the spatial and temporal variability in the data. We extract the residuals from our linear model using the function `residuals`.

```
Tmax_no_14$residuals <- residuals(Tmax_July_lm)
```

Now let us plot the residuals of the last eight days. Notice how these residuals, shown in the bottom panel of Figure 3.9, are strongly spatially correlated. The triangles in the image correspond to the two stations whose time series of residuals we shall analyze later.

```
g <- ggplot(filter(Tmax_no_14, day %in% 24:31)) +
  geom_point(aes(lon, lat, colour = residuals)) +
  facet_wrap(~ day, ncol=4) +
  col_scale(name = "degF") +
  geom_point(data = filter(Tmax_no_14,day %in% 24:31 &
                                     id %in% c(3810, 3889)),
              aes(lon, lat),  colour = "black",
              pch = 2, size = 2.5) +
  theme_bw()
```

```
print(g)
```

We can use Moran's *I* test, described in Technical Note 3.2, to test for spatial dependence in the residuals on each day. In the following code, we take each day in our data set, compute the distances, form the weight matrix, and carry out Moran's *I* test using the function `Moran.I` from the package **ape**.

```
P <- list()                              # init list
days <- c(1:13, 15:31)                   # set of days
for(i in seq_along(days)) {              # for each day
  Tmax_day <- filter(Tmax_no_14,
                     day == days[i])     # filter by day
  station.dists <- Tmax_day %>%          # take the data
    select(lon, lat) %>%                 # extract coords.
    dist() %>%                           # comp. dists.
    as.matrix()                          # conv. to matrix
  station.dists.inv <- 1/station.dists   # weight matrix
  diag(station.dists.inv) <- 0           # 0 on diag
  P[[i]] <- Moran.I(Tmax_day$residuals,  # run Moran's I
                    station.dists.inv) %>%
            do.call("cbind", .)          # conv. to df
}
```

The object P is a list of single-row data frames that can be collapsed into a single data frame by calling `do.call` and proceeding to row-bind the elements of each list item together. We print the first six records of the resulting data frame below.

```
do.call("rbind", P) %>% head()
```

```
##      observed expected      sd p.value
## [1,]    0.272 -0.00758 0.0124       0
## [2,]    0.226 -0.00758 0.0124       0
## [3,]    0.211 -0.00758 0.0124       0
## [4,]    0.163 -0.00758 0.0124       0
```

```
## [5,]     0.258 -0.00758 0.0124          0
## [6,]     0.122 -0.00758 0.0123          0
```

The maximum p-value from the 30 tests is 8.04×10^{-6}, which is very small. Since we are in a multiple-hypothesis setting, we need to control the familywise error rate and, for a level of significance α, reject the null hypothesis of no correlation only if the p-value is less than $c(\alpha)$ $(< \alpha)$, where $c(\cdot)$ is a correction function. In this case, even the very conservative Bonferroni correction (for which $c(\alpha) = \alpha/T$, where T is the number of time points) will result in rejecting the null hypothesis at each time point.

It is straightforward to extend Moran's I test to the spatio-temporal setting, as one need only extend the concept of "spatial distance" to "spatio-temporal distance." We are faced with the usual problem of how to appropriately scale time to make a Euclidean distance across space and time have a realistic interpretation. One way to do this is to fit a dependence model that allows for scaling in time, and subsequently scale time by an estimate of the scaling factor prior to computing the Euclidean distance. We shall work with one such model, which uses an anisotropic covariance function, in Chapter 4. For now, as we did with IDW, we do not scale time and compute distances on the spatio-temporal domain (which happens to be reasonable for these data).

```
station.dists <- Tmax_no_14 %>%    # take the data
  select(lon, lat, day) %>%        # extract coordinates
  dist() %>%                       # compute distances
  as.matrix()                      # convert to matrix
```

We now need to compute the weights from the distances, set the diagonal to zero and call `Moran.I`.

```
station.dists.inv <- 1/station.dists
diag(station.dists.inv) <- 0
Moran.I(Tmax_no_14$residuals, station.dists.inv)$p.value
```

```
## [1] 0
```

Unsurprisingly, given what we saw when analyzing individual time slices, the p-value is very small, strongly suggesting that there is spatio-temporal dependence in the data.

When the data are regularly spaced in time, as is the case here, one may also look at the "temporal" residuals at some location and test for temporal correlation in these residuals using the Durbin–Watson test. For example, consider the maximum temperature (`Tmax`) residuals at stations 3810 and 3889.

```
TS1 <- filter(Tmax_no_14, id == 3810)$residuals
TS2 <- filter(Tmax_no_14, id == 3889)$residuals
```

These residuals can be easily plotted using base R graphics as follows; see the top panel of Figure 3.9.

```
par(mar=c(4, 4, 1, 1))
plot(TS1,                                # Station 3810 residuals
     xlab = "day of July 1993",
     ylab = "residuals (degF)",
     type = 'o', ylim = c(-8, 7))
lines(TS2,                               # Station 3889 residuals
      xlab = "day of July 1993",
      ylab = "residuals (degF)",
      type = 'o', col = "red")
```

Note that there is clear temporal correlation in the residuals; that is, residuals close to each other in time tend to be more similar than residuals further apart. It is also interesting to note that the residuals are correlated between the stations; that is, at the same time point, the residuals at both stations are more similar than at different time points. This is due to the spatial correlation in the residuals that was tested for above (these two stations happen to be quite close to each other; recall the previous image of the spatial residuals). One may also look at the *correlogram* (the empirical autocorrelation function) of the residuals by typing `acf(TS1)` and `acf(TS2)`, respectively. From these plots it can be clearly seen that there is significant lag-1 correlation in both these residual time series.

Now let us proceed with carrying out a Durbin–Watson test for the residuals at every station. This can be done using a `for` loop as we did with Moran's *I* test; however, we shall now introduce a more elaborate way of carrying out multiple tests and predictions on groups of data within a data frame, using the packages **tidyr**, **purrr**, and **broom**, which will also be used in Lab 3.3.

In Lab 2.1 we investigated data wrangling techniques for putting data that are in a data frame into groups using `group_by`, and then we performed an operation on each of those groups using `summarise`. The grouped data frame returned by `group_by` is simply the original frame with each row associated with a group. A more elaborate representation of these data is in a *nested data frame*, where we have a data frame containing one row *for each group*. The "nested" property comes from the fact that we may have a data frame, conventionally under the field name "data," for each group. For example, if we group Tmax_no_14 by `lon` and `lat`, we obtain the following first three records.

```
nested_Tmax_no_14 <- group_by(Tmax_no_14, lon, lat) %>% nest()
head(nested_Tmax_no_14, 3)

## # A tibble: 3 x 3
##    lon   lat data
##   <dbl> <dbl> <list>
## 1 -81.4  39.3 <tibble [30 x 16]>
```

```
## 2 -81.4  35.7 <tibble [30 x 16]>
## 3 -88.9  35.6 <tibble [30 x 16]>
```

Note the third column, `data`, which is a column of *tibbles* (which, for the purposes of this book, should be treated as sophisticated data frames). Next we define a function that takes the data frame associated with a single group, carries out the test (in this case the Durbin–Watson test), and returns the results. The function `dwtest` takes an R formula as the first argument and the data as the second argument. In this case, we test for autocorrelation in the residuals after removing a temporal (constant) trend and by using the formula `residuals ~ 1`.

```
dwtest_one_station <- function(data)
                        dwtest(residuals ~ 1, data = data)
```

Calling `dwtest_one_station` for the data in the first record will carry out the test at the first station, in the second record at the second station, and so on. For example,

```
dwtest_one_station(nested_Tmax_no_14$data[[1]])
```

carries out the Durbin–Watson test on the residuals at the first station.

To carry out the test on each record in the nested data frame, we use the function `map` from the package **purrr**. For example, the command

```
map(nested_Tmax_no_14$data, dwtest_one_station) %>% head()
```

shows the test results for the first six stations. These results can be assigned to another column within our nested data frame using `mutate`. These results are of class `htest` and not easy to analyze or visualize in their native form. We therefore use the function `tidy` from the package **broom** to extract the key information from the test (in this case the statistic, the *p*-value, the method, and the hypothesis) and put it into a data frame. For example,

```
dwtest_one_station_tidy <- nested_Tmax_no_14$data[[1]] %>%
                              dwtest_one_station() %>%
                              tidy()
```

tidies up the results at the first station. The first three columns of the returned data are

```
dwtest_one_station_tidy[, 1:3]
```

```
## # A tibble: 1 x 3
##   statistic p.value method
##       <dbl>   <dbl> <chr>
## 1     0.982 0.00122 Durbin-Watson test
```

To assign the test results to each record in the nested data frame as added fields (instead of as another data frame), we then use the function **unnest**. In summary, the code

```
Tmax_DW_no_14 <- nested_Tmax_no_14 %>%
    mutate(dwtest = map(data, dwtest_one_station)) %>%
    mutate(test_df = map(dwtest, tidy)) %>%
    unnest(test_df)
```

provides all the information we need. The first three records, excluding the last two columns, are

```
Tmax_DW_no_14 %>% select(-method, -alternative) %>% head(3)
```

```
## # A tibble: 3 x 6
##     lon   lat data          dwtest     statistic p.value
##   <dbl> <dbl> <list>        <list>         <dbl>   <dbl>
## 1 -81.4  39.3 <tibble [30~  <S3: hte~      0.982 1.22e-3
## 2 -81.4  35.7 <tibble [30~  <S3: hte~      0.921 5.86e-4
## 3 -88.9  35.6 <tibble [30~  <S3: hte~      1.59  1.29e-1
```

The proportion of p-values *below* the 5% level of significance divided by the number of tests (Bonferroni correction) is

```
mean(Tmax_DW_no_14$p.value < 0.05/nrow(Tmax_DW_no_14)) * 100
```

```
## [1] 21.8
```

This proportion of 21.8% is reasonably high and provides evidence that there is considerable temporal autocorrelation in the residuals, as expected.

Finally, we can also compute and visualize the empirical spatio-temporal semivariogram of the residuals. Recall that in Lab 2.1 we put the maximum temperature data in the NOAA data set into an STFDF object that we labeled STObj3. We now load these data and subset the month of July 1993.

```
data("STObj3", package = "STRbook")
STObj4 <- STObj3[, "1993-07-01::1993-07-31"]
```

All we need to do is merge Tmax_no_14, which contains the residuals, with the STFDF object STObj4, so that the empirical semivariogram of the residuals can be computed. This can be done quickly, and safely, using the function **left_join**.

```
STObj4@data <- left_join(STObj4@data, Tmax_no_14)
```

As in Lab 2.3, we now compute the empirical semivariogram with the function **variogram**.

```
vv <- variogram(object = residuals ~ 1, # fixed effect component
                data = STObj4,      # July data
                width = 80,         # spatial bin (80 km)
                cutoff = 1000,      # consider pts < 1000 km apart
                tlags = 0.01:6.01)  # 0 days to 6 days
```

The command `plot` (vv) displays the empirical semivariogram of the residuals, which is shown in Figure 3.8. This empirical semivariogram is clearly different from that of the data (Figure 2.17) and has a lower sill, but it suggests that there is still spatial and temporal correlation in the residuals.

Predictions

Prediction from linear or generalized linear models in R is carried out using the function `predict`. As in Lab 3.1, we use the following prediction grid.

```
pred_grid <- expand.grid(lon = seq(-100, -80, length = 20),
                         lat = seq(32, 46, length = 20),
                         day = seq(4, 29, length = 6))
```

We require all the covariate values at all the prediction locations. Hence, the 12 basis functions need to be evaluated on this grid. As above, we do this by calling `eval_basis` and converting the result to a matrix, which we then attach to our prediction grid.

```
Spred <- eval_basis(basis = G,                      # basis functs
                    s = pred_grid[,c("lon","lat")] %>%  # pred locs
                      as.matrix()) %>%              # conv. to matrix
    as.matrix()                                     # results as matrix
colnames(Spred) <- paste0("B", 1:ncol(Spred)) # assign col names
pred_grid <- cbind(pred_grid, Spred)          # attach to grid
```

Now that we have all the covariates in place, we can call `predict`. We supply `predict` with the model Tmax_July_lm, the prediction grid, and the argument interval = "prediction", so that `predict` returns the prediction intervals.

```
linreg_pred <- predict(Tmax_July_lm,
                       newdata = pred_grid,
                       interval = "prediction")
```

When `predict` is called as above, it returns a matrix containing three columns with names fit, lwr, and upr, which contain the prediction and the lower and upper bounds of the 95% prediction interval, respectively. Since in this case the prediction interval is the prediction $\pm$ 1.96 times the prediction standard error, we can calculate the prediction standard error from the given interval as follows.

```
## Assign prediction and prediction s.e. to the prediction grid
pred_grid$z_pred <- linreg_pred[,1]
pred_grid$z_err <- (linreg_pred[,3] - linreg_pred[,2]) / (2*1.96)
```

Plotting the prediction and prediction standard error proceeds in a straightforward fashion using **ggplot2**; see Figure 3.7. This is left as an exercise for the reader.

Lab 3.3: Regression Models for Forecasting

In this Lab we fit a simple linear model to every pixel in the SST data set, and we use these models to predict SST for a month in which we have no SST data. The models will simply contain an intercept and a single covariate, namely the Southern Oscillation Index (SOI). The SOI data we use here are supplied with **STRbook** and were retrieved from `https://www.esrl.noaa.gov/psd/gcos_wgsp/Timeseries/SOI/`.

For this Lab we need the usual data-wrangling and plotting packages, as well as the packages **broom** and **purrr** for fitting and predicting with multiple models simultaneously.

```
library("broom")
library("dplyr")
library("ggplot2")
library("STRbook")
library("purrr")
library("tidyr")
```

In the first section of this Lab we tidy up the data to obtain the SST data frame from the raw data. You may also skip this section by loading `SST_df` from **STRbook**, and fast-forwarding to the section that is concerned with fitting the data.

```
data("SST_df", package = "STRbook")
```

Tidying Up the Data

The first task in this Lab is to wrangle the SST data into a long-format data frame that is amenable to linear fitting and plotting. Recall from Lab 2.3 that the SST data are provided in three data frames, one describing the land mask, one containing the SST values in wide format, and one containing the coordinates.

```
data("SSTlandmask", package = "STRbook")
data("SSTdata", package = "STRbook")
data("SSTlonlat", package = "STRbook")
```

We first combine the land mask data with the coordinates data frame.

```
lonlatmask_df <- data.frame(cbind(SSTlonlat, SSTlandmask))
names(lonlatmask_df) <- c("lon", "lat", "mask")
```

Then we form our SST data frame in wide format by attaching `SSTdata` to the coordinate-mask data frame.

```
SSTdata <- cbind(lonlatmask_df, SSTdata)
```

Finally, we use `gather` to put the data frame into long format.

```
SST_df <- gather(SSTdata, date, sst, -lon, -lat, -mask)
```

Our data frame now contains the SST data, but the `date` field contains as entries V1, V2, . . . , which were the names of the columns in `SSTdata`.

```
SST_df %>% head(3)
```

```
##   lon lat mask date       sst
## 1 124 -29    1   V1 -0.36289
## 2 126 -29    1   V1 -0.28461
## 3 128 -29    1   V1 -0.19195
```

We replace this `date` field with two fields, one containing the month and one containing the year. We can do this by first creating a mapping table that links V1 to January 1970, V2 to February 1970, and so on, and then merging using `left_join`.

```
date_grid <- expand.grid(Month = c("Jan", "Feb", "Mar", "Apr",
                                   "May", "Jun", "Jul", "Aug",
                                   "Sep", "Oct", "Nov", "Dec"),
                         Year = 1970:2002,
                         stringsAsFactors =  FALSE)
date_grid$date <- paste0("V", 1:396)
SST_df <- left_join(SST_df, date_grid) %>%
          select(-date)
```

For good measure, we also add in the `date` field again but this time in month–year format.

```
SST_df$date <- paste(SST_df$Month, SST_df$Year)
SST_df %>% head(3)
```

```
##   lon lat mask      sst Month Year     date
## 1 124 -29    1 -0.36289   Jan 1970 Jan 1970
## 2 126 -29    1 -0.28461   Jan 1970 Jan 1970
## 3 128 -29    1 -0.19195   Jan 1970 Jan 1970
```

Next, we set SST data that are coincident with land locations to `NA`:

```
SST_df$sst<- ifelse(SST_df$mask == 0, SST_df$sst, NA)
```

Our SST data frame is now in place. The following code plots a series of SSTs leading up to the 1997 El Niño event; see Figure 3.10.

```
g <- ggplot(filter(SST_df, Year == 1997 &   # subset by month/year
                    Month %in% c("Apr","Aug","Jun","Oct"))) +
  geom_tile(aes(lon, lat,
                fill = pmin(sst, 4))) +   # clamp SST at 4deg
    facet_wrap(~date, dir = "v") +        # facet by date
    fill_scale(limits = c(-4, 4),         # color limits
               name = "degC") +           # legend title
    theme_bw() + coord_fixed()            # fix scale and theme
```

Now we need to add the SOI data to the SST data frame. The SOI time series is available as a 14-column data frame, with the first column containing the year, the next 12 columns containing the SOI for each month in the respective year, and the last column containing the mean SOI for that year. In the following, we remove the annual average from the data frame, which is in wide format, and then put it into long format using `gather`.

```
data("SOI", package = "STRbook")
SOI_df <- select(SOI, -Ann) %>%
        gather(Month, soi, -Year)
```

Finally, we use `left_join` to merge the SOI data and the SST data.

```
SST_df <- left_join(SST_df, SOI_df,
                    by = c("Month", "Year"))
```

Fitting the Models Pixelwise

In this section we fit linear time-series models to the SSTs in each pixel using data up to April 1997. We first create a data frame containing the SST data between January 1970 and April 1997.

```
SST_pre_May <- filter(SST_df, Year <= 1997) %>%
            filter(!(Year == 1997 &
                    Month %in% c("May", "Jun", "Jul",
                                 "Aug", "Sep", "Oct",
                                 "Nov", "Dec")))
```

Next, as in Lab 3.2, we use **purrr** and **broom** to construct a nested data frame that contains a linear model fitted to every pixel. We name the function that fits the linear model at a single pixel to the data over time as `fit_one_pixel`.

```
fit_one_pixel <- function(data)
               mod <- lm(sst ~ 1 + soi, data = data)

pixel_lms <- SST_pre_May %>%
            filter(!is.na(sst)) %>%
            group_by(lon, lat) %>%
            nest() %>%
            mutate(model = map(data, fit_one_pixel)) %>%
            mutate(model_df = map(model, tidy))
```

The string of commands above describes an operation that is practically identical to what we did in Lab 3.2. We take the data, filter them to remove missing data, group by pixel, create a nested data frame, fit a model to each pixel, and extract a data frame containing information on the linear fit by pixel. The first three records of the nested data frame are as follows.

```
pixel_lms %>% head(3)

## # A tibble: 3 x 5
##     lon   lat data                  model      model_df
##   <dbl> <dbl> <list>                <list>     <list>
## 1   154   -29 <tibble [328 x 6]> <S3: lm> <tibble [2 x 5]>
## 2   156   -29 <tibble [328 x 6]> <S3: lm> <tibble [2 x 5]>
## 3   158   -29 <tibble [328 x 6]> <S3: lm> <tibble [2 x 5]>
```

To extract the model parameters from the linear-fit data frames, we use **unnest**:

```
lm_pars <- pixel_lms %>%
          unnest(model_df)
```

For each pixel, we now have an estimate of the intercept and the effect associated with the covariate `soi`, as well as other information such as the p-values.

```
head(lm_pars, 3)

## # A tibble: 3 x 7
##     lon   lat term    estimate std.error statistic  p.value
##   <dbl> <dbl> <chr>      <dbl>     <dbl>     <dbl>    <dbl>
## 1   154   -29 (Inter~   0.132    0.0266      4.96  1.13e-6
## 2   154   -29 soi      0.0277    0.0223      1.24  2.16e-1
## 3   156   -29 (Inter~  0.0365    0.0262      1.40  1.64e-1
```

We can plot spatial maps of the intercept and the regression coefficient associated with `soi` directly. We first merge this data frame with the coordinates data frame using **left_join**, which also contains land pixels. In this way, regression coefficients over land pixels are marked as `NA`, which is appropriate.

```
lm_pars <- left_join(lonlatmask_df, lm_pars)
```

The following code plots the spatial intercept and the spatial regression coefficient associated with `soi`; see the top panels of Figure 3.11.

```
g2 <- ggplot(filter(lm_pars, term == "(Intercept)" | mask == 1)) +
    geom_tile(aes(lon, lat, fill = estimate)) +
    fill_scale() +
    theme_bw() + coord_fixed()

g3 <- ggplot(filter(lm_pars, term == "soi" | mask == 1)) +
    geom_tile(aes(lon, lat, fill = estimate)) +
    fill_scale() +
    theme_bw() + coord_fixed()
```

Predicting SST Pixelwise

We now use the linear models at the pixel level to predict the SST in October 1997 using the SOI index for that month. The SOI for that month is extracted from `SOI_df` as follows.

```
soi_pred <- filter(SOI_df, Month == "Oct" & Year == "1997") %>%
            select(soi)
```

We next define the function that carries out the prediction at the pixel level. The function takes a linear model `lm` and the SOI at the prediction date `soi_pred`, runs the **predict** function for this date, and returns a data frame containing the prediction and the prediction standard error.

```
predict_one_pixel <- function(lm, soi_pred) {
    predict(lm,                          # linear model
            newdata = soi_pred,          # pred. covariates
            interval = "prediction") %>% # output intervals
    data.frame() %>%                     # convert to df
    mutate(se = (upr-lwr)/(2 * 1.96)) %>% # comp pred. se
    select(fit, se)                      # return fit & se
  }
```

Prediction proceeds at each pixel by calling **predict_one_pixel** on each row in our nested data frame `pixel_lms`.

```
SST_Oct_1997 <- pixel_lms %>%
                mutate(preds = map(model,
                                   predict_one_pixel,
                                   soi_pred = soi_pred)) %>%
                unnest(preds)
```

We have unnested the `preds` data frame above to save the fit and prediction standard
error as fields in the `SST_Oct_1997` data frame. You can type `SST_Oct_1997 %>%`
`head`(3) to have a look at the first three records. It is straightforward to plot the prediction
and prediction standard error from `SST_Oct_1997`; see the middle and bottom panels of
Figure 3.11. This is left as an exercise for the reader.

Lab 3.4: Generalized Linear Spatio-Temporal Regression

In this Lab we fit a generalized linear spatio-temporal model to yearly counts of Carolina
wren in and around the state of Missouri between 1994 and 2014. These counts are part
of the BBS data set. We need **gstat**, **sp**, and **spacetime** for fitting an empirical semivariogram to the residuals, **FRK** to construct the basis functions (as in Lab 3.2), **ape** for running
Moran's *I* test, and the usual packages for wrangling and plotting.

```
library("ape")
library("dplyr")
library("FRK")
library("ggplot2")
library("gstat")
library("sp")
library("spacetime")
library("STRbook")
library("tidyr")
```

Fitting the Model

The Carolina wren counts in the BBS data set, in both wide and long format, are supplied
with **STRbook**. Here we load the data directly in long format and remove any records that
contain missing observations.

```
data("MOcarolinawren_long", package = "STRbook")
MOcarolinawren_long <- MOcarolinawren_long %>%
                       filter(!is.na(cnt))
```

We use the same covariates to fit these data as we did to fit the maximum temperature,
`Tmax`, in Lab 3.2. Twelve of these covariates were basis functions constructed using

`auto_basis` from the package **FRK**; see Lab 3.2 for details. The matrix `S` below then contains the basis functions evaluated at the Carolina wren observation locations.

```
G <- auto_basis(data = MOcarolinawren_long[,c("lon","lat")] %>%
                     SpatialPoints(),              # To sp obj
               nres = 1,                           # One resolution
               type = "Gaussian")                  # Gaussian BFs

S <- eval_basis(basis = G,                         # basis functions
               s = MOcarolinawren_long[,c("lon","lat")] %>%
                     as.matrix()) %>%              # conv. to matrix
    as.matrix()                                    # conv. to matrix
colnames(S) <- paste0("B", 1:ncol(S)) # assign column names
```

Next, we attach the basis-function covariate information to the data frame containing the counts, and remove the fields `loc.ID` and `t`, which we will not explicitly use when fitting the model. We list the first five columns of the first three records of our constructed data frame `Wren_df` as follows.

```
Wren_df <- cbind(MOcarolinawren_long,S) %>%
  select(-loc.ID, -t)
Wren_df[1:3, 1:5]

##   cnt  lat   lon year       B1
## 1   4 36.8 -89.2 1994 0.00258
## 2   2 36.6 -90.7 1994 0.03551
## 3   8 36.9 -91.7 1994 0.11588
```

Generalized linear models (GLMs) are fitted in R using the function `glm`. The function works similarly to `lm`, but in addition it requires one to specify the exponential-family model that is used (in this first instance we consider the Poisson family), as well as the link function (here we use the log function, which is the canonical link). The `glm` function is called as follows (note that we have used the same formula as in Lab 3.2).

```
Wren_GLM <- glm(cnt ~ (lon + lat + year)^2 + ., # formula
               family = poisson("log"),         # Poisson + log link
               data = Wren_df)                  # data set
```

The mean and variance of a random variable that has a Poisson distribution are the same. In cases where the variance in the data is greater than that suggested by this model, the data are said to exhibit "over-dispersion." An estimate of the dispersion is given by the ratio of the deviance to the total degrees of freedom (the number of data points minus the number of covariates). In this case the dispersion estimate is

```
Wren_GLM$deviance / Wren_GLM$df.residual
```

```
## [1] 3.78
```

which is greater than 1, a sign of over-dispersion.

Another way to obtain an estimate of the disperson parameter (and, to account for it if present) is to replace `poisson` with `quasipoisson` when calling `glm`, and then type `summary`(Wren_GLM). The quasi-Poisson model assumes that the variance is proportional to the mean, and that the constant of the proportionality is the over-dispersion parameter. Note from the output of `summary` that the dispersion parameter is 3.9, which is close to what we estimated above.

It can be shown that under the null hypothesis of no over-dispersion, the deviance is approximately chi-squared distributed with degrees of freedom equal to $m - p - 1$.

```
Wren_GLM$df.residual
```

```
## [1] 764
```

The observed deviance is

```
Wren_GLM$deviance
```

```
## [1] 2890
```

The probability of observing such a large or larger deviance under the null hypothesis of no over-dispersion (i.e., the p-value) is

```
1 - pchisq(q = Wren_GLM$deviance, df = Wren_GLM$df.residual)
```

```
## [1] 0
```

Therefore, we reject the null hypothesis of no over-dispersion at the usual levels of significance (10%, 5%, and 1%). One may use other models in the exponential family, such as the negative-binomial distribution, to account explicitly for the over-dispersion. For convenience, in this Lab we proceed with the Poisson family. We use the negative-binomial distribution in Lab 4.4.

Prediction

As in the other Labs, prediction proceeds through use of the function `predict`. We first generate our space-time prediction grid, which is an $80 \times 80 \times 21$ grid in degrees $\times$ degrees $\times$ years, covering the observations in space and in time.

```
pred_grid <- expand.grid(lon = seq(
                         min(MOcarolinawren_long$lon) - 0.2,
                         max(MOcarolinawren_long$lon) + 0.2,
                         length.out = 80),
                    lat = seq(
                         min(MOcarolinawren_long$lat) - 0.2,
                         max(MOcarolinawren_long$lat) + 0.2,
                         length.out = 80),
                    year = 1994:2014)
```

As in Lab 3.2, we now evaluate the basis functions at the prediction locations.

```
S_pred <- eval_basis(basis = G,                          # basis functs
               s = pred_grid[,c("lon","lat")] %>%         # pred locs
                   as.matrix()) %>%                        # conv. to matrix
     as.matrix()                                           # as matrix
colnames(S_pred) <- paste0("B", 1:ncol(S_pred))          # assign  names
pred_grid <- cbind(pred_grid, S_pred)                     # attach to grid
```

In the call to **predict** below, we specify type = "link" to indicate that we predict the link function of the response and not the response (analogous to the log-intensity of the process).

```
wren_preds <- predict(Wren_GLM,
                      newdata = pred_grid,
                      type = "link",
                      se.fit = TRUE)
```

The predictions and prediction standard errors of the link function of the response are then attached to our prediction grid for plotting; see Figure 3.12. Plotting to obtain Figure 3.12 is left as an exercise for the reader.

```
pred_grid <- pred_grid %>%
             mutate(log_cnt = wren_preds$fit,
                    se = wren_preds$se.fit)
```

When fitting GLMs, it is good practice to check the deviance residuals and inspect them for any residual correlation. The default GLM residuals returned by **residuals** are deviance residuals.

```
Wren_df$residuals <- residuals(Wren_GLM)
```

Interestingly, the plot of the deviance residuals in Figure 3.13 is "noisy," indicating a lack of spatial correlation.

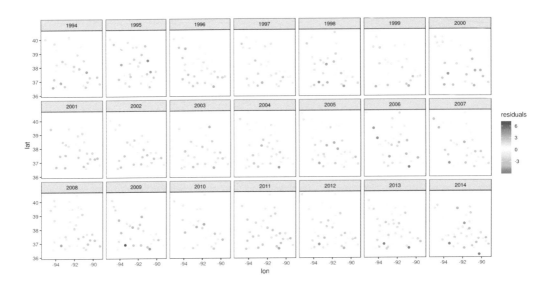

Figure 3.13: The deviance residuals from the fitted GLM between $t = 1$ (the year 1994) and $t = 21$ (2014).

```
g2 <- ggplot(Wren_df) +
    geom_point(aes(lon, lat, colour = residuals)) +
    col_scale(name = "residuals") +
    facet_wrap(~year, nrow = 3) + theme_bw()
```

We can test for spatial correlation of the deviance residuals by running Moran's I test on the spatial deviance residuals for each year. The code below follows closely that for Moran's I test in Lab 3.2 and then summarizes the p-values obtained for each year.

```
P <- list()                          # init list
years <- 1994:2014
for(i in seq_along(years)) {         # for each day
  Wren_year <- filter(Wren_df,
                  year == years[i])  # filter by year
  obs_dists <- Wren_year %>%         # take the data
    select(lon,lat) %>%              # extract coords.
    dist() %>%                       # comp. dists.
    as.matrix()                      # conv. to matrix
  obs_dists.inv <- 1/obs_dists       # weight matrix
  diag(obs_dists.inv) <- 0           # 0 on diag
  P[[i]] <- Moran.I(Wren_year$residuals,  # run Moran's I
                  obs_dists.inv) %>%
          do.call("cbind", .)        # conv. to df
```

```
}
do.call("rbind",P) %>% summary(digits = 2)
```

```
##      observed               expected              sd                p.value
##   Min.    :-0.084     Min.    :-0.040     Min.    :0.025     Min.    :0.06
##   1st Qu.:-0.059     1st Qu.:-0.029     1st Qu.:0.028     1st Qu.:0.24
##   Median :-0.044     Median :-0.029     Median :0.030     Median :0.42
##   Mean    :-0.041     Mean    :-0.028     Mean    :0.031     Mean    :0.47
##   3rd Qu.:-0.022     3rd Qu.:-0.025     3rd Qu.:0.033     3rd Qu.:0.68
##   Max.    : 0.010     Max.    :-0.023     Max.    :0.041     Max.    :0.94
```

Hence, at the 5% level of significance, the null hypothesis (of no spatial correlation in these deviance residuals) is not rejected. This was expected from the visualization in Figure 3.13.

More insight can be obtained by looking at the empirical semivariogram of the deviance residuals. To do this we first construct an STIDF, thereby casting the irregular space-time data into a **spacetime** object.

```
Wren_STIDF <- STIDF(sp = SpatialPoints(
                        Wren_df[,c("lon","lat")],
                        proj4string = CRS("+proj=longlat")),
                    time = as.Date(Wren_df[, "year"] %>%
                                    as.character(),
                                format = "%Y"),
                    data = Wren_df)
```

Then we compute the empirical semivariogram using **variogram**. We consider time bins of width 1 year (i.e., of width 52.1429 weeks). Bins specified in units of weeks are required, as this is the largest temporal unit recognized by **variogram** .

```
tlags <- seq(0.01, 52.1429*6 + 0.01, by = 52.1429)
vv <- variogram(object = residuals ~ 1,  # fixed effect component
                data = Wren_STIDF,        # data set
                tlags = tlags,            # temp. bins
                width = 25,               # spatial bin (25 km)
                cutoff = 150,             # use pts < 150 km apart
                tunit = "weeks")          # time unit
```

The empirical semivariogram can be plotted using `plot(vv)`. Notice how there is little evidence of spatial correlation but ample evidence of temporal correlation in the residuals. (The variance of the differences over a large range of time lags at the same spatial location is small.) This is a clear sign that a more sophisticated spatio-temporal random-effects model should be considered for these data.

Chapter 4

Descriptive Spatio-Temporal Statistical Models

Chapter 3 is the linchpin for the "two Ds" of spatio-temporal statistical modeling, which are now upon us in this chapter (the first "D," namely "descriptive") and the next chapter (the second "D," namely "dynamic"). We hope to have eased you from the free form of spatio-temporal exploratory data analysis presented in Chapter 2 into the "rigor" needed to build a coherent statistical model. The independent probability structure assumed in Chapter 3 was a place-holder for the sorts of probability structures that respect Tobler's law, discussed in the previous chapters: in our context, this says that a set of values at nearby spatio-temporal locations should *not* be assumed independent. As we shall see, there is a "descriptive" way (this chapter, Chapter 4) and a "dynamic" way (Chapter 5) to incorporate spatio-temporal statistical dependence into models.

In this chapter we focus on two of the goals of spatio-temporal modeling given in Chapter 3: prediction at some location in space within the time span of the observations and, to a lesser extent, parameter inference for spatio-temporal covariates. For both goals we assume that our observations can be decomposed into a true (latent) spatio-temporal process plus observation error. We then assume that the true process can be written in terms of spatio-temporal fixed effects due to covariates plus a spatio-temporally dependent random process. We call this a *descriptive* approach because its main concern is to specify (or describe) the dependence structure in the random process. This is in contrast to the *dynamic* approach presented in Chapter 5 that models the evolution of the dependent random process through time. To implement the prediction and inference approaches discussed herein we must perform estimation. We mention the most popular and relevant estimation approaches and algorithms as they come up, but omit most of the details. The interested reader can explore these details in the references given in Section 4.6. Finally, we note that these discussions require a bit more statistical formality and mathematical notation, and so the presentations in this and the next chapter are at a higher technical level than those in Chapter 3.

4.1   Additive Measurement Error and Process Models

In this section we describe more formally a two-stage model that considers additive measurement error in a data (observation) model, and a process model that is decomposed into a fixed- (covariate-) effect term and a random-process term. This general decomposition is the basis for the models that we present in this and the next chapter.

Recall that at each time $t \in \{t_1, \ldots, t_T\}$ we have m_{t_j} observations. With a slight abuse of notation, we write the number of observations at time t_j as m_j. The vector of all observations is then given by

$$\mathbf{Z} = (Z(\mathbf{s}_{11}; t_1), Z(\mathbf{s}_{21}; t_1), \ldots, Z(\mathbf{s}_{m_11}; t_1), \ldots, Z(\mathbf{s}_{1T}; t_T), \ldots, Z(\mathbf{s}_{m_TT}; t_T))'.$$

That is, different numbers of irregular spatial observations are allowed for each time (note that if there are no observations at a given time, t_j, the set of spatial locations is empty for that time and $m_j = 0$). We seek a prediction at some spatio-temporal location $(\mathbf{s}_0; t_0)$. As described in Chapter 1, if $t_0 < t_T$, so that we have all data available to us, then we are in a smoothing situation; if we only have data up to time t_0 then we are in a filtering situation; and if $t_0 > t_T$ then we are in a forecasting situation. We seek statistically optimal predictions for an underlying latent (i.e., hidden) random spatio-temporal process. We denote this process by $\{Y(\mathbf{s}; t) : \mathbf{s} \in D_s, \ t \in D_t\}$, for spatial location $\mathbf{s}$ in spatial domain D_s (a subset of d-dimensional Euclidean space), and time index t in temporal domain D_t (along the one-dimensional real line).

More specifically, suppose we represent the data in terms of the latent spatio-temporal process of interest plus a measurement error. For example,

$$Z(\mathbf{s}_{ij}; t_j) = Y(\mathbf{s}_{ij}; t_j) + \epsilon(\mathbf{s}_{ij}; t_j), \quad i = 1, \ldots, m_j; \ j = 1, \ldots, T, \tag{4.1}$$

where the errors $\{\epsilon(\mathbf{s}_{ij}; t_j)\}$ represent *iid* mean-zero measurement error that is independent of $Y(\cdot; \cdot)$ and has variance σ_ϵ^2. So, in the simple data model (4.1) we assume that the data are noisy observations of the *latent* process Y at a finite collection of locations in the space-time domain, where typically we have not observed data at all locations of interest. Consequently, we would like to predict the latent value $Y(\mathbf{s}_0; t_0)$ at a spatio-temporal location $(\mathbf{s}_0; t_0)$ as a function of the data vector represented by $\mathbf{Z}$ (or some subset of these observations), which is of dimension $\sum_{j=1}^{T} m_j$. To simplify the notation that follows, we shall sometimes assume that data were observed at the same set of m locations for each of the T times, in which case $\mathbf{Z}$ is of length mT.

Now suppose that the latent process follows the model

$$Y(\mathbf{s}; t) = \mu(\mathbf{s}; t) + \eta(\mathbf{s}; t), \tag{4.2}$$

for all $(\mathbf{s}; t)$ in our space-time domain of interest (e.g., $D_s \times D_t$), where each component in (4.2) has a special role to play. In (4.2), $\mu(\mathbf{s}; t)$ represents the process mean, which is not random, and $\eta(\mathbf{s}; t)$ represents a mean-zero random process with spatial and temporal statistical dependence. Our goal here is to find the optimal linear predictor in the sense that it minimizes the mean squared prediction error between $Y(\mathbf{s}_0; t_0)$ and our prediction, which we write as $\widehat{Y}(\mathbf{s}_0; t_0)$. Depending on the problem at hand, we may choose to let $\mu(\mathbf{s}; t)$ be: (i) known, (ii) constant but unknown, or (iii) modeled in terms of p covariates, $\mu(\mathbf{s}; t) = \mathbf{x}(\mathbf{s}; t)'\boldsymbol{\beta}$, where the p-dimensional vector of parameters $\boldsymbol{\beta}$ is unknown. In the context of the descriptive methods considered in this chapter, these choices result in spatio-temporal (S-T) (i) simple, (ii) ordinary, and (iii) universal kriging, respectively.

4.2 Prediction for Gaussian Data and Processes

Recall from Chapter 3 that when we interpolate with spatio-temporal data we specify that the value of the process at some location is simply a weighted combination of nearby observations. We described a couple of deterministic methods to obtain such weights (inverse distance weighting and kernel smoothing). Here we are concerned with determining the statistically "optimal" weights in this linear combination. At this point, it is worth taking a step back and looking at the big picture.

In the case of predicting statistically within the domain of our space-time observation locations (smoothing), we are just interpolating our observations $\mathbf{Z}$ to the location $(\mathbf{s}_0; t_0)$ in a way that respects that we have observational uncertainty. For example, in the special case where $(\mathbf{s}_0; t_0)$ corresponds to an observation location, we are simply smoothing out this observation uncertainty. Unlike the deterministic approaches to spatio-temporal prediction in Chapter 3, we seek the weights in a linear predictor that minimize the interpolation error on average. This optimization criterion is $E(Y(\mathbf{s}_0; t_0) - \widehat{Y}(\mathbf{s}_0; t_0))^2$, the mean square prediction error (MSPE). The best linear unbiased predictor that minimizes the MSPE is referred to as the *kriging predictor*. As we shall see, the kriging weights are determined by the statistical dependence (i.e., covariances) between observation locations (roughly, the greater the covariability, the greater the weight), yet respect the measurement uncertainty.

There are several different approaches to deriving the form of the optimal linear predictor, which we henceforth call *S-T kriging*. Given that we are just focusing on the first two moments in the descriptive approach (i.e., the means, variances, and covariances of $Y(\cdot; \cdot)$), it is convenient to assume that the underlying process is a *Gaussian process* and the measurement error has a Gaussian distribution. We take this approach in this book.

What is a Gaussian process? Consider a stochastic process denoted by $\{Y(\mathbf{r}) : \mathbf{r} \in D\}$, where $\mathbf{r}$ is a spatial, temporal, or spatio-temporal location in D, a subset of d-dimensional space. This process is said to be a Gaussian process, often denoted $Y(\mathbf{r}) \sim GP(\mu(\mathbf{r}), c(\cdot; \cdot))$, if the process has all its finite-dimensional distributions Gaussian, determined by a mean function $\mu(\mathbf{r})$ and a covariance function $c(\mathbf{r}, \mathbf{r}') = \text{cov}(Y(\mathbf{r}), Y(\mathbf{r}'))$ for *any* location $\{\mathbf{r}, \mathbf{r}'\} \in D$. (Note that in spatio-temporal statistics it is common to use $Gau(\cdot, \cdot)$ instead of $GP(\cdot, \cdot)$, and we follow that convention in this book.) There are two important points to make about the Gaussian process. First, because the Gaussian process determines a probability distribution over functions, it exists *everywhere* in the domain of interest D; so, if the mean and covariance functions are known, the process can be described anywhere in the domain. Second, only finite distributions need to be considered in practice because of the fundamental property that any finite collection of Gaussian process random variables $\{Y(\mathbf{r}_i)\}$ has a joint multivariate normal (Gaussian) distribution. This allows the use of traditional machinery of multivariate normal distributions when performing prediction and inference. Gaussian processes are fundamental to the theoretical and practical foundation of spatial and spatio-temporal statistics and, since the first decade of the twenty-first century, have become increasingly important and popular modeling tools in the machine-learning community (e.g., Rasmussen and Williams, 2006).

In the context of S-T kriging, time is implicitly treated as another dimension, and we consider covariance functions that describe covariability between any two space-time locations (where in general we should use covariance functions that respect that durations in time are different from distances in space). We can write the data model in terms of vectors,

$$\mathbf{Z} = \mathbf{Y} + \boldsymbol{\varepsilon}, \tag{4.3}$$

where $\mathbf{Y} \equiv (Y(\mathbf{s}_{11}; t_1), \ldots, Y(\mathbf{s}_{m_T T}; t_T))'$ and $\boldsymbol{\varepsilon} \equiv (\epsilon(\mathbf{s}_{11}; t_1), \ldots, \epsilon(\mathbf{s}_{m_T T}; t_T))'$. Similarly, the vector form of the process model for $\mathbf{Y}$ is written

$$\mathbf{Y} = \boldsymbol{\mu} + \boldsymbol{\eta}, \tag{4.4}$$

where $\boldsymbol{\mu} \equiv (\mu(\mathbf{s}_{11}; t_1), \ldots, \mu(\mathbf{s}_{m_T T}; t_T))' = \mathbf{X}\boldsymbol{\beta}$, and $\boldsymbol{\eta} \equiv (\eta(\mathbf{s}_{11}; t_1), \ldots, \eta(\mathbf{s}_{m_T T}; t_T))'$. Note that $\text{cov}(\mathbf{Y}) \equiv \mathbf{C}_y = \mathbf{C}_\eta$, $\text{cov}(\boldsymbol{\varepsilon}) \equiv \mathbf{C}_\epsilon$, and $\text{cov}(\mathbf{Z}) \equiv \mathbf{C}_z = \mathbf{C}_y + \mathbf{C}_\epsilon$.

Now, defining $\mathbf{c}_0' \equiv \text{cov}(Y(\mathbf{s}_0; t_0), \mathbf{Z})$, $c_{0,0} \equiv \text{var}(Y(\mathbf{s}_0; t_0))$, and $\mathbf{X}$ the $(\sum_{j=1}^{T} m_j) \times (p+1)$ matrix given by $\mathbf{X} \equiv [\mathbf{x}(\mathbf{s}_{ij}; t_j)' : i = 1, \ldots, m_j; \ j = 1, \ldots, T]$, consider the joint Gaussian distribution,

$$\begin{bmatrix} Y(\mathbf{s}_0; t_0) \\ \mathbf{Z} \end{bmatrix} \sim Gau \left(\begin{bmatrix} \mathbf{x}(\mathbf{s}_0; t_0)' \\ \mathbf{X} \end{bmatrix} \boldsymbol{\beta} \,, \begin{bmatrix} c_{0,0} & \mathbf{c}_0' \\ \mathbf{c}_0 & \mathbf{C}_z \end{bmatrix} \right).$$

Using well-known results for conditional distributions from a joint multivariate normal (Gaussian) distribution (e.g., Johnson and Wichern, 1992), and assuming (for the moment) that $\boldsymbol{\beta}$ is known (recall that this is called *S-T simple kriging*), one can obtain the conditional distribution,

$$Y(\mathbf{s}_0; t_0) \mid \mathbf{Z} \sim Gau(\mathbf{x}(\mathbf{s}_0; t_0)'\boldsymbol{\beta} + \mathbf{c}_0'\mathbf{C}_z^{-1}(\mathbf{Z} - \mathbf{X}\boldsymbol{\beta}) \,, c_{0,0} - \mathbf{c}_0'\mathbf{C}_z^{-1}\mathbf{c}_0), \tag{4.5}$$

for which the mean is the S-T simple kriging predictor,

$$\widehat{Y}(\mathbf{s}_0; t_0) = \mathbf{x}(\mathbf{s}_0; t_0)'\boldsymbol{\beta} + \mathbf{c}_0'\mathbf{C}_z^{-1}(\mathbf{Z} - \mathbf{X}\boldsymbol{\beta}), \tag{4.6}$$

and the variance is the S-T simple kriging variance,

$$\sigma_{Y,sk}^2(\mathbf{s}_0; t_0) = c_{0,0} - \mathbf{c}_0'\mathbf{C}_z^{-1}\mathbf{c}_0. \tag{4.7}$$

Note that we call $\sigma_{Y,sk}(\mathbf{s}_0; t_0)$ the S-T simple kriging *prediction standard error*, and it has the same units as $\widehat{Y}(\mathbf{s}_0; t_0)$.

It is fundamentally important in kriging that one be able to specify the covariance between the process at *any* two locations in the domain of interest (i.e., $\mathbf{c}_0$). That is, we assume that the process is defined for an uncountable set of locations and the data correspond to a partial realization of this process. As mentioned above, this is the benefit of considering S-T kriging from the Gaussian-process perspective. That is, if we assume we have a Gaussian process, then we can specify a valid finite-dimensional Gaussian distribution for *any* finite subset of locations.

Another important observation to make here is that (4.6) is a predictor of the hidden value, $Y(\mathbf{s}_0; t_0)$, not of $Z(\mathbf{s}_0; t_0)$. The form of the conditional distribution given by (4.5) helps clarify the intuition behind S-T kriging. In particular, note that the conditional mean takes the residuals between the observations and their marginal means (i.e., $\mathbf{Z} - \mathbf{X}\boldsymbol{\beta}$), weights them according to $\mathbf{w}' \equiv \mathbf{c}_0' \mathbf{C}_z^{-1}$, and adds the result back onto the marginal mean corresponding to the prediction location (i.e., $\mathbf{x}(\mathbf{s}_0; t_0)'\boldsymbol{\beta}$). Furthermore, the weights, $\mathbf{w}$, are only a function of the covariances and the measurement-error variance. Another way to think of this is that the trend term $\mathbf{x}(\mathbf{s}_0; t_0)'\boldsymbol{\beta}$ is the mean of $Y(\mathbf{s}_0; t_0)$ *prior to* considering the observations; then the simple S-T kriging predictor combines this prior mean with a weighted average of the mean-corrected observations to get a new, conditional, mean. Similarly, if one interprets $c_{0,0}$ as the variance prior to considering the observations, then the conditional (on the data) variance reduces this initial variance by an amount given by $\mathbf{c}_0' \mathbf{C}_z^{-1} \mathbf{c}_0$. Consider the following numerical example.

Example: Simple S-T Kriging

Suppose we have four observations in a one-dimensional space and a one-dimensional time domain: $Z(2; 0.2) = 15$, $Z(2; 1.0) = 22$, $Z(6; 0.2) = 17$, and $Z(6; 0.9) = 23$. We seek an S-T simple kriging prediction for $Y(\mathbf{s}_0; t_0) = Y(3; 0.5)$. The data locations and prediction location are shown in Figure 4.1. Let $x(s; t) = 1$ for all s and t, $\beta = 20$, and $\mathrm{var}(Y(s; t)) = 2$ for all s and t. Using the spatio-temporal covariance function (4.13) discussed below (with parameters $a = 2$, $b = 0.2$, $\sigma^2 = c_{0,0} = 2.0$, and $d = 1$), the covariance (between data) matrix $\mathbf{C}_z$, the covariance (between the data and the latent $Y(\cdot; \cdot)$ at the prediction location) vector $\mathbf{c}_0$, and the weights $\mathbf{w}' = \mathbf{c}_0' \mathbf{C}_z^{-1}$ are given by

$$\mathbf{C}_z = \begin{bmatrix} 2.0000 & 1.0600 & 1.0546 & 0.9364 \\ 1.0600 & 2.0000 & 0.8856 & 1.0599 \\ 1.0546 & 0.8856 & 2.0000 & 1.1625 \\ 0.9364 & 1.0599 & 1.1625 & 2.0000 \end{bmatrix}, \quad \mathbf{c}_0 = \begin{bmatrix} 1.6653 \\ 1.3862 \\ 1.3161 \\ 1.2539 \end{bmatrix}, \quad \mathbf{w} = \begin{bmatrix} 0.5377 \\ 0.2565 \\ 0.1841 \\ 0.1323 \end{bmatrix}.$$

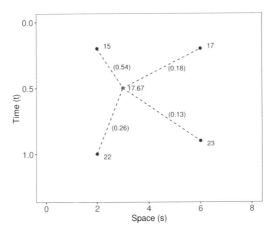

Figure 4.1: Data locations (blue dots) and prediction location (red dot) in a (one-dimensional) space-time domain for an example of S-T simple kriging. The data values and the S-T simple kriging prediction are given next to the locations. The S-T simple kriging weights associated with each data location are given in parentheses next to the dashed lines connecting the data locations to the prediction location.

Substituting these matrices, vectors, and the data vector, $\mathbf{Z} = (15, 22, 17, 23)'$, into the formulas for the S-T kriging predictor (4.6) and prediction variance (4.7), we obtain

$$
\begin{aligned}
\widehat{Y}(3; 0.5) &= 17.67, \\
\widehat{\sigma}^2_{Y,sk} &= 0.34.
\end{aligned}
$$

Note that the S-T simple kriging prediction (17.67) is substantially smaller than the prior mean (20), mainly because the highest weights are associated with the earlier times, which have smaller values. In addition, the S-T simple kriging prediction variance (0.34) is much less than the prior variance (2), as expected when there is strong spatio-temporal dependence.

In most real-world problems, one would not know β. In this case, our optimal prediction problem is analogous to the estimation of effects in a linear mixed model, that is, in a model that considers the response in terms of both fixed effects (e.g., regression terms) and random effects, η. It is straightforward to show that the optimal linear unbiased predictor, or S-T *universal kriging* predictor of $Y(\mathbf{s}_0; t_0)$ is

$$
\widehat{Y}(\mathbf{s}_0; t_0) = \mathbf{x}(\mathbf{s}_0; t_0)'\widehat{\boldsymbol{\beta}}_{\text{gls}} + \mathbf{c}_0'\mathbf{C}_z^{-1}(\mathbf{Z} - \mathbf{X}\widehat{\boldsymbol{\beta}}_{\text{gls}}), \tag{4.8}
$$

where the generalized least squares (gls) estimator of β is given by

$$
\widehat{\boldsymbol{\beta}}_{\text{gls}} \equiv (\mathbf{X}'\mathbf{C}_z^{-1}\mathbf{X})^{-1}\mathbf{X}'\mathbf{C}_z^{-1}\mathbf{Z}. \tag{4.9}
$$

The associated S-T universal kriging variance is given by

$$\sigma^2_{Y,\text{uk}}(\mathbf{s}_0; t_0) = c_{0,0} - \mathbf{c}'_0 \mathbf{C}_z^{-1} \mathbf{c}_0 + \kappa, \tag{4.10}$$

where

$$\kappa \equiv (\mathbf{x}(\mathbf{s}_0; t_0) - \mathbf{X}' \mathbf{C}_z^{-1} \mathbf{c}_0)' (\mathbf{X}' \mathbf{C}_z^{-1} \mathbf{X})^{-1} (\mathbf{x}(\mathbf{s}_0; t_0) - \mathbf{X}' \mathbf{C}_z^{-1} \mathbf{c}_0)$$

represents the additional uncertainty brought to the prediction (relative to S-T simple kriging) due to the estimation of $\boldsymbol{\beta}$. We call $\sigma_{Y,\text{uk}}(\mathbf{s}_0; t_0)$ the S-T universal kriging *prediction standard error*.

Both the S-T simple and universal kriging equations can be extended easily to accommodate prediction at many locations in space and time, including those at which we have observations. For example, in Figure 4.2, we show predictions of maximum temperature from data in the NOAA data set in July 1993 on a space-time grid (using a separable spatio-temporal covariance function, defined in Section 4.2.1), with 14 July deliberately omitted from the data set. The respective prediction standard errors are shown in Figure 4.2, where those for 14 July are substantially larger. We produce these figures in Lab 4.1.

For readers who have some experience with spatial statistics, particularly geostatistics, the development given above in the spatio-temporal context will look very familiar. S-T simple, ordinary, and universal kriging are the same as their spatial counterparts, but now in space *and* time.

So far, we have assumed that we know the variances and covariances that make up $\mathbf{C}_y$, $\mathbf{C}_\epsilon$ (recall that $\mathbf{C}_z = \mathbf{C}_y + \mathbf{C}_\epsilon$), $\mathbf{c}_0$, and $c_{0,0}$. Of course, in reality we would rarely (if ever) know these. The seemingly simple solution is to parameterize them, say in terms of parameters $\boldsymbol{\theta}$, and then estimate them through maximum likelihood, restricted maximum likelihood (see Technical Note 4.2) as in the classical linear mixed model, or perhaps through a fully Bayesian implementation, in which case one specifies prior distributions for the elements of $\boldsymbol{\theta}$ (see Section 4.2.3). As in spatial statistics, the parameterization of these covariance functions is one of the most challenging problems in spatio-temporal statistics.

4.2.1 Spatio-Temporal Covariance Functions

We saw in the previous section that S-T kriging predictors require that we know $\mathbf{C}_z$ and $\mathbf{c}_0$, and hence we need to know the spatio-temporal covariances between the hidden random process evaluated at any two locations in space and time. It is important to note that *not any function can be used as a covariance function*. Let a general *spatio-temporal covariance function* be denoted by

$$c_*(\mathbf{s}, \mathbf{s}'; t, t') \equiv \text{cov}(Y(\mathbf{s}; t), Y(\mathbf{s}'; t')), \tag{4.11}$$

which is appropriate only if the function is *valid* (i.e., non-negative-definite, which guarantees that the kriging variances are non-negative). (Note that in (4.11) the primes are not transposes, but are used to denote different spatio-temporal locations.)

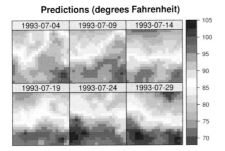
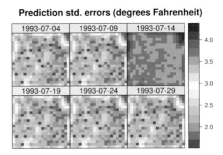

Figure 4.2: (Left) S-T universal kriging predictions and (right) prediction standard errors of maximum temperature (in degrees Fahrenheit) within a square lat-lon box enclosing the domain of interest for six days (each 5 days apart) in July 1993 using the R package **gstat**. Data for 14 July 1993 were omitted from the original data set.

   In practice, classical-kriging implementations assume second-order stationarity: the random process is said to be *second-order* (or *weakly*) *stationary* if it has a constant expectation μ (say) and a covariance function that can be expressed in terms of spatial and temporal lags:

$$c_*(\mathbf{s}, \mathbf{s}'; t, t') = c(\mathbf{s}' - \mathbf{s}; t' - t) = c(\mathbf{h}; \tau),$$

where $\mathbf{h} \equiv \mathbf{s}' - \mathbf{s}$ and $\tau \equiv t - t'$ are the spatial and temporal lags, respectively. Recall from Chapter 2 that if the dependence on spatial lag is only a function of $\|\mathbf{h}\|$, we say there is spatial *isotropy*. Arguably, the two biggest benefits of the second-order stationarity assumption are that it allows for more parsimonious parameterizations of the covariance function, and that it provides pseudo-replication of dependencies at given lags in space and time, both of which facilitate estimation of the covariance function's parameters. (In practice, it is unlikely that the spatio-temporal stationary covariance function is completely known and it is usually specified in terms of some parameters $\boldsymbol{\theta}$.)

   The next question is how to obtain valid stationary (or non-stationary) spatio-temporal covariance functions. Mathematically speaking, how do we ensure that the functions we choose are non-negative-definite?

Separable (in Space and Time) Covariance Functions

Separable classes of spatio-temporal covariance functions have often been used in spatio-temporal modeling because they offer a convenient way to guarantee validity. The separable class is given by

$$c(\mathbf{h}; \tau) \equiv c^{(s)}(\mathbf{h}) \cdot c^{(t)}(\tau),$$

which is valid if both the spatial covariance function, $c^{(s)}(\mathbf{h})$, and the temporal covariance function, $c^{(t)}(\tau)$, are valid. There are a large number of classes of valid spatial and

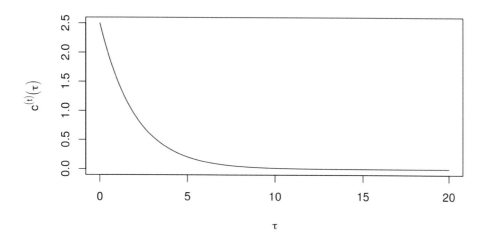

Figure 4.3: Exponential covariance function for time lag τ, $\sigma_t^2 = 2.5$, and $a_t = 2$.

valid temporal covariance functions in the literature (e.g., the Matérn, power exponential, and Gaussian classes, to name a few). For example, the exponential covariance function (which is a special case of both the Matérn covariance function and the power exponential covariance function) is given by

$$c^{(s)}(\mathbf{h}) = \sigma_s^2 \exp\left\{ -\frac{||\mathbf{h}||}{a_s} \right\},$$

where σ_s^2 is the variance parameter and a_s is the spatial-dependence (or scale) parameter in units of distance. The larger a_s is, the more dependent the spatial process is. Similarly, $c^{(t)}(\tau) = \sigma_t^2 \exp\{-|\tau|/a_t\}$ is a valid temporal covariance function (see Figure 4.3 for an example).

A consequence of separability is that the resulting spatio-temporal correlation function, $\rho(\mathbf{h}; \tau) \equiv c(\mathbf{h}; \tau)/c(\mathbf{0}; 0)$, is given by

$$\rho(\mathbf{h}; \tau) = \rho^{(s)}(\mathbf{h}; 0) \cdot \rho^{(t)}(\mathbf{0}; \tau),$$

where $\rho^{(s)}(\mathbf{h}; 0)$ and $\rho^{(t)}(\mathbf{0}; \tau)$ are the corresponding marginal spatial and temporal correlation functions, respectively. Thus, one only needs the marginal spatial and temporal correlation functions to obtain the joint spatio-temporal correlation function under separability. In addition, separable models facilitate computation. Notice (e.g., from (4.6) and (4.7)) that the inverse $\mathbf{C}_z^{-1}$ is ubiquitous in S-T kriging equations. Separability can allow one to consider the inverse of the spatial and temporal components separately. For example, assume that $Z(\mathbf{s}_{ij}; t_j)$ is observed at the same $i = 1, \ldots, m_j = m$ locations at each time point, $j = 1, \ldots, T$. In this case, one can write $\mathbf{C}_z = \mathbf{C}_z^{(t)} \otimes \mathbf{C}_z^{(s)}$, where $\otimes$ is the Kronecker product (see Technical Note 4.1), $\mathbf{C}_z^{(t)}$ is the $T \times T$ temporal covariance matrix,

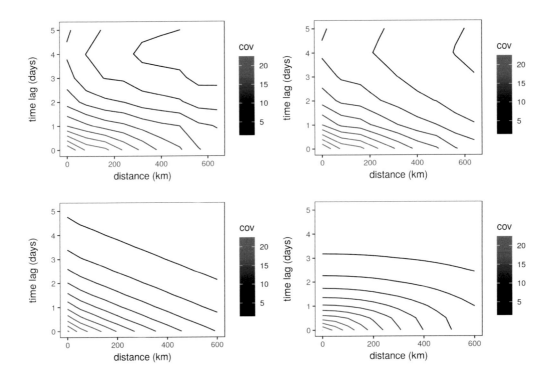

Figure 4.4: Contour plot of the empirical covariance function (top left), fitted separable covariance function obtained by taking the product of $\hat{c}^{(s)}(\|\mathbf{h}\|)$ and $\hat{c}^{(t)}(|\tau|)$ (top right), fitted separable covariance function using the spatio-temporal separable model given in equation (4.18) and (4.19) (bottom left) and fitted covariance function using the non-separable model given in equation (4.20) (bottom right).

and $\mathbf{C}_z^{(s)}$ is the $m \times m$ spatial covariance matrix. Taking advantage of a useful property of Kronecker products (see Technical Note 4.1), $\mathbf{C}_z^{-1} = (\mathbf{C}_z^{(t)})^{-1} \otimes (\mathbf{C}_z^{(s)})^{-1}$, which shows that to take the inverse of the $mT \times mT$ matrix $\mathbf{C}_z$, one only has to take the inverses of $T \times T$ and $m \times m$ matrices.

Consider the maximum-temperature observations (Tmax) from the NOAA data set presented in Chapter 2. After removing the obvious linear trend in latitude, we consider the empirical isotropic spatio-temporal covariance function (discussed in Section 2.4.2) calculated for the residuals, shown in Figure 4.4 (top left panel), and we compare that to the empirical separable model in Figure 4.4 (top right panel). That is, we are simply considering the product of $\hat{c}(0; |\tau|)$ and $\hat{c}(\|\mathbf{h}\|; 0)$. Note that these two plots are remarkably similar, giving visual support for a separable model in this case. We shall discuss the lower two panels of this figure in Section 4.2.3. See Crujeiras et al. (2010) and references therein for formal tests of separability.

A consequence of the separability property is that the temporal evolution of the process at a given spatial location does not depend directly on the process' temporal evolution at other locations. As we discuss in Chapter 5, this is very seldom the case for real-world processes as it implies no interaction across space and time. The question then becomes, "how can we obtain other classes of spatio-temporal covariance functions?" Several approaches that have been developed in the literature: (i) sums-and-products formulation; (ii) construction by a spectral representation through Bochner's theorem (which formally relates the spectral representation to the covariance representation; e.g., the inverse Fourier transform is a special case); and (iii) covariance functions from the solution of stochastic partial differential equations (SPDEs). We discuss these briefly below.

Technical Note 4.1: Kronecker Products

Consider two matrices, an $n_a \times m_a$ matrix, $\mathbf{A}$, and an $n_b \times m_b$ matrix, $\mathbf{B}$. The Kronecker product is given by the $n_a n_b \times m_a m_b$ matrix $\mathbf{A} \otimes \mathbf{B}$ defined as

$$
\mathbf{A} \otimes \mathbf{B} =
\begin{bmatrix}
a_{11}\mathbf{B} & \cdots & a_{1m_a}\mathbf{B} \\
\vdots & \vdots & \vdots \\
a_{n_a 1}\mathbf{B} & \cdots & a_{n_a m_a}\mathbf{B}
\end{bmatrix}.
$$

The Kronecker product has some nice properties that facilitate matrix representations. For example, if $\mathbf{A}$ is $n_a \times n_a$ and $\mathbf{B}$ is $n_b \times n_b$, the inverse and determinants can be expressed in terms of the individual matrices:

$$
(\mathbf{A} \otimes \mathbf{B})^{-1} = \mathbf{A}^{-1} \otimes \mathbf{B}^{-1},
$$
$$
|\mathbf{A} \otimes \mathbf{B}| = |\mathbf{A}|^{n_b} |\mathbf{B}|^{n_a}.
$$

In the context of spatio-temporal processes, Kronecker products are useful in at least two ways. First, they provide a convenient way to represent spatio-temporal covariance matrices for separable processes. That is, consider $\{Y(\mathbf{s}_i; t_j) : i = 1, \ldots, m; \ j = 1, \ldots, T\}$ and define $\mathbf{C}_y^{(s)}$ to be the $m \times m$ matrix of purely spatial covariances and $\mathbf{C}_y^{(t)}$ to be the $T \times T$ matrix of purely temporal covariances. Then the $mT \times mT$ spatio-temporal covariance matrix can be written as, $\mathbf{C}_y = \mathbf{C}_y^{(t)} \otimes \mathbf{C}_y^{(s)}$ if the process is separable. Although this may not be realistic for many processes, it is advantageous because of the inverse property, $\mathbf{C}_y^{-1} = (\mathbf{C}_y^{(t)})^{-1} \otimes (\mathbf{C}_y^{(s)})^{-1}$; see Section 4.2.1.

The second way that Kronecker products are useful for spatio-temporal modeling is for forming spatio-temporal basis functions, which we discuss in Section 4.4. In particular, if we construct an $m \times n_{\alpha,s}$ matrix $\mathbf{\Phi}$ by evaluating $n_{\alpha,s}$ spatial basis functions at m spatial locations, and a $T \times n_{\alpha,t}$ matrix $\mathbf{\Psi}$ by evaluating $n_{\alpha,t}$ temporal basis functions at T temporal locations, then the matrix constructed from spatio-temporal basis functions

> formed through the *tensor product* of the spatial and temporal basis functions and evaluated at all combinations of spatial and temporal locations is given by the $mT \times n_{\alpha,s}n_{\alpha,t}$ matrix $\mathbf{B} = \boldsymbol{\Psi} \otimes \boldsymbol{\Phi}$. Basis functions can be used to construct spatio-temporal covariance functions. Note that using a set of basis functions constructed through the tensor product yields a class of *spatio-temporal covariance functions* that are in general not separable.

Sums-and-Products Formulation

There is a useful result in mathematics that states that, as well as the product, the *sum* of two non-negative-definite functions is non-negative-definite. This allows us to construct valid spatio-temporal covariance functions as the product and/or sum of valid covariance functions. For example,

$$c(\mathbf{h}; \tau) \equiv p \, c_1^{(s)}(\mathbf{h}) \cdot c_1^{(t)}(\tau) + q \, c_2^{(s)}(\mathbf{h}) + r \, c_2^{(t)}(\tau) \tag{4.12}$$

is a valid spatio-temporal covariance function when $p > 0$, $q \geq 0$, $r \geq 0$; $c_1^{(s)}(\mathbf{h})$ and $c_2^{(s)}(\mathbf{h})$ are valid spatial covariance functions; and $c_1^{(t)}(\tau)$ and $c_2^{(t)}(\tau)$ are valid temporal covariance functions. Of course, (4.12) can be extended to include the sum of many terms and the result is non-negative definite if each component covariance function is non-negative-definite.

The sums-and-products formulation above points to connections between separable covariance functions and other special cases. For example, consider the *fully symmetric* spatio-temporal covariance functions: a spatio-temporal random process $\{Y(\mathbf{s}; t)\}$ is said to have a fully symmetric spatio-temporal covariance function if, for all spatial locations $\mathbf{s}, \mathbf{s}'$ in the spatial domain of interest and time points t, t' in the temporal domain of interest, we can write

$$\mathrm{cov}(Y(\mathbf{s}; t), Y(\mathbf{s}'; t')) = \mathrm{cov}(Y(\mathbf{s}; t'), Y(\mathbf{s}'; t)).$$

Using such covariances to model spatio-temporal dependence is not always reasonable for real-world processes. For example, is it reasonable that the covariance between yesterday's temperature in London and today's temperature in Paris is the same as that between yesterday's temperature in Paris and today's temperature in London? Such a relationship might be appropriate under certain meteorological conditions, but not in general (imagine a weather system moving from northwest to southeast across Europe). So, for scientific reasons or as a result of an exploratory data analysis, the fully symmetric covariance function may not be an appropriate choice.

Now, note that the covariance given by (4.12) is an example of a fully symmetric covariance, but it is only separable if $q = r = 0$. In general, separable covariance functions are always fully symmetric, while the converse is not true.

Construction via a Spectral Representation

An important example of the construction approach to spatio-temporal covariance function development was given by Cressie and Huang (1999). They were able to cast the problem in the spectral domain so that one only needs to choose a one-dimensional positive-definite function of time lag in order to obtain a class of valid non-separable spatio-temporal covariance functions. In their Example 1, they construct the stationary spatio-temporal covariance function,

$$c(\mathbf{h}; \tau) = \sigma^2 \exp\{-b^2 ||\mathbf{h}||^2 / (a^2 \tau^2 + 1)\} / (a^2 \tau^2 + 1)^{d/2}, \qquad (4.13)$$

where $\sigma^2 = c(\mathbf{0}; 0)$, d corresponds to the spatial dimension (often $d = 2$), and $a \geq 0$ and $b \geq 0$ are scale parameters in space and time, respectively. There are other classes of such spatio-temporal models, and this has been an active area of research in the past few decades (see the overview in Montero et al., 2015).

> **R tip:** In this book we limit our focus to **gstat** when doing S-T kriging. However, there are numerous other packages in R that could be used. Among these **CompRandFld** and **RandomFields** are worth noting because of the large selection of non-separable spatio-temporal covariance functions they make available to the user.

Stochastic Partial Differential Equation (SPDE) Approach

The SPDE approach to deriving spatio-temporal covariance functions was originally inspired by statistical physics, where physical equations forced by random processes that describe advective, diffusive, and decay behavior were used to describe the second moments of macro-scale processes, at least in principle. A famous example of this approach in spatial statistics resulted in the ubiquitous Matérn spatial covariance function, which was originally derived as the solution to a fractional stochastic diffusion equation and has been extended by several authors (e.g., Montero et al., 2015).

Although such an approach can suggest non-separable spatio-temporal covariance functions, only a few special (simple) cases lead to closed-form functions (see, for example, Cressie and Wikle, 2011, p. 300). Perhaps more importantly, although these models appear to have a physical basis through the SPDE, macro-scale real-world processes of interest are seldom this simple (e.g., linear and stationary in space and/or time). That is, the spatio-temporal covariance functions that can be obtained in closed form from SPDEs are seldom directly appropriate models for physical processes (but may still provide good fits to data).

4.2.2   Spatio-Temporal Semivariograms

Historically, it has been common in the area of spatial statistics known as geostatistics to consider dependence through the variogram. In the context of a spatio-temporal random process $\{Y(\mathbf{s};t)\}$, the *spatio-temporal variogram* is defined as

$$\text{var}(Y(\mathbf{s};t) - Y(\mathbf{s}';t')) \equiv 2\gamma(\mathbf{s},\mathbf{s}';t,t'), \tag{4.14}$$

where $\gamma(\cdot)$ is called the *semivariogram* (see Technical Note 2.1). The stationary version of the spatio-temporal variogram is denoted by $2\gamma(\mathbf{h};\tau)$, where $\mathbf{h} = \mathbf{s}' - \mathbf{s}$ and $\tau = t' - t$, analogous to the stationary-covariance representation given previously. The underlying process Y is considered to be *intrinsically stationary* if it has a constant expectation and a stationary variogram. When the process is second-order stationary (second-order stationarity is a stronger restriction than intrinsic stationarity), there is a useful and simple relationship between the spatio-temporal semivariogram and the covariance function, namely,

$$\gamma(\mathbf{h};\tau) = c(\mathbf{0};0) - c(\mathbf{h};\tau). \tag{4.15}$$

Notice that strong spatio-temporal dependence corresponds to small values of the semivariogram. Thus, contour plots of $\{\gamma(\mathbf{h};\tau)\}$ in (4.15) start near zero close to the origin $(\mathbf{h};\tau) = (\mathbf{0},0)$, and they rise to a constant value (the "sill") as both $\mathbf{h}$ and τ move away from the origin.

Although there has been a preference to consider dependence through the variogram in geostatistics, this has not been the case in more mainstream spatio-temporal statistical analyses. The primary reason for this is that most real-world processes are best characterized in the context of *local* second-order stationarity. The difference between intrinsic stationarity and second-order stationarity is most appreciated when the lags $\mathbf{h}$ and τ are large. If only local stationarity is expected and modeled, the extra generality given by the variogram is not needed. Still, the empirical semivariogram offers a useful way to summarize the spatio-temporal dependence in the data and to fit a spatio-temporal covariance function.

On a theoretical level, the stationary variogram allows S-T kriging for a larger class of processes (i.e., intrinsically stationary processes) than the second-order stationary processes. A price to pay for this extra generality is the extreme caution needed when using the variogram to find optimal kriging coefficients. Cressie and Wikle (2011, p. 148) point out that the universal-kriging weights may not sum to 1 and, in situations where they do not, the resulting variogram-based kriging predictor will not be optimal. However, when using the covariance-based kriging predictor, there are no such issues and it is always optimal.

In addition, on a more practical level, most spatio-temporal analyses consider models that are specified from a likelihood perspective or a Bayesian perspective, where covariance matrices are needed. The variogram by itself does not specify the covariance matrix, since one also needs to model the variance function $\sigma^2(\mathbf{s};t) \equiv \text{var}(Y(\mathbf{s};t))$, which is usually impractical unless it is stationary and does not depend on $\mathbf{s}$ and t. Some software packages that perform S-T kriging, such as **gstat**, fit variogram functions to data, mainly for historical reasons and because of the implicit assumption in (4.14) that a constant mean need not be assumed when estimating the variogram. (This is generally a good thing because the constant mean assumption is tenuous in practice, since the mean for real-world processes typically depends on exogenous covariates that vary with space and time.)

4.2.3 Gaussian Spatio-Temporal Model Estimation

The spatio-temporal covariance and variogram functions presented above depend on unknown parameters. These are almost never known in practice and must be estimated from the data. There is a history in spatial statistics of fitting covariance functions (or semivariograms) directly to the empirical estimates – for example, by using a least squares or weighted least squares approach (see Cressie, 1993, for an overview). However, in the spatio-temporal context we prefer to consider fully parameterized covariance models and infer the parameters through likelihood-based methods or through fully Bayesian methods. This follows closely the approaches in mixed-linear-model parameter estimation; for an overview, see McCulloch and Searle (2001). We briefly describe the likelihood-based approach and the Bayesian approach below.

Likelihood Estimation

Given the data model (4.3), note that $\mathbf{C}_z = \mathbf{C}_y + \mathbf{C}_\epsilon$. Then, in obvious notation, $\mathbf{C}_z$ depends on parameters $\boldsymbol{\theta} \equiv \{\boldsymbol{\theta}_y, \boldsymbol{\theta}_\epsilon\}$ for the covariance functions of the hidden process Y and the measurement-error process ϵ, respectively. The likelihood can then be written as

$$L(\boldsymbol{\beta}, \boldsymbol{\theta}; \mathbf{Z}) \propto |\mathbf{C}_z(\boldsymbol{\theta})|^{-1/2} \exp\left\{-\frac{1}{2}(\mathbf{Z} - \mathbf{X}\boldsymbol{\beta})'(\mathbf{C}_z(\boldsymbol{\theta}))^{-1}(\mathbf{Z} - \mathbf{X}\boldsymbol{\beta})\right\}, \qquad (4.16)$$

and we maximize this with respect to $\{\boldsymbol{\beta}, \boldsymbol{\theta}\}$, thus obtaining the maximum likelihood estimates (MLEs), $\{\widehat{\boldsymbol{\beta}}_{\mathrm{mle}}, \widehat{\boldsymbol{\theta}}_{\mathrm{mle}}\}$. Because the covariance parameters appear in the matrix inverse and determinant in (4.16), analytical maximization for most parametric covariance models is not possible, but numerical methods can be used. To reduce the number of parameters in this maximization, we often consider "profiling," where we replace $\boldsymbol{\beta}$ in (4.16) with the generalized least squares estimator, $\boldsymbol{\beta}_{\mathrm{gls}} = (\mathbf{X}'\mathbf{C}_z(\boldsymbol{\theta})^{-1}\mathbf{X})^{-1}\mathbf{X}'\mathbf{C}_z(\boldsymbol{\theta})^{-1}\mathbf{Z}$ (which depends only on $\boldsymbol{\theta}$). Then the profile likelihood is just a function of the unknown parameters $\boldsymbol{\theta}$. Using a numerical optimization method (e.g., Newton–Raphson) to obtain $\widehat{\boldsymbol{\theta}}_{\mathrm{mle}}$, we then obtain $\widehat{\boldsymbol{\beta}}_{\mathrm{mle}} = (\mathbf{X}'\mathbf{C}_z(\widehat{\boldsymbol{\theta}}_{\mathrm{mle}})^{-1}\mathbf{X})^{-1}\mathbf{X}'\mathbf{C}_z(\widehat{\boldsymbol{\theta}}_{\mathrm{mle}})^{-1}\mathbf{Z}$, which is the MLE of $\boldsymbol{\beta}$. The parameter estimates $\{\widehat{\boldsymbol{\beta}}_{\mathrm{mle}}, \widehat{\boldsymbol{\theta}}_{\mathrm{mle}}\}$ are then substituted into the kriging equations above (e.g., (4.8) and (4.10)) to obtain the empirical best linear unbiased predictor (EBLUP) and the associated empirical prediction variance.

R tip: Maximizing the log-likelihood (i.e., the log of (4.16)) in R can be done in a number of ways. Among the most popular functions in base R are **nlm**, which implements a Newton-type algorithm, and **optim**, which contains a number of general-purpose routines, some of which are gradient-based. When a simple covariance function is used, the gradient can be found analytically, and gradient information may then be used to facilitate optimization. Many of the parameters in our models (such as the variance or dependence-scale parameters) need to be positive to ensure positive-definite covariance matrices. This can be easily achieved by finding the MLEs of the log of the parameters, instead of the parameters themselves. Then the MLE of the parameter on the original scale is obtained by exponentiating the MLE on the log scale. In this case, one typically uses the *delta method* to obtain the variance of the transformed parameter estimates (see, for example, Kendall and Stuart, 1969).

As described in Technical Note 4.2, restricted maximum likelihood (REML) considers the likelihood of a linear transformation of the data vector such that the errors are orthogonal to the $\mathbf{X}$s that make up the mean function. Numerical maximization of the associated likelihood, which is only a function of the parameters $\boldsymbol{\theta}$ (i.e., not of $\boldsymbol{\beta}$), gives $\widehat{\boldsymbol{\theta}}_{\text{reml}}$. These estimates are substituted into (4.9), the GLS formula for $\boldsymbol{\beta}$, to obtain $\widehat{\boldsymbol{\beta}}_{\text{reml}}$ as well as the kriging equations (4.8) and (4.10).

Both the MLE and REML approaches have the advantage that they are based on the "likelihood principle" and, assuming that the Gaussian distributional assumptions are correct, they have desirable properties such as sufficiency, invariance, consistency, efficiency, and asymptotic normality. In mixed-effects models and in spatial statistics, REML is usually preferred over MLE for estimation of covariance parameters because REML typically has less bias in small samples (see, for example, the overview in Wu et al., 2001).

Technical Note 4.2: Restricted Maximum Likelihood
Consider a contrast matrix $\mathbf{K}$ such that $E(\mathbf{KZ}) = 0$. For example, let $\mathbf{K}$ be an $(m - p - 1) \times m$ matrix orthogonal to the column space of the $m \times (p + 1)$ design matrix $\mathbf{X}$. That is, let $\mathbf{K}$ correspond to the $m - p - 1$ linearly independent rows of $(\mathbf{I} - \mathbf{X}(\mathbf{X}'\mathbf{X})^{-1}\mathbf{X}')$. Because $\mathbf{KX} = 0$, it follows that $E(\mathbf{KZ}) = \mathbf{KX}\boldsymbol{\beta} = 0$, and $\text{var}(\mathbf{KZ}) = \mathbf{KC}_z(\boldsymbol{\theta})\mathbf{K}'$. In this case, the likelihood based on $\mathbf{KZ}$ is not a function of the mean parameters $\boldsymbol{\beta}$ and is given by

$$L_{\text{reml}}(\boldsymbol{\theta}; \mathbf{Z}) \propto |\mathbf{KC}_z(\boldsymbol{\theta})\mathbf{K}'|^{-1/2} \exp\left\{-\frac{1}{2}(\mathbf{KZ})'(\mathbf{KC}_z(\boldsymbol{\theta})\mathbf{K}')^{-1}(\mathbf{KZ})\right\}. \quad (4.17)$$

Then (4.17) is maximized numerically to obtain $\widehat{\boldsymbol{\theta}}_{\text{reml}}$. Note that parameter estimation and statistical inference with REML do not depend on the specific choice of $\mathbf{K}$, so long as it is a contrast matrix that leads to $E(\mathbf{KZ}) = \mathbf{0}$ (Patterson and Thompson, 1971). One can then use these estimates in a GLS estimate of β: $\widehat{\beta}_{\mathrm{reml}} \equiv (\mathbf{X}'\mathbf{C}_z(\widehat{\boldsymbol{\theta}}_{\mathrm{reml}})^{-1}\mathbf{X})^{-1}\mathbf{X}'\mathbf{C}_z(\widehat{\boldsymbol{\theta}}_{\mathrm{reml}})^{-1}\mathbf{Z}$.

Bayesian Inference

Instead of treating $\boldsymbol{\beta}$ and $\boldsymbol{\theta}$ as fixed, unknown, and to be estimated (e.g., from the likelihood), prior distributions $[\boldsymbol{\beta}]$ and $[\boldsymbol{\theta}]$ (often assumed independent) could be posited for the mean parameters $\boldsymbol{\beta}$ and the covariance parameters $\boldsymbol{\theta}$, respectively. Typical choices for $[\boldsymbol{\theta}]$ do not admit closed-form posterior distributions for $[Y(\mathbf{s}_0)|\mathbf{Z}]$, which means that the predictor $E(Y(\mathbf{s}_0; t_0)|\mathbf{Z})$ and the associated uncertainty, $\mathrm{var}(Y(\mathbf{s}_0; t_0)|\mathbf{Z})$, are not available in closed form and must be obtained through numerical evaluation of the posterior distribution (for more details, see Section 4.5.2 below; Cressie and Wikle, 2011; Banerjee et al., 2015).

Example: S-T Kriging

Consider the maximum-temperature observations in the NOAA data set (Tmax). The empirical covariogram of these data is shown in the top left panel of Figure 4.4. Consider two spatio-temporal covariance functions fitted to the residuals from a model with a regression component that includes an intercept and latitude as a covariate. The first of these covariance functions is given by an isotropic and stationary separable model of the form

$$c^{(\mathrm{sep})}(\|\mathbf{h}\|; |\tau|) \equiv c^{(s)}(\|\mathbf{h}\|) \cdot c^{(t)}(|\tau|), \tag{4.18}$$

in which we let both covariance functions, $c^{(s)}(\cdot)$ and $c^{(t)}(\cdot)$, take the form

$$c^{(\cdot)}(h) = b_1 \exp(-\phi h) + b_2 I(h = 0), \tag{4.19}$$

where ϕ, b_1, and b_2 are parameters that are different for $c^{(s)}(\cdot)$ and $c^{(t)}(\cdot)$ and need to be estimated; and $I(\cdot)$ is the indicator function that is used to represent the so-called *nugget effect*, made up of the measurement-error variance plus the micro-scale variation. The fitted model is shown in the bottom left panel of Figure 4.4.

The second model we fit is a non-separable spatio-temporal covariance function, in which the temporal lag is scaled to account for the different nature of space and time. This model is given by

$$c^{(\mathrm{st})}(\|\mathbf{v}_a\|) \equiv b_1 \exp(-\phi\|\mathbf{v}_a\|) + b_2 I(\|\mathbf{v}_a\| = 0), \tag{4.20}$$

where $\mathbf{v}_a \equiv (\mathbf{h}', a\tau)'$, and recall that $\|\mathbf{v}_a\| = (\mathbf{h}'\mathbf{h} + a^2\tau^2)^{1/2}$. Here, a is the scaling factor used for generating the space-time anisotropy. The fitted model is shown in the bottom right panel of Figure 4.4.

The non-separable spatio-temporal covariance function (4.20) allows for space-time anisotropy, but it is otherwise relatively inflexible. It only contains one parameter (a) to account for the different scaling needed for space and time, one parameter (ϕ) for the length scale, and two parameters to specify the variance (the nugget effect, b_2, and the variance of the smooth component, b_1). Thus, (4.20) has a total of four parameters, in contrast to the six parameters in (4.18). This results in a relatively poor fit to the `Tmax` data from the NOAA data set. In this case, the separable model is able to provide a better reconstruction of the empirical covariance function despite its lack of space-time interaction, which is not surprising given that the fitted separable covariance function (Figure 4.4, top right) is visually similar to the empirical spatio-temporal covariance function (Figure 4.4, top left). We note that although the separable model fits better in this case, it is still a rather unrealistic model for most processes of interest.

4.3   Random-Effects Parameterizations

As discussed previously, it can be difficult to specify realistic valid spatio-temporal co-variance functions and to work with large spatio-temporal covariance matrices (e.g., $\mathbf{C}_z$) in situations with large numbers of prediction or observation locations. One way to miti-gate these problems is to take advantage of *conditional* specifications that the hierarchical modeling framework allows.

We can consider classical linear mixed models from either a conditional perspective, where we condition the response on the random effects, or from a marginal perspective, where the random effects have been averaged (integrated) out (see Technical Note 4.3), and it is this marginal distribution that is modeled. We digress briefly from the spatio-temporal context to illustrate the conditional versus marginal approach in a simple longitudinal-data-analysis setting. Longitudinal data are collected over time, often in a clinical trial where the response to drug treatments and controls is measured on the same subjects at different follow-up times. Here, one might allow there to be subject-specific intercepts or slopes corresponding to the treatment effect over time.

Figure 4.5 shows simulated data for a longitudinal study in which 90 individuals are assigned randomly to three treatment groups (control, treatment 1, and treatment 2), 30 per group. Their responses are then plotted through time (20 time points). In each case, the re-sponse is generally linear with time, with individual-specific random intercepts and slopes. These responses can be modeled in terms of a linear mixed model, with fixed effects corre-sponding to the treatment (control, treatment 1, and treatment 2), individual random effects for the slope and intercept, and a random effect for the error. The random effects correspond to a situation where individuals have somewhat different baseline responses (intercept), and their response with time to the treatment is also subject to individual variation (slope).

For the simulated data shown in Figure 4.5, we might consider a longitudinal model such as (see, for example, Verbeke and Molenberghs, 2009, Section 3.3):

$$
Z_{ij} = \begin{cases}
(\beta_0 + \alpha_{0i}) + (\beta_1 + \alpha_{1i})t_j + \epsilon_{ij}, & \text{if the subject receives the control,} \\
(\beta_0 + \alpha_{0i}) + (\beta_2 + \alpha_{1i})t_j + \epsilon_{ij}, & \text{if the subject receives treatment 1,} \\
(\beta_0 + \alpha_{0i}) + (\beta_3 + \alpha_{1i})t_j + \epsilon_{ij}, & \text{if the subject recieves treatment 2,}
\end{cases}
$$

where Z_{ij} is the response for the ith subject ($i = 1, \ldots, n = 90$) at time $j = 1, \ldots, T = 20$; β_0 is an overall fixed intercept; $\beta_1, \beta_2, \beta_3$ are fixed time-trend effects; and $\alpha_{0i} \sim iid\, Gau(0, \sigma_1^2)$ and $\alpha_{1i} \sim iid\, Gau(0, \sigma_2^2)$ are individual-specific random intercept and slope effects, respectively. We can write this model in the classical linear mixed-model notation as

$$
\mathbf{Z}_i = \mathbf{X}_i \boldsymbol{\beta} + \boldsymbol{\Phi}\boldsymbol{\alpha}_i + \boldsymbol{\varepsilon}_i,
$$

where $\mathbf{Z}_i$ is a 20-dimensional vector of responses for the ith individual; $\mathbf{X}_i$ is a 20×4 matrix consisting of a column vector of 1s (intercept) and three columns indicating the treatment group of the ith individual; $\boldsymbol{\beta}$ is a four-dimensional vector of fixed effects; $\boldsymbol{\Phi}$ is a 20×2 matrix with a vector of 1s in the first column and the second column consists of the vector of times, $(1, 2, \ldots, 20)'$; the associated random-effect vector is $\boldsymbol{\alpha}_i \equiv (\alpha_{0i}, \alpha_{1i})' \sim Gau(\mathbf{0}, \mathbf{C}_\alpha)$, where $\mathbf{C}_\alpha = \text{diag}(\sigma_1^2, \sigma_2^2)$; and $\boldsymbol{\varepsilon}_i \sim Gau(\mathbf{0}, \sigma_\epsilon^2 \mathbf{I})$ is a 20-dimensional error vector. We assume that the elements of $\{\boldsymbol{\alpha}_i\}$ and $\{\boldsymbol{\varepsilon}_i\}$ are all mutually independent.

Because the variation in the individuals' intercepts and slopes is specified by random effects, this formulation allows one to consider inference at the subject (individual) level (e.g., predictions of an individual's true values). However, if interest is in the fixed treatment effects $\boldsymbol{\beta}$, one might consider the marginal distribution of the responses in which these individual random effects have been removed through averaging (integration). Responses that share common random effects exhibit marginal dependence through the marginal covariance matrix, and so the inference on the fixed effects (e.g., via generalized least squares) then accounts for this more complicated marginal dependence. For the example presented here, one can show that the marginal covariance for an individual's response at time t_j and t_k is given by $\text{cov}(Z_{ij}, Z_{ik}) = \sigma_1^2 + t_j t_k \sigma_2^2 + \sigma_\epsilon^2 I(j = k)$, which says that the marginal variance is time-varying, whereas the conditional covariance (conditioned on $\boldsymbol{\alpha}$) is simply $\sigma_\epsilon^2 I(j = k)$.

In the context of spatial or spatio-temporal modeling, the same considerations as for the classical linear mixed-effects model apply. That is, we can also write the process of interest conditional on random effects, where the random effects might be spatial, temporal, or spatio-temporal. Why is this important? As we show in the next section, it allows us to *build* spatio-temporal dependence *conditionally*, in such a way that the implied marginal spatio-temporal covariance function is *always valid*, and it provides some computational advantages.

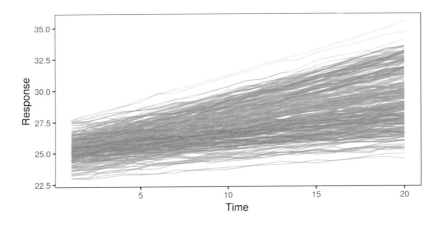

Figure 4.5: Simulated longitudinal data showing the response of individuals through time. The red lines are the simulated responses for a control group, the green lines are the simulated responses for treatment 1, and the blue lines are the simulated responses for treatment 2.

Technical Note 4.3: Marginal and Conditional Linear Mixed Models
Consider the conditional representation of a classic general linear mixed-effects model (Laird and Ware, 1982) for response vector $\mathbf{Z}$ and fixed and random effects vectors, $\boldsymbol{\beta}$ and $\boldsymbol{\alpha}$, respectively. Specifically, consider

$$\mathbf{Z}|\boldsymbol{\alpha} \sim Gau(\mathbf{X}\boldsymbol{\beta} + \boldsymbol{\Phi}\boldsymbol{\alpha}, \mathbf{C}_\epsilon), \qquad (4.21)$$
$$\boldsymbol{\alpha} \sim Gau(\mathbf{0}, \mathbf{C}_\alpha),$$

where $\mathbf{X}$ and $\boldsymbol{\Phi}$ are assumed to be known matrices, and $\mathbf{C}_\epsilon$ and $\mathbf{C}_\alpha$ are known covariance matrices. The marginal distribution of $\mathbf{Z}$ is then given by integrating out the random effects:

$$[\mathbf{Z}] = \int [\mathbf{Z} \mid \boldsymbol{\alpha}][\boldsymbol{\alpha}]d\boldsymbol{\alpha}. \qquad (4.22)$$

Note that dependence on $\boldsymbol{\theta}$, which recall are the covariance parameters in $\mathbf{C}_z$ and $\mathbf{C}_\alpha$, has been suppressed in (4.22), although the (implicit) presence of $\boldsymbol{\theta}$ can be seen in (4.23)–(4.26) below. We can obtain this distribution by making use of iterated conditional expectation and variance formulas. In particular, note that we can write the model associated with (4.21) as

$$\mathbf{Z} = \mathbf{X}\boldsymbol{\beta} + \boldsymbol{\Phi}\boldsymbol{\alpha} + \boldsymbol{\varepsilon}, \quad \boldsymbol{\varepsilon} \sim Gau(\mathbf{0}, \mathbf{C}_\epsilon), \qquad (4.23)$$

and then

$$E(\mathbf{Z}) = E_\alpha\{E(\mathbf{Z}|\boldsymbol{\alpha})\} = E_\alpha\{\mathbf{X}\boldsymbol{\beta} + \boldsymbol{\Phi}\boldsymbol{\alpha}\} = \mathbf{X}\boldsymbol{\beta}, \tag{4.24}$$

$$\mathrm{var}(\mathbf{Z}) = \mathrm{var}_\alpha\{E(\mathbf{Z}|\boldsymbol{\alpha})\} + E_\alpha\{\mathrm{var}(\mathbf{Z}|\boldsymbol{\alpha})\} = \boldsymbol{\Phi}\mathbf{C}_\alpha\boldsymbol{\Phi}' + \mathbf{C}_\epsilon. \tag{4.25}$$

Then, since (4.23) shows that $\mathbf{Z}$ is a linear combination of normally distributed random variables, it is also normally distributed and the marginal distribution is given by

$$\mathbf{Z} \sim Gau(\mathbf{X}\boldsymbol{\beta}, \boldsymbol{\Phi}\mathbf{C}_\alpha\boldsymbol{\Phi}' + \mathbf{C}_\epsilon). \tag{4.26}$$

Thus, we can see that the integration over the common random effects $\boldsymbol{\alpha}$ in (4.22) induces a more complicated error covariance structure in the marginal distribution (i.e., compare the marginal covariance matrix, $\boldsymbol{\Phi}\mathbf{C}_\alpha\boldsymbol{\Phi}' + \mathbf{C}_\epsilon$, to the conditional covariance matrix, $\mathbf{C}_\epsilon$). This idea of conditioning on random effects and inducing dependence through integration is fundamentally important to hierarchical statistical modeling. That is, it is typically easier to model means than it is to model covariances, and so we put our modeling effort into the conditional mean and then let the integration induce the more complicated marginal dependence rather than specifying it directly.

4.4 Basis-Function Representations

By themselves, the conditional specifications discussed in Section 4.3 are often not enough to help us deal with the problem of specifying realistic spatio-temporal covariance structures and deal with the "curse of dimensionality," which is endemic in spatio-temporal statistics. We also need to pay particular attention to our choice of $\boldsymbol{\Phi}$, and we often do this through basis-function expansions (recall that we introduced basis functions in Chapter 1 and in more detail in Chapter 3).

Basis functions, like covariates, can be nonlinear functions of $(\mathbf{s}; t)$; however, the expansion is a linear function of the basis functions' coefficients. We assume that these coefficients are the objects of inference in a statistical additive model. If the coefficients are fixed but unknown and to be estimated, then we have a regression model and the basis functions act as covariates (see, for example, Section 3.2). If the coefficients are random, then we have a random-effects model (or, if covariates are also present, a mixed-effects model) and we can perform inference on the moments of those random effects. More importantly, as we have shown in Section 4.3, this framework allows us to build complexity through marginalization. This often simplifies the model specification, particularly if we consider the random effects to be associated with spatial, temporal, or spatio-temporal basis functions. In the following subsections, we consider spatio-temporal models that involve these three types of basis functions.

4.4.1   Random Effects with Spatio-Temporal Basis Functions

Assuming the same data model (4.3) as above, we rewrite the process model (4.2) in terms of fixed and random effects, $\boldsymbol{\beta}$ and $\{\alpha_i : i = 1, \ldots, n_\alpha\}$, respectively:

$$Y(\mathbf{s};t) = \mathbf{x}(\mathbf{s};t)'\boldsymbol{\beta} + \eta(\mathbf{s};t) = \mathbf{x}(\mathbf{s};t)'\boldsymbol{\beta} + \sum_{i=1}^{n_\alpha} \phi_i(\mathbf{s};t)\alpha_i + \nu(\mathbf{s};t), \qquad (4.27)$$

where $\{\phi_i(\mathbf{s};t) : i = 1, \ldots, n_\alpha\}$ are specified *spatio-temporal* basis functions corresponding to location $(\mathbf{s};t)$, $\{\alpha_i\}$ are *random effects*, and $\nu(\mathbf{s};t)$ is sometimes needed to represent small-scale spatio-temporal random effects not captured by the basis functions. So, in (4.27) we are just decomposing the spatio-temporal random process, $\eta(\mathbf{s};t)$, into a linear combination of random effects and a "residual" error term.

Let $\boldsymbol{\alpha} \sim Gau(\mathbf{0}, \mathbf{C}_\alpha)$, where $\boldsymbol{\alpha} \equiv (\alpha_1, \ldots, \alpha_{n_\alpha})'$. Suppppose we are interested in making inference on the process Y at n_y spatio-temporal locations, which we denote by the n_y-dimensional vector $\mathbf{Y}$. The process model then becomes

$$\mathbf{Y} = \mathbf{X}\boldsymbol{\beta} + \boldsymbol{\Phi}\boldsymbol{\alpha} + \boldsymbol{\nu}, \qquad (4.28)$$

where the ith column of the $n_y \times n_\alpha$ matrix $\boldsymbol{\Phi}$ corresponds to the ith basis function, $\phi_i(\cdot; \cdot)$, at all of the n_y spatio-temporal locations, and in the same order as that used to construct $\mathbf{Y}$. The vector $\boldsymbol{\nu}$ also corresponds to the spatio-temporal ordering given in $\mathbf{Y}$, and $\boldsymbol{\nu} \sim Gau(\mathbf{0}, \mathbf{C}_\nu)$. In this case, one can see (Technical Note 4.3) that the marginal distribution of $\mathbf{Y}$ is given by $\mathbf{Y} \sim Gau(\mathbf{X}\boldsymbol{\beta}, \boldsymbol{\Phi}\mathbf{C}_\alpha\boldsymbol{\Phi}' + \mathbf{C}_\nu)$, so that $\mathbf{C}_y = \boldsymbol{\Phi}\mathbf{C}_\alpha\boldsymbol{\Phi}' + \mathbf{C}_\nu$. Now the vector of covariance parameters $\boldsymbol{\theta}$ is augmented to include parameters in $\mathbf{C}_\nu$. The spatio-temporal dependence is accounted for by the spatio-temporal basis functions, $\boldsymbol{\Phi}$, and in general this could accommodate non-separable dependence. A benefit of this approach is that the spatio-temporal modeling effort focuses on the fixed number n_α of random effects. In this case, note that the random effects $\boldsymbol{\alpha}$ are not indexed by space and time, so it should be easier to specify a model for them. For example, we can specify a *covariance matrix* to describe their dependence, which is easier than specifying a *covariance function*.

In situations where $n_\alpha \ll n_y$ (i.e., a *low-rank representation*), an additional benefit comes from being able to perform matrix inverses in terms of n_α-dimensional matrices (through well-known matrix-algebra relationships). Specifically, under model (4.28) we note that we can write $\mathbf{C}_z = \boldsymbol{\Phi}\mathbf{C}_\alpha\boldsymbol{\Phi}' + \mathbf{V}$, where we define $\mathbf{V} \equiv \mathbf{C}_\nu + \mathbf{C}_\epsilon$. Then, using the well-known Sherman–Morrison–Woodbury matrix identities (e.g., Searle, 1982), we can write

$$\mathbf{C}_z^{-1} = \mathbf{V}^{-1} - \mathbf{V}^{-1}\boldsymbol{\Phi}(\boldsymbol{\Phi}'\mathbf{V}^{-1}\boldsymbol{\Phi} + \mathbf{C}_\alpha^{-1})^{-1}\boldsymbol{\Phi}'\mathbf{V}^{-1}.$$

Importantly, if $\mathbf{V}^{-1}$ has simple structure (e.g., is sparse or diagonal) and $n_\alpha \ll n_y$, then this inverse is easy to calculate because it is a function of a simple high-dimensional matrix $\mathbf{V}^{-1}$ and a low-dimensional matrix inverse $\mathbf{C}_\alpha^{-1}$.

It is important to note that even in the *full-rank* $(n_\alpha = n_y)$ and *over-complete* $(n_\alpha > n_y)$ cases there can still be computational benefits through induced sparsity in $\mathbf{C}_\alpha$ and the use of efficient matrix-multiplication routines that use multiresolution algorithms, orthogonality, and/or sparse precision matrices. In addition, basis-function implementations may assume that $\boldsymbol{\nu} = \mathbf{0}$ and often that $\boldsymbol{\Phi}$ is orthogonal, so that $\boldsymbol{\Phi}\boldsymbol{\Phi}' = \mathbf{I}$; in those cases, one can reduce the computational burden significantly. Finally, we note that specific basis functions and methodologies are devised to take advantage of other properties of various matrices (e.g., sparse structure on the random-effects covariance matrix, $\mathbf{C}_\alpha$, or on the random-effects precision matrix, $\mathbf{C}_\alpha^{-1}$).

> **R tip:** Sparse matrices can be used in R using definitions in the packages **Matrix** or **spam**. For both these packages, arithmetic operations, decompositions (e.g., the Cholesky decomposition), back-solves and forward-solves, and other important matrix operations, can be done seamlessly using standard R commands. With **Matrix**, a sparse matrix can be constructed using the function `sparseMatrix`, while a sparse diagonal matrix can be constructed using the function `Diagonal`. With the former, the argument `symmetric = TRUE` can be used to specify a sparse symmetric matrix.

The definition of "basis function" in our spatio-temporal context is pretty liberal; the matrix $\boldsymbol{\Phi}$ in the product $\boldsymbol{\Phi}\boldsymbol{\alpha}$ is a spatio-temporal basis-function matrix so long as its coefficients $\boldsymbol{\alpha}$ are random and the columns of $\boldsymbol{\Phi}$ are spatio-temporally referenced. One decision associated with fitting model (4.27) concerns the choice of basis functions. For spatial processes, the decisions one makes with regard to the choice of basis functions are usually not that critical, as there are multiple types of bases that can accommodate the same spatial variability. However, as one starts considering spatio-temporal processes, the choice of basis functions can make a difference, especially for the dynamical formulations presented in Chapter 5.

In general, one can use (i) fixed or parameterized basis functions, (ii) local or global basis functions, (iii) reduced-rank, complete, or over-complete bases, and (iv) basis functions with expansion coefficients possibly indexed by space, time, or space-time. Further, the choice is affected by the presence and type of residual structure and the distribution of the random effects. Historically, it has been fairly challenging to come up with good spatio-temporal basis functions (for the same reason it has been difficult to come up with truly realistic spatio-temporal covariance functions). One simplification is to consider tensor-product basis functions (mentioned in Section 3.2 and Technical Note 4.1), where we define the spatio-temporal basis function as the product of a spatial basis function and a temporal basis function. Note that this does *not* yield a separable spatio-temporal model, in general. It is also quite common to see spatio-temporal-dependence models for Y, where the statistical dependence comes from spatial-only basis functions whose coefficients are temporal stochastic processes (Section 4.4.2).

Predictions (degrees Fahrenheit)

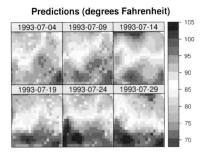

Prediction std. errors (degrees Fahrenheit)

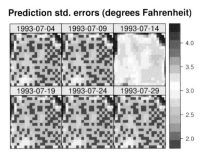

Figure 4.6: (Left) Predictions of `Tmax` and (right) prediction standard errors in degrees Fahrenheit within a square box enclosing the domain of interest for six days (each 5 days apart) spanning the temporal window of the data, 01 July 1993–20 July 2003, using bisquare spatio-temporal basis functions and the R package **FRK**. Data for 14 July 1993 were omitted from the original data set.

Example: Fixed Rank Kriging

A widely adopted method for rank reduction is *fixed rank kriging* (FRK), implemented in R through the package **FRK**. Lab 4.2 demonstrates how FRK can be applied to the maximum temperature (`Tmax`) in the NOAA data set using $n_\alpha = 1880$ space-time tensor-product basis functions (see Technical Note 4.1) at two resolutions for $\{\phi_i(\mathbf{s}; t) : i = 1, \ldots, n_\alpha\}$. In particular, *bisquare* basis functions are used (see Lab 4.2 for details). FRK also considers a fine-scale-variation component ν such that $\mathbf{C}_\nu$ is diagonal. The matrix $\mathbf{C}_\alpha$ is constructed such that the coefficients $\boldsymbol{\alpha}$ at each resolution are independent, and such that the covariances between these coefficients within a resolution decay exponentially with the distance between the centers of the basis functions. Parameters are estimated using an EM algorithm for computing maximum likelihood estimates (see Algorithm 4.1).

Figure 4.6 shows the predictions and prediction standard errors obtained using FRK; as is typical for kriging, the computations are made with $\widehat{\boldsymbol{\theta}}$ substituted in for the unknown covariance parameters $\boldsymbol{\theta}$. Although the uncertainty in $\widehat{\boldsymbol{\theta}}$ is not accounted for in this setting, it is typically thought to be a fairly minor component of the variation in the spatio-temporal prediction. The predictions are similar to those obtained using S-T kriging in Figure 4.2 of Section 4.2, but they are also a bit "noisier" because of the assumed uncorrelated fine-scale variation term; see (4.27). The prediction standard errors show similar patterns to those obtained earlier (Figure 4.2), although there are notable differences upon visual examination. This is commonly observed when using reduced-rank methods, and it is particularly evident with very-low-rank implementations (e.g., with EOFs) accompanied with spatially uncorrelated fine-scale variation. In such cases, the prediction-standard-error maps can have prediction standard errors related more to the shapes of the basis functions and less to the prediction location's proximity to an observation.

Algorithm 4.1: Basic EM Algorithm

In some cases, it can be computationally more efficient to perform maximum likelihood estimation using the expectation-maximization (EM) algorithm rather than through direct optimization of the likelihood function. The basic idea is that one defines *complete data* to be a combination of actual observations and missing observations. Let W denote these complete data made up of observations (W_{obs}) and "missing" observations (W_{mis}), and θ represents the unknown parameters in the model, so that the complete-data log-likelihood is given by $\log(L(\theta|W))$. The basic EM algorithm is given below.

Choose starting values for the parameter, $\hat{\theta}^{(0)}$
repeat $i = 1, 2, \ldots$
    1. E-Step: Obtain $Q(\theta|\hat{\theta}^{(i-1)}) = E\{\log(L(\theta \mid W)) \mid W_{\text{obs}}, \hat{\theta}^{(i-1)}\}$
    2. M-Step: Obtain $\hat{\theta}^{(i)} = \max_{\theta}\{Q(\theta \mid \hat{\theta}^{(i-1)})\}$
until Convergence either in $\hat{\theta}^{(i)}$ or in $\log(L(\theta \mid W))$

In Section 4.4, W_{obs} corresponds to the data $\mathbf{Z}$, while W_{mis} corresponds to the coefficients $\boldsymbol{\alpha}$.

4.4.2 Random Effects with Spatial Basis Functions

Consider the case where the basis functions of the spatio-temporal process are functions of space only and their random coefficients are indexed by time:

$$Y(\mathbf{s}; t_j) = \mathbf{x}(\mathbf{s}; t_j)'\boldsymbol{\beta} + \sum_{i=1}^{n_\alpha} \phi_i(\mathbf{s})\alpha_i(t_j) + \nu(\mathbf{s}; t_j), \quad j = 1, \ldots, T, \qquad (4.29)$$

where $\{\phi_i(\mathbf{s}) : i = 1, \ldots, n_\alpha; \mathbf{s} \in D_s\}$ are known spatial basis functions, $\alpha_i(t_j)$ are temporal random processes, and the other model components are defined as above. We can consider a wide variety of spatial basis functions for this model, and again these might be of reduced rank, of full rank, or over-complete. For example, we might consider complete global basis functions (e.g., Fourier), or reduced-rank empirically defined basis functions (e.g., EOFs), or a variety of non-orthogonal bases (e.g., Gaussian functions, wavelets, bisquare functions, or Wendland functions). We illustrate a few of these in one dimension in Figure 4.7 (see also Section 3.2). It is often not important which basis function is used; still, one has to be careful to ensure that the type and number of basis functions are flexible and large enough to model the true dependence in Y (and the data $\mathbf{Z}$). This requires some experimentation and model diagnostics (see, for example, Chapter 6).

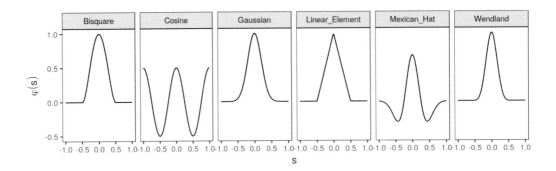

Figure 4.7: Some spatial basis functions that can be employed in spatio-temporal modeling, depicted in one-dimensional space. From left to right: bisquare, cosine, Gaussian, linear element, Mexican-hat wavelet, and first-order Wendland functions.

Assuming interest in the spatio-temporal dependence at n spatial locations $\{\mathbf{s}_1, \ldots, \mathbf{s}_n\}$ and at times $\{t_j : j = 1, 2, \ldots, T\}$, we can write model (4.29) in vector form as

$$\mathbf{Y}_{t_j} = \mathbf{X}_{t_j}\boldsymbol{\beta} + \boldsymbol{\Phi}\boldsymbol{\alpha}_{t_j} + \boldsymbol{\nu}_{t_j}, \tag{4.30}$$

where $\mathbf{Y}_{t_j} = (Y(\mathbf{s}_1; t_j), \ldots, Y(\mathbf{s}_n; t_j))'$ is the n-dimensional process vector, $\boldsymbol{\nu}_{t_j} \sim Gau(\mathbf{0}, \mathbf{C}_\nu)$, $\boldsymbol{\alpha}_{t_j} \equiv (\alpha_1(t_j), \ldots, \alpha_{n_\alpha}(t_j))'$, $\boldsymbol{\Phi} \equiv (\phi(\mathbf{s}_1), \ldots, \phi(\mathbf{s}_n))'$, and $\phi(\mathbf{s}_i) \equiv (\phi_1(\mathbf{s}_i), \ldots, \phi_{n_\alpha}(\mathbf{s}_i))'$, $i = 1, \ldots, n$. An important question is then what the preferred distribution for $\boldsymbol{\alpha}_{t_j}$ is.

It can be shown that if $\boldsymbol{\alpha}_{t_1}, \boldsymbol{\alpha}_{t_2}, \ldots$ are independent in time, where $\boldsymbol{\alpha}_{t_j} \sim iid\ Gau(\mathbf{0}, \mathbf{C}_\alpha)$, then the marginal distribution of $\mathbf{Y}_{t_j}$ is $Gau(\mathbf{X}_{t_j}\boldsymbol{\beta}, \boldsymbol{\Phi}\mathbf{C}_\alpha\boldsymbol{\Phi}' + \mathbf{C}_\nu)$, and $\mathbf{Y}_{t_1}, \mathbf{Y}_{t_2}, \ldots$ are independent. Hence, the $nT \times nT$ joint spatio-temporal covariance matrix is given by the Kronecker product, $\mathbf{C}_Y = \mathbf{I}_T \otimes (\boldsymbol{\Phi}\mathbf{C}_\alpha\boldsymbol{\Phi}' + \mathbf{C}_\nu)$, where $\mathbf{I}_T$ is the T-dimensional identity matrix (see Technical Note 4.1). So the independence-in-time assumption implies a simple separable spatio-temporal dependence structure. To model more complex spatio-temporal dependence structure using spatial-only basis functions, one must specify the model for the random coefficients such that $\{\boldsymbol{\alpha}_{t_j} : j = 1, \ldots, T\}$ are *dependent in time*. This is simplified by assuming *conditional* temporal dependence (dynamics) as discussed in Chapter 5.

4.4.3 Random Effects with Temporal Basis Functions

We can also express the spatio-temporal random process in terms of temporal basis functions and spatially indexed random effects:

$$Y(\mathbf{s}; t) = \mathbf{x}(\mathbf{s}; t)'\boldsymbol{\beta} + \sum_{i=1}^{n_\alpha} \phi_i(t)\alpha_i(\mathbf{s}) + \nu(\mathbf{s}; t), \tag{4.31}$$

where $\{\phi_i(t) : i = 1, \ldots, n_\alpha; \ t \in D_t\}$ are temporal basis functions and $\{\alpha_i(\mathbf{s})\}$ are their spatially indexed random coefficients. In this case, one could model $\{\alpha_i(\mathbf{s}) : \mathbf{s} \in D_s; \ i = 1, \ldots, n_\alpha\}$ using multivariate geostatistics. The temporal-basis-function representation given in (4.31) is not as common in spatio-temporal statistics as the spatial-basis-function representation given in (4.29). This is probably because most spatio-temporal processes have a scientific interpretation of spatial processes evolving in time. However, this need not be the case, and temporal basis functions are increasingly being used to model non-stationary-in-time processes (e.g., complex seasonal or high-frequency time behavior) that vary across space.

Example Using Temporal Basis Functions

Spatio-temporal modeling and prediction using temporal basis functions can be carried out with the package **SpatioTemporal** (see Lab 4.3). In the top panel of Figure 4.8 we show the three temporal basis functions used to model maximum temperature in the NOAA data set. These basis functions were obtained following a procedure similar to EOF analysis, which is described in Technical Note 2.2. Note that the basis function $\phi_1(t) = 1$ is time-invariant.

Once $\phi_1(t), \phi_2(t)$, and $\phi_3(t)$ are selected, estimates (e.g., ordinary least squares) of $\alpha_1(\mathbf{s}), \alpha_2(\mathbf{s})$, and $\alpha_3(\mathbf{s})$ can be found and used to indicate how they might be modeled. For example, in Lab 4.3 we see that while both $\alpha_1(\mathbf{s})$ and $\alpha_2(\mathbf{s})$ have a latitudinal trend, $\alpha_3(\mathbf{s})$ does not. Assigning these fields exponential covariance functions, we obtain the models:

$$E(\alpha_1(\mathbf{s})) = \alpha_{11} + \alpha_{12}s_2, \qquad \text{cov}(\alpha_1(\mathbf{s}), \alpha_1(\mathbf{s}+\mathbf{h})) = \sigma_1^2 \exp(-\|\mathbf{h}\|/r_1), \quad (4.32)$$

$$E(\alpha_2(\mathbf{s})) = \alpha_{21} + \alpha_{22}s_2, \qquad \text{cov}(\alpha_2(\mathbf{s}), \alpha_2(\mathbf{s}+\mathbf{h})) = \sigma_2^2 \exp(-\|\mathbf{h}\|/r_2), \quad (4.33)$$

$$E(\alpha_3(\mathbf{s})) = \alpha_{31}, \qquad \text{cov}(\alpha_3(\mathbf{s}), \alpha_3(\mathbf{s}+\mathbf{h})) = \sigma_3^2 \exp(-\|\mathbf{h}\|/r_3), \quad (4.34)$$

where s_2 denotes the latitude coordinate at $\mathbf{s} = (s_1, s_2)'$, r_1, r_2, and r_3 are scale parameters, and σ_1^2, σ_2^2, and σ_3^2 are stationary variances. We further assume that $\text{cov}(\alpha_k(\mathbf{s}), \alpha_\ell(\mathbf{s}')) = 0$ for $k \neq \ell$, which is a strong assumption.

Using maximum likelihood to estimate all unknown parameters and "plugging" the estimates in, the resulting prediction is the spatio-temporal smoothed map, $E(Y(\cdot; \cdot) \mid \mathbf{Z})$, obtained from maps of $E(\alpha_1(\cdot) \mid \mathbf{Z}), E(\alpha_2(\cdot) \mid \mathbf{Z})$, and $E(\alpha_3(\cdot) \mid \mathbf{Z})$, which can all be written in closed form. We show the first three basis-function times series in the top panel of Figure 4.8 and the predicted spatial maps (i.e., the basis-function coefficients), corresponding to these three basis functions in the bottom panel. Note how $E(\alpha_1(\cdot) \mid \mathbf{Z})$ picks up the latitude component evident in the NOAA maximum-temperature data. On the other hand, the fields $E(\alpha_2(\cdot) \mid \mathbf{Z})$ and $E(\alpha_3(\cdot) \mid \mathbf{Z})$ appear to capture oblique and longitudinal trends that have not been considered up to now, but with much smaller magnitudes. Although not shown here, these predictions of the basis-function coefficients $\alpha_1(\cdot), \alpha_2(\cdot)$, and $\alpha_3(\cdot)$ have associated uncertainties and those can be plotted as prediction standard-error maps as well.

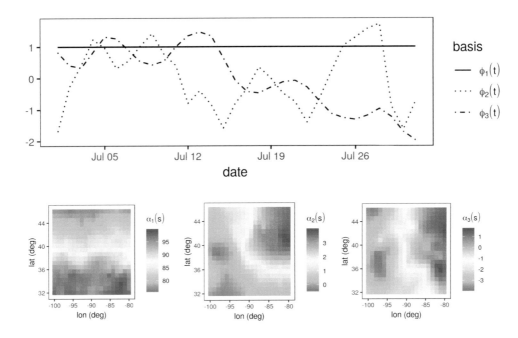

Figure 4.8: Top: Basis functions $\phi_1(t), \phi_2(t)$, and $\phi_3(t)$, where the latter two were obtained from the left-singular vectors following a singular value decomposition of the data matrix. Bottom: $E(\alpha_1(\mathbf{s}) \mid \mathbf{Z}), E(\alpha_2(\mathbf{s}) \mid \mathbf{Z})$, and $E(\alpha_3(\mathbf{s}) \mid \mathbf{Z})$.

4.4.4   Confounding of Fixed Effects and Random Effects

Consider the general mixed-effects representation given in (4.28):

$$\mathbf{Y} = \mathbf{X}\boldsymbol{\beta} + \boldsymbol{\Phi}\boldsymbol{\alpha} + \boldsymbol{\nu}, \quad \boldsymbol{\nu} \sim Gau(\mathbf{0}, \mathbf{C}_{\nu}),$$

and recall that $\mathbf{Z} = \mathbf{Y} + \boldsymbol{\varepsilon}$. Although the columns of $\boldsymbol{\Phi}$ are basis functions, they are indexed in space and time in the same way that the columns of $\mathbf{X}$ are. Then, depending on the structure of the columns in these two matrices, it is quite possible that the random effects can be confounded with the fixed effects, similarly to the way extreme collinearity can affect the estimation of fixed effects in traditional regression (recall Section 3.2.2). This suggests that if primary interest is in inference on the fixed-effect parameters ($\boldsymbol{\beta}$), then one should mitigate potential confounding associated with the random effects. As with collinearity, if the columns of $\boldsymbol{\Phi}$ and $\mathbf{X}$ are linearly independent, then there is no concern about confounding. This has led to mitigation strategies that tend to restrict the random effects by selecting basis functions in $\boldsymbol{\Phi}$ that are orthogonal to the column space of $\mathbf{X}$ (or approximately so). If prediction of the hidden process $\mathbf{Y}$ is the primary goal, one is typically much less concerned about potential confounding.

4.5 Non-Gaussian Data Models with Latent Gaussian Processes

There is only one way to be Gaussian, but an infinite number of ways to be non-Gaussian! This is a challenge that we address in this section through the use of hierarchical statistical models. The modeling paradigm that we follow is to find a Gaussian process, possibly deep in the hierarchy, that describes the spatio-temporal behavior of a hidden process or of parameters that vary with space and time. The marginal distribution of the data is then non-Gaussian, but somewhere there is a Gaussian process that results in spatio-temporal dependence in the data through marginalization.

The examples presented thus far in this chapter have all assumed additive Gaussian error and random-effects distributions. Many spatio-temporal problems of interest deal with distinctly non-Gaussian data (e.g., counts, binary responses, extreme values). One of the most useful aspects of the hierarchical-modeling paradigm is that it allows one to accommodate fairly easily non-Gaussian data models, so long as the observations are *conditionally independent*, conditional on latent dependent Gaussian processes. This is the spatio-temporal manifestation of traditional generalized linear mixed models (GLMMs) and generalized additive mixed models (GAMMs) in statistics. That is, the likelihood assumes that the observations are conditionally independent given a spatio-temporal mean response that is some transformation of an additive mixed model. Our situation is a bit more flexible than the GLMM and GAMM in that our data model does not necessarily have to be from the exponential family, so long as we can allow conditional independence in the observations conditioned on spatio-temporal structure in the hidden process (and/or the associated process parameters).

As an example, consider a data model from the exponential family as in Section 3.4.1, such that

$$Z(\mathbf{s};t) \mid Y(\mathbf{s};t), \gamma \sim indep. \; EF(Y(\mathbf{s};t), \gamma), \quad \mathbf{s} \in D_s, \, t \in D_t, \qquad (4.35)$$

where EF corresponds to a distribution from the exponential family with scale parameter γ and mean $Y(\mathbf{s};t)$. In Section 3.4.1, we modeled a transformation of the mean response in terms of additive fixed effects (e.g., a linear combination of covariates). Here, we extend that and model the transformed mean response in terms of additive fixed effects and random effects,

$$g(Y(\mathbf{s};t)) = \mathbf{x}(\mathbf{s};t)'\boldsymbol{\beta} + \eta(\mathbf{s};t), \quad \mathbf{s} \in D_s, t \in D_t,$$

where $g(\cdot)$ is a specified monotonic link function, $\mathbf{x}(\mathbf{s};t)$ is a p-dimensional vector of covariates for spatial location $\mathbf{s}$ and time t, and $\eta(\mathbf{s};t)$ is a spatio-temporal Gaussian random process that can be modeled either in terms of spatio-temporal covariances (as in Section 4.2), a special case of which uses a basis-function expansion (Section 4.3), or as a dynamic spatio-temporal process (Chapter 5). The same modeling issues associated with this latent Gaussian spatio-temporal process are present here as with the Gaussian-data case, but estimation of parameters and prediction of $Y(\mathbf{s}_0;t_0)$ are typically more involved given the non-Gaussian data model.

As an illustration, a simple model involving spatio-temporal count data could be represented by

$$\mathbf{Z}_t \mid \mathbf{Y}_t \sim indep. \; Poi(\mathbf{Y}_t),$$

$$\log(\mathbf{Y}_t) = \mathbf{X}_t \boldsymbol{\beta} + \boldsymbol{\Phi}_t \boldsymbol{\alpha}_t + \boldsymbol{\nu}_t,$$

where $\mathbf{Z}_t$ is an m_t-dimensional data vector of counts at m_t spatial locations, $\mathbf{Y}_t$ represents the latent spatio-temporal mean process at m_t locations, $\boldsymbol{\Phi}_t$ is an $m_t \times n_\alpha$ matrix of n_α spatial basis functions, and the associated random coefficients are modeled as $\boldsymbol{\alpha}_t \sim Gau(\mathbf{0}, \mathbf{C}_\alpha)$, independent in time, with micro-scale error term $\boldsymbol{\nu}_t \sim indep. \; Gau(\mathbf{0}, \sigma_\nu^2 \mathbf{I})$; that is, $\mathbf{C}_\nu = \sigma_\nu^2 \mathbf{I}$. As discussed in Section 4.4.2, it is often more realistic to consider temporal dependence through a dynamic model on $\{\boldsymbol{\alpha}_t\}$, which will be explored in Chapter 5. As was the case for the Gaussian data models in Sections 4.1–4.4, the parameters $\boldsymbol{\beta}$ and $\boldsymbol{\theta}$ (in $\mathbf{C}_\alpha$ and $\mathbf{C}_\nu$) could be estimated or a prior distribution could be put on them.

4.5.1   Generalized Additive Models (GAMs)

We often seek more flexible models that can accommodate nonlinear structure in the mean function. Recall from Section 3.4.1 that one successful approach to this problem has been through the use of GAMs. In general, these models consider a transformation of the mean response to have an additive form in which the additive components are smooth functions (e.g., splines) of the covariates, where generally the functions themselves are expressed as basis-function expansions. In practical applications, the basis coefficients are treated as random coefficients in the estimation procedure. However, just as one can add random effects to generalized linear models (GLMs) to get generalized linear mixed models (GLMMs), one can also add (additional) random effects to GAMs to get generalized additive mixed models (GAMMs).

For example, consider data model (4.35). Similarly to (3.12), we can write the transformed mean response additively as

$$g(Y(\mathbf{s};t)) = \mathbf{x}(\mathbf{s};t)'\boldsymbol{\beta} + \sum_{i=1}^{n_f} f_i(\mathbf{x}(\mathbf{s};t); \mathbf{s};t) + \nu(\mathbf{s};t), \qquad (4.36)$$

where again $g(\cdot)$ is a specified monotonic link function; $\mathbf{x}(\mathbf{s};t)$ is a p-dimensional vector of covariates for spatial location $\mathbf{s}$ and time t; $f_i(\cdot)$ are functions of the covariates, the spatial locations, and the time index; and $\nu(\mathbf{s};t)$ is a spatio-temporal random effect. Typically, the functions $f_i(\cdot)$ are modeled in terms of a truncated basis-function expansion; for example, $f_i(x_1(\mathbf{s};t); \mathbf{s};t) = \sum_{k=1}^{q_i} \phi_k(x_1(\mathbf{s};t); \mathbf{s};t)\alpha_{ik}$. Thus, we can see that the basis-function expansions with random coefficients given in (4.27), (4.29), and (4.31) are essentially GAMMs. But, whereas in those models the smooth functions are typically only a function of spatio-temporal location, spatial location, or time, respectively, it is more common in the GAM/GAMM setting to allow the basis functions to also depend nonlinearly on covariates. On the other hand, GAM/GAMMs typically assume that the basis functions are smooth functions, whereas there is no such requirement for spatio-temporal-basis-function models. GAM/GAMMs can easily be implemented in R (e.g., we provide an example with the **mgcv** package in Lab 4.4).

4.5.2   Inference for Spatio-Temporal Hierarchical Models

Implicit in the estimation associated with the linear Gaussian spatio-temporal model discussed in Section 4.2.3 is that the covariance and fixed-effects parameters can be estimated more easily when we marginalize (integrate) out the latent Gaussian spatio-temporal process. In general, the likelihood is

$$[\mathbf{Z} \mid \boldsymbol{\theta}, \boldsymbol{\beta}] = \int [\mathbf{Z} \mid \mathbf{Y}, \boldsymbol{\theta}][\mathbf{Y} \mid \boldsymbol{\theta}, \boldsymbol{\beta}] \mathrm{d}\mathbf{Y}, \tag{4.37}$$

viewed as a function of $\boldsymbol{\theta}$ and $\boldsymbol{\beta}$. For linear mixed models (Sections 4.1–4.4), we assumed that the two distributions inside the integral in (4.37) were Gaussian with linear relationships; this implied that the marginal likelihood function was in the form of a Gaussian density (e.g., Technical Note 4.3), and thus can be written in closed form. More generally, we can relax the Gaussian assumption for the data model and, in the models presented here, the latent Gaussian spatio-temporal process $\mathbf{Y}$ is transformed through a nonlinear link function. This lack of Gaussianity and the presence of nonlinearity complicates the analysis, as generally the likelihood (4.37) cannot be obtained in closed form.

The integral in (4.37) can in principle be evaluated numerically, from which one can estimate the relatively few fixed effects and covariance parameters $\{\boldsymbol{\beta}, \boldsymbol{\theta}\}$ through numerical optimization. In spatio-temporal models this is complicated by the high dimensionality of the integral; recall that $\mathbf{Y}$ is a $(\sum_{t=1}^{T} m_t)$-dimensional vector. Traditional approaches to this problem are facilitated by the usual conditional-independence assumption in the data model and by exploiting the latent Gaussian nature of the random effects. These approaches include methods such as Laplace approximation, quasi-likelihood, generalized estimating equations, pseudo-likelihood, and penalized quasi-likelihood. For example, recent advances in automatic differentiation have led to very efficient Laplace approximation approaches for performing inference with such likelihoods, even when there are a very large number of random effects (see, for example, the Template Model Builder (**TMB**) R package). Although these methods are increasingly being used successfully in the spatial context, there has tended to be more focus on Bayesian estimation approaches for spatio-temporal models in the literature. Either way, some type of approximation is needed (approximating the integrals, approximating the models using linearization, or approximating the posterior distribution through various Bayesian computational methods).

Bayesian Hierarchical Modeling

The Bayesian hierarchical model (BHM) paradigm provides the estimation and inferential framework for many complex spatio-temporal models in the literature. Recall from Technical Note 1.1 that we can decompose an arbitrary joint distribution in terms of a hierarchical sequence of conditional distributions and a marginal distribution; for example,

$$[A, B, C] = [A \mid B, C][B \mid C][C].$$

In the context of our general geostatistical spatio-temporal model given in Section 4.2,

$$
\begin{aligned}
[\mathbf{Z}, \mathbf{Y}, \boldsymbol{\beta}, \boldsymbol{\theta}] &= [\mathbf{Z} \mid \mathbf{Y}, \boldsymbol{\beta}, \boldsymbol{\theta}][\mathbf{Y} \mid \boldsymbol{\beta}, \boldsymbol{\theta}][\boldsymbol{\beta} \mid \boldsymbol{\theta}][\boldsymbol{\theta}] \\
&= [\mathbf{Z} \mid \mathbf{Y}, \boldsymbol{\theta}_\epsilon][\mathbf{Y} \mid \boldsymbol{\beta}, \boldsymbol{\theta}_y][\boldsymbol{\theta}_\epsilon][\boldsymbol{\theta}_y][\boldsymbol{\beta}],
\end{aligned}
$$

where $\boldsymbol{\theta}$ contains all of the variance and covariance parameters from the data model and the process model. Note that the first equality is based on the probability decomposition and the second equality is based on writing $\boldsymbol{\theta} = \{\boldsymbol{\theta}_\epsilon, \boldsymbol{\theta}_y\}$ and assuming that $\boldsymbol{\beta}$, $\boldsymbol{\theta}_\epsilon$, and $\boldsymbol{\theta}_y$ are independent *a priori*. Now, Bayes' Rule implies that

$$[\mathbf{Y}, \boldsymbol{\beta}, \boldsymbol{\theta} \mid \mathbf{Z}] \propto [\mathbf{Z} \mid \mathbf{Y}, \boldsymbol{\theta}_\epsilon][\mathbf{Y} \mid \boldsymbol{\beta}, \boldsymbol{\theta}_y][\boldsymbol{\beta}][\boldsymbol{\theta}_\epsilon][\boldsymbol{\theta}_y]. \qquad (4.38)$$

For example, in the linear Gaussian case, $[\mathbf{Z} \mid \mathbf{Y}, \boldsymbol{\theta}_\epsilon]$ is given by (4.3) and $[\mathbf{Y} \mid \boldsymbol{\beta}, \boldsymbol{\theta}_y]$ is given by (4.4). The prior distributions $[\boldsymbol{\beta}]$, $[\boldsymbol{\theta}_\epsilon]$, and $[\boldsymbol{\theta}_y]$ are then specified according to the particular modeling choices made.

If we are interested in inference on the parameters, then we focus on the posterior distribution, $[\boldsymbol{\beta}, \boldsymbol{\theta} \mid \mathbf{Z}]$; if our interest is in prediction, we focus on the predictive distribution, $[\mathbf{Y} \mid \mathbf{Z}]$. In principle, we can obtain these posterior distributions if we can evaluate the normalizing constant in (4.38), which is a function of the data $\mathbf{Z}$, specifically, the marginal distribution $[\mathbf{Z}]$. However, in the general spatio-temporal case (and in most hierarchical models) there is no analytical form for this normalizing constant, and one must use numerical approximations. A common and useful approach is to use Markov chain Monte Carlo (MCMC) techniques to obtain (Markov dependent) Monte Carlo (MC) samples from the posterior distribution and then to perform inference on the parameters and prediction of the hidden process by summarizing these MC samples (see Algorithm 4.2 for a basic Gibbs sampler MCMC algorithm). The advantage of the BHM approach is that parameter uncertainty is accounted for directly. But, there is no "free lunch," and this usually comes at a cost of greater computational complexity.

In cases where the BHM computational complexity is formidable one can sometimes find approximations that help simplify the computational burden. For example, just as penalized-quasi-likelihood methods use Laplace approximations to deal with the integral in (4.37), the integrated nested Laplace approximation (INLA) approach is sometimes well suited for latent Gaussian spatial and spatio-temporal processes. The method exploits the Laplace approximation in Bayesian latent-Gaussian models and does not require generating samples from the posterior distribution. Hence, it can often be used for quite large data sets at reasonable computational expense. We use INLA to fit a latent separable spatio-temporal model in Lab 4.5.

Another way to mitigate the computational burden of a BHM is to obtain estimates $\widehat{\theta}$ of the parameters θ outside of the fully Bayesian model as in *empirical Bayesian estimation* (e.g., Carlin and Louis, 2010). As mentioned in Chapter 1, Cressie and Wikle (2011, pp. 23–24) call this approach *empirical hierarchical modeling* in the spatio-temporal context. In this case, one focuses on the "empirical predictive distribution," $[\mathbf{Y} \mid \mathbf{Z}, \widehat{\theta}]$. The primary example of this in spatio-temporal statistics is S-T kriging as discussed in Section 4.2. That is, rather than assigning prior distributions to the parameters, they are estimated and the estimates are "plugged in" to the closed-form kriging formulas. This typically has the advantage of substantially less computational burden but at a cost of overly liberal uncertainty quantification. Ideally, one should take additional steps to account for the uncertainty associated with using these plug-in estimates (e.g., via the bootstrap).

Algorithm 4.2: Basic Gibbs Sampler MCMC Algorithm

Consider the joint posterior distribution of K random variables, $w_1, \ldots, w_K$, given data, $\mathbf{Z}$, which we denote as $[w_1, \ldots, w_K \mid \mathbf{Z}]$. As is typical, assume that we do not know the normalizing constant for this posterior distribution. Markov chain Monte Carlo (MCMC) approaches can be used to obtain samples from such distributions indirectly. Specifically, rather than compute the posterior distribution directly, one computes successive simulations from a Markov chain constructed so that samples from the stationary distribution of this chain are equivalent to samples from the target posterior distribution. That is, after some "burn-in" time, samples of the chain are viewed as samples simulated from the posterior distribution. Note that these samples are statistically dependent. The posterior distribution can be explored by various Monte Carlo summaries of the MCMC samples.

One of the simplest MCMC algorithms is the *Gibbs sampler*, which is most appropriate when the distributions of each of the random variables conditioned on all of the others and the data (the "full-conditional" distributions) are available in closed form. For a basic overview, see Gelman et al. (2014). A generic Gibbs sampler algorithm is given below.

An initial step in the Gibbs sampler algorithm is to derive all of the full conditional distributions in closed form. That is, derive

$$[w_1 | w_2, \ldots, w_K, \mathbf{Z}], \ [w_2 | w_1, w_3, \ldots, w_K, \mathbf{Z}], \ldots, [w_K | w_1, w_2, \ldots, w_{K-1}, \mathbf{Z}].$$

Obtain starting values: $\{w_1^{(0)}, \ldots, w_K^{(0)}\}$

for $i = 1, 2, \ldots, N_{\text{gibbs}}$ **do**

1. Sample $w_1^{(i)} \sim [w_1|w_2^{(i-1)}, \ldots, w_K^{(i-1)}, \mathbf{Z}]$
2. Sample $w_2^{(i)} \sim [w_2|w_1^{(i)}, w_3^{(i-1)}, \ldots, w_K^{(i-1)}, \mathbf{Z}]$

   $\vdots$

K. Sample $w_K^{(i)} \sim [w_K|w_1^{(i)}, \ldots, w_{K-1}^{(i)}, \mathbf{Z}]$

end for

Discard the first b "burn-in" samples and use the remaining $b+1, \ldots, N_{\text{gibbs}}$ samples as though they are coming from the posterior distribution $[w_1, \ldots, w_K|\mathbf{Z}]$.

Note that this is one of the most basic MCMC algorithms. Many modifications exist to improve efficiency and deal with the common case where the full conditional distributions are not available in closed form (see, for example, Gelman et al., 2014, for an overview).

4.6   Chapter 4 Wrap-Up

Time marches forward, but it can be valuable to look back at a changing landscape over a period of time. We can describe how space and time interact using spatio-temporal mean and covariance functions, without having to commit to a mechanistic model that expresses the interaction dynamically. Hence, in this chapter we considered spatio-temporal modeling using what we have called the "descriptive" approach. Importantly, we made a clear distinction between the data and the underlying latent process that represents the real-world process upon which measurements were taken. That is, we need to think conditionally! Thus, we considered a data model where the conditional distribution was Gaussian and where the conditional distribution was non-Gaussian. In both cases, we conditioned on a latent Gaussian spatio-temporal process.

We also considered the latent spatio-temporal Gaussian process by specifying the first-order (mean) structure in terms of exogenous covariates (including functions of locations of space or time) and the second-order dependence in terms of spatio-temporal covariance functions. We discussed various assumptions for such models related to stationarity, separability, and full symmetry. These sorts of representations are ideally suited for problems where there are not too many observations or locations in time and space at which one wants to predict, and where either we feel comfortable that we know the dependence structure (and can represent it by covariance functions), or we just want to account for dependence and do not care so much that the model is not all that realistic. In situations with large data sets and/or large numbers of prediction locations, it is often more efficient computationally to consider random-effects representations of the second-order structure using basis-function expansions. The basis-function construction also frees the modeler from having to develop valid spatio-temporal covariance functions, as our conditional basis-function random effects *induce* a valid marginal covariance function. We considered this from the perspective of basis functions that are defined in space and time, in space only, and in time only. The descriptive-modeling framework is similar for each. In addition, we briefly showed how these spatio-temporal mixed models using basis functions are related to GAM/GAMMs, depending on the choice of basis functions and the estimation approach. An overview of GAMs can be found in Wood (2017).

A potential issue with performing parameter inference in descriptive models with spatial or spatio-temporal random effects is the problem of confounding. Traditionally, this has not been as big a concern in spatial and spatio-temporal statistics because the focus has been on prediction. But, as these methods have increasingly been used to account for dependence when *interpreting* fixed effects, confounding has received much more attention (e.g., Hodges and Reich, 2010; Hughes and Haran, 2013; Hanks et al., 2015).

An overview of Bayesian computation for spatial and spatio-temporal descriptive models is presented in Diggle and Ribeiro Jr. (2007) and Banerjee et al. (2015). The INLA approximate-Bayesian methodology is discussed in Rue et al. (2009), Lindgren et al. (2011), Blangiardo and Cameletti (2015), and Krainski et al. (2019). Descriptive models that can be formulated using simple dynamic equations in a Bayesian framework can also be implemented using **spTimer** (Bakar and Sahu, 2015) and the function `spDynLM` in **spBayes** (Finley et al., 2007). Computational methods for non-Bayesian approaches to non-Gaussian spatial data can be found in Schabenberger and Gotway (2005). An overview on using `R` to perform some exploratory and geostatistical modeling for spatio-temporal data can be found in RESSTE Network et al. (2017).

There are a number of informative books on spatio-temporal statistical methodology. These include Le and Zidek (2006),Cressie and Wikle (2011), Sherman (2011), Blangiardo and Cameletti (2015), Diggle (2013), Mateu and Müller (2013), Baddeley et al. (2015), Banerjee et al. (2015), Montero et al. (2015), Shaddick and Zidek (2015), and Christakos (2017).

One of the most challenging aspects of characterizing the spatio-temporal dependence structure, from either the marginal-covariance-model perspective or the conditional-basis-function perspective, is the ability to model real-world interactions that occur across time and space. In that case, the underlying processes are often best described by spatial fields that evolve through time according to "rules" that govern the spatio-temporal variability. That is, they represent a dynamical system. As we shall see in Chapter 5, spatio-temporal models that explicitly account for these dynamics offer the benefit of providing more realistic models in general, and they can simplify model construction and estimation through conditioning.

Lab 4.1: Spatio-Temporal Kriging with gstat

In this Lab we go through the process of carrying out spatio-temporal universal kriging using the semivariogram with the package **gstat**. We focus on the maximum temperature data in the NOAA data set (`Tmax`) in July 1993. In addition to the packages used in Chapter 2 for data wrangling, we need **RColorBrewer** to color some of the surfaces that will be produced.

```
library("sp")
library("spacetime")
library("ggplot2")
library("dplyr")
library("gstat")
library("RColorBrewer")
library("STRbook")
library("tidyr")
```

For S-T kriging of the maximum-temperature data set in July 1993, we need to fit a parametric function to the empirical semivariogram `vv` computed in Lab 2.3. The code is reproduced below for completeness.

```
data("STObj3", package = "STRbook")
STObj4 <- STObj3[, "1993-07-01::1993-07-31"]
vv <- variogram(object = z ~ 1 + lat, # fixed effect component
                data = STObj4,        # July data
                width = 80,           # spatial bin (80 km)
                cutoff = 1000,        # consider pts < 1000 km apart
                tlags = 0.01:6.01)    # 0 days to 6 days
```

A number of covariance-function models are available with the package **gstat**; see the **gstat** vignette "spatio-temporal-kriging" for details by typing

```
vignette("spatio-temporal-kriging")
```

The first semivariogram we consider here corresponds to the spatio-temporal separable covariance function in (4.18) and (4.19). Observe from the vignette that a separable covariance function (4.18) corresponds to a semivariogram of the form

$$\gamma^{\mathrm{sep}}(\mathbf{h}; \tau) = \mathrm{sill} \cdot \left(\bar{\gamma}^{(s)}(\|\mathbf{h}\|) + \bar{\gamma}^{(t)}(|\tau|) - \bar{\gamma}^{(s)}(\|\mathbf{h}\|)\bar{\gamma}^{(t)}(|\tau|) \right),$$

where the "standardized" semivariograms $\bar{\gamma}^{(s)}$ and $\bar{\gamma}^{(t)}$ have separate nugget effects and sills equal to 1.

A spatio-temporal semivariogram is constructed with **gstat** using the function **vgmST**. The argument `stModel = "separable"` is used to define a separable model, while the function **vgm** is used to construct the individual semivariograms (one for space and one for time). Several arguments can be passed to **vgm**. The first four, which we use below, correspond to the partial sill, the model type, the range, and the nugget, respectively. The argument `sill` that is supplied to **vgmST** defines the joint spatio-temporal sill. The numbers used in their definition are initial values supplied to the optimization routine used for fitting in the function `fit.StVariogram`, which fits `sepVgm` to `vv`. These initial values should be reasonable – for example, the length scale ϕ can be set to a value that spans 10% of the spatial/temporal domain, and the variances/sills can be set such that they have similar orders of magnitude to the total variance of the measurements.

```
sepVgm <- vgmST(stModel = "separable",
                space = vgm(10, "Exp", 400, nugget = 0.1),
                time = vgm(10, "Exp", 1, nugget = 0.1),
                sill = 20)
sepVgm <- fit.StVariogram(vv, sepVgm)
```

The second model we fit has the covariance function given in (4.20). For this model, the function **vgmST** takes the `joint` semivariogram as an argument, as well as the sill (`sill`) and the scaling factor (`stAni`), denoted by a in **v**, defined just below (4.20). This parameter can be initially set by considering orders of magnitudes – if the spatial field is evolving on scales of the order of hundreds of kilometers and the temporal evolution has a scale on the order of days, then an initial value of `stAni = 100` is reasonable.

```
metricVgm <- vgmST(stModel = "metric",
                   joint = vgm(100, "Exp", 400, nugget = 0.1),
                   sill = 10,
                   stAni = 100)
metricVgm <- fit.StVariogram(vv, metricVgm)
```

We can compare the fits of the two semivariograms by checking the mean squared error of the fits. These can be found by directly accessing the final function value of the optimizer used by `fit.StVariogram`.

```
metricMSE <- attr(metricVgm, "optim")$value
sepMSE <- attr(sepVgm, "optim")$value
```

Here the variable `metricMSE` is 2.1 while `sepMSE` is 1.4, indicating that the separable semivariogram gives a better fit to the empirical semivariogram in this case. The fitted semivariograms can be plotted using the standard **plot** function.

```
plot(vv, list(sepVgm, metricVgm), main = "Semi-variance")
```

Contour plots of the the fitted variograms are shown in the bottom panels of Figure 4.4. The corresponding stationary S-T covariance function is obtained from (4.15).

Next, we use the fitted S-T covariance models for prediction using S-T kriging, in this case *universal* S-T kriging since we are treating the latitude coordinate as a covariate. First, we need to create a space-time prediction grid. For our spatial grid, we consider 20 spatial locations between 100°W and 80°W, and 20 spatial locations between 32°N and 46°N. In the code below, when converting to `SpatialPoints`, we ensure that the coordinate reference system (CRS) of the prediction grid is the same as that of the observations.

```
spat_pred_grid <- expand.grid(
                      lon = seq(-100, -80, length = 20),
                      lat = seq(32, 46, length = 20)) %>%
             SpatialPoints(proj4string = CRS(proj4string(STObj3)))
gridded(spat_pred_grid) <- TRUE
```

For our temporal grid, we consider six equally spaced days in July 1993.

```
temp_pred_grid <- as.Date("1993-07-01") + seq(3, 28, length = 6)
```

We can then combine `spat_pred_grid` and `temp_pred_grid` to construct an `STF` object for our space-time prediction grid.

```
DE_pred <- STF(sp = spat_pred_grid,      # spatial part
               time = temp_pred_grid)    # temporal part
```

Since there are missing observations in `STObj4`, we first need to cast `STObj4` into either an `STSDF` or an `STIDF`, and remove the data recording missing observations. For simplicity here, we consider the `STIDF` (considering `STSDF` would be around twice as fast). Also, in order to show the capability of S-T kriging to predict across time, we omitted data on 14 July 1993 from the data set.

```
STObj5 <- as(STObj4[, -14], "STIDF")       # convert to STIDF
STObj5 <- subset(STObj5, !is.na(STObj5$z))  # remove missing data
```

Now we can call **krigeST** using `STObj5` as our data.

```
pred_kriged <- krigeST(z ~ 1 + lat,        # latitude trend
                       data = STObj5,       # data set w/o 14 July
                       newdata = DE_pred,   # prediction grid
                       modelList = sepVgm,  # semivariogram
                       computeVar = TRUE)   # compute variances
```

To plot the predictions and accompanying prediction standard errors, it is straightforward to use the function **stplot**. First, we define our color palette using the function **brewer.pal** and the function **colorRampPalette** (see help files for details on what these functions do).

```
color_pal <- rev(colorRampPalette(brewer.pal(11, "Spectral"))(16))
```

Second, we call the `stplot` function with the object containing the results.

```
stplot(pred_kriged,
       main = "Predictions (degrees Fahrenheit)",
       layout = c(3, 2),
       col.regions = color_pal)
```

The prediction (kriging) standard errors can be plotted in a similar way.

```
pred_kriged$se <- sqrt(pred_kriged$var1.var)
stplot(pred_kriged[, , "se"],
       main = "Prediction std. errors (degrees Fahrenheit)",
       layout = c(3, 2),
       col.regions = color_pal)
```

Spatio-temporal kriging as shown in this Lab is relatively quick and easy to implement for small data sets, but it starts to become prohibitive as data sets grow in size, unless some approximation is used. For example, the function `krigeST` allows one to use the argument `nmax` to determine the maximum number of observations to use when doing prediction. The predictor is no longer optimal, but it is close enough to the optimal predictor in many cases of practical interest.

Lab 4.2: Spatio-Temporal Basis Functions with FRK

In this Lab we shall focus on modeling the maximum temperature in July 1993 from data in the NOAA data set using spatio-temporal basis functions. The packages we need are the following:

```
library("dplyr")
library("FRK")
library("ggplot2")
library("gstat")
library("RColorBrewer")
library("sp")
library("spacetime")
library("STRbook")
library("tidyr")
```

The package **FRK** implements a low-rank approach to spatial and spatio-temporal modeling known as *fixed rank kriging* (FRK). FRK considers the random-effects model (4.27), sometimes known as the spatio-temporal random-effects model (Cressie et al., 2010), and provides functionality to the user for choosing the basis functions $\{\phi_i(\mathbf{s};t) : i = 1, \ldots, n_\alpha\}$ from the data.

A key difference between **FRK** and other geostatistical packages is that, in **FRK**, modeling and prediction are carried out on a fine, regular discretization of the spatio-temporal domain. The small grid cells are known as *basic areal units* (BAUs), and their primary utility is to account for problems of change of support (varying measurement footprint), which we do not consider in this Lab. The package is loaded by typing in the console

```
library("FRK")
```

For spatio-temporal modeling and prediction, **FRK** requires the user to provide the point-level data as objects of class `STIDF` (see p. 23). Hence, for this exercise, we use `STObj5` from Lab 3.1, which we reconstruct below (for completeness) from `STObj3`.

```
data("STObj3", package = "STRbook")         # load STObj3
STObj4 <- STObj3[, "1993-07-01::1993-07-31"] # subset time
STObj5 <- as(STObj4[, -14], "STIDF")         # omit t = 14
STObj5 <- subset(STObj5, !is.na(STObj5$z))   # remove NAs
```

The spatio-temporal BAUs are constructed using the function **auto_BAUs** which takes several arguments, as shown below and detailed using the in-line comments. For more details see **help**(auto_BAUs). Note that as cellsize we chose **c**(1, 0.75, 1) which indicates a BAU size of 1 degree longitude × 0.75 degrees latitude × 1 day – this choice ensures that the BAUs are similar to the prediction grid used in Lab 3.1. The argument convex is an "extension radius" used in domain construction via the package **INLA**. See the help file of **inla.nonconvex.hull** for details.

```
BAUs <- auto_BAUs(manifold = STplane(),    # ST field on the plane
                  type = "grid",           # gridded (not "hex")
                  data = STObj5,           # data
                  cellsize = c(1, 0.75, 1), # BAU cell size
                  convex = -0.12,          # hull extension
                  tunit = "days")          # time unit is "days"
```

The BAUs are of class `STFDF` since they are three-dimensional pixels arranged regularly in both space and in time. To plot the spatial BAUs overlaid with the data locations, we run

```
plot(as(BAUs[, 1], "SpatialPixels"))    # plot pixel BAUs
plot(SpatialPoints(STObj5),
     add = TRUE, col = "red")            # plot data points
```

This generates the left panel of Figure 4.9. The BAUs, which we will also use as our prediction grid, overlap all the data points. The user has other options in BAU construction; for example, the following code generates *hexagonal* BAUs using a convex hull for a boundary.

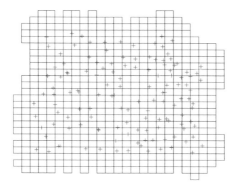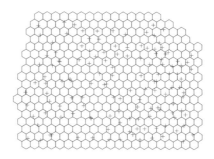

Figure 4.9: BAUs constructed for modeling and predicting maximum temperature from data in the NOAA data set. Left: Gridded BAUs arranged within a non-convex hull enclosing the data. Right: Hexagonal BAUs arranged within a convex hull enclosing the data.

```
BAUs_hex <- auto_BAUs(manifold = STplane(), # model on the plane
                      type = "hex",            # hex (not "grid")
                      data = STObj5,           # data
                      cellsize = c(1, 0.75, 1), # BAU cell size
                      nonconvex_hull = FALSE,  # convex hull
                      tunit = "days")          # time unit is "days"
```

Plotting proceeds in a similar fashion, except that the first line in the code chunk above now becomes

```
plot(as(BAUs_hex[, 1], "SpatialPolygons"))
```

This allows for the fact the the BAUs are now (hexagonal) polygons and not rectangular pixels. The resulting plot is shown in the right panel of Figure 4.9.

Next we construct the basis functions $\{\phi_i(\mathbf{s}; t) : i = 1, \ldots, n_\alpha\}$. In **FRK**, these are constructed by taking the tensor product of spatial basis functions with temporal basis functions. Specifically, consider a set of r_s spatial basis functions $\{\phi_p(\mathbf{s}) : p = 1, \ldots, r_s\}$, and a set of r_t temporal basis functions $\{\psi_q(t) : q = 1, \ldots, r_t\}$. Then we construct the set of spatio-temporal basis functions as $\{\phi_{st,u}(s, t) : u = 1, \ldots, r_s r_t\} = \{\phi_p(\mathbf{s})\psi_q(t) : p = 1, \ldots, r_s; \ q = 1, \ldots, r_t\}$.

The generic basis function that **FRK** uses by default is the bisquare function (see Figure 4.7) given by

$$b(\mathbf{s}, \mathbf{v}) \equiv \begin{cases} \{1 - (\|\mathbf{v} - \mathbf{s}\|/r)^2\}^2, & \|\mathbf{v} - \mathbf{s}\| \leq r, \\ 0, & \text{otherwise,} \end{cases}$$

where r is the aperture parameter. Basis functions can be either regularly placed, or irregularly placed, and they are often multiresolutional. We choose two resolutions below, yielding $r_s = 94$ spatial basis functions in total, and place them irregularly in the domain. (Note that r_s and the bisquare apertures are determined automatically by `auto_basis`.)

```
G_spatial <- auto_basis(manifold = plane(),          # fns on plane
                        data = as(STObj5, "Spatial"), # project
                        nres = 2,                     # 2 res.
                        type = "bisquare",            # bisquare.
                        regular = 0)                  # irregular
```

Temporal basis functions also need to be defined. We use the function `local_basis` below to construct a regular sequence of $r_t = 20$ bisquare basis functions between day 1 and day 31 of the month. Each of these bisquare basis functions is assigned an aperture of 2 days; that is, the support of each bisquare function is 4 days. The temporal grid is defined through

```
t_grid <- matrix(seq(1, 31, length = 20))
```

The basis functions are constructed using the following commands.

```
G_temporal <- local_basis(manifold = real_line(),   # fns on R1
                          type = "bisquare",          # bisquare
                          loc = t_grid,               # centroids
                          scale = rep(2, 20))         # aperture par.
```

Finally, we construct the $r_s r_t = 1880$ spatio-temporal basis functions by taking the tensor product of the spatial and the temporal ones, using the function `TensorP`.

```
G <- TensorP(G_spatial, G_temporal)       # take the tensor product
```

The basis functions `G_spatial` and `G_temporal` can be visualized using the plotting function `show_basis`; see Figure 4.10. While the basis functions are of tensor-product form, the resulting S-T covariance function obtained from the spatio-temporal random effects model is not separable in space and time.

In **FRK**, the fine-scale variation term at the BAU level, (4.28), is assumed to be Gaussian with covariance matrix proportional to $\mathrm{diag}(\{\sigma_{\nu,i}^2\})$, where $\{\sigma_{\nu,i}^2 : i = 1, \ldots, n_y\}$ are pre-specified at the BAU level (the constant of proportionality is then estimated by **FRK**). Typically, these are related to some geographically related quantity such as surface roughness. In our case, we simply set $\sigma_{\nu,i}^2 = 1$ for all i.

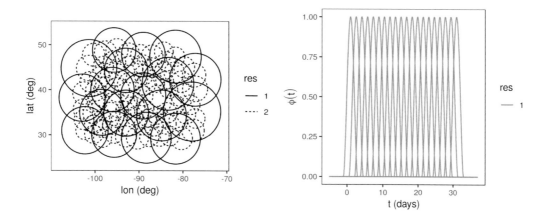

Figure 4.10: Spatial and temporal basis functions used to construct the spatio-temporal basis functions. Left: Locations of spatial basis functions (circles denote spatial support). Right: Temporal basis functions.

```
BAUs$fs = 1
```

The fine-scale variance at the BAU level is confounded with the measurement-error variance. In some cases, the measurement-error variance is known; when it is not (as in this case), one can carry out a simple analysis to estimate the value of the semivariogram at the origin. In this case, we simply assume that the nugget effect estimated when fitting the separable covariance function in Lab 4.1 is the measurement-error variance – any residual nugget component is then assumed to be the fine-scale variance introduced as a consequence of the low-rank approximation to the process. The measurement-error variance is specified in the `std` field in the data `ST` object.

```
STObj5$std <- sqrt(0.049)
```

The response variable and covariates are identified through a standard R formula. In this case we use latitude as a covariate and set

```
f <- z ~ lat + 1
```

We are now ready to call the main function **FRK**, which estimates all the unknown parameters in the models, including the covariance matrix of the basis-function coefficients and the fine-scale variance. We need to supply the formula, the data, the basis functions, the BAUs, and any other parameters configuring the expectation-maximization (EM) algorithm used for finding the maximum likelihood estimates. To reduce processing time, we have set the number of EM-algorithm steps to 3. Convergence of the EM algorithm can be assessed visually by setting `print_lik = TRUE` below.

```
S <- FRK(f = f,                 # formula
         data = list(STObj5),   # (list of) data
         basis = G,             # basis functions
         BAUs = BAUs,           # BAUs
         n_EM = 3,              # max. no. of EM iterations
         tol = 0.01)            # tol. on change in log-likelihood
```

Once the model is fitted, prediction proceeds via the function **predict**. If the argument `newdata` is not specified, then prediction is done at all the BAUs.

```
grid_BAUs <- predict(S)
```

The resulting object, `grid_BAUs`, is also of class `STFDF`, and plotting proceeds as per Lab 4.1 using the **stplot** function. The resulting predictions and prediction standard errors are illustrated in Figure 4.6.

Lab 4.3: Temporal Basis Functions with SpatioTemporal

In this Lab we model the maximum temperature in the NOAA data set (`Tmax`) using temporal basis functions and spatial random fields. Specifically, we use the model

$$Y(\mathbf{s};t) = \mathbf{x}(\mathbf{s};t)'\boldsymbol{\beta} + \sum_{i=1}^{n_\alpha} \phi_i(t)\alpha_i(\mathbf{s}) + \nu(\mathbf{s};t), \qquad (4.39)$$

where $\mathbf{x}(\mathbf{s};t)$ are the covariates; $\boldsymbol{\beta}$ are the regression coefficients; $\{\phi_i(t)\}$ are the temporal basis functions; $\{\alpha_i(\mathbf{s})\}$ are coefficients of the temporal basis functions, modeled as multivariate (spatial) random fields; and $\nu(\mathbf{s};t)$ is a spatially correlated, but temporally independent, random process.

Spatio-temporal modeling using temporal basis functions can be carried out using the package **SpatioTemporal**. For this Lab we require the following packages.

```
library("dplyr")
library("ggplot2")
library("gstat")
library("RColorBrewer")
library("sp")
library("spacetime")
library("SpatioTemporal")
library("STRbook")
library("tidyr")
```

The space-time object used by **SpatioTemporal** is of class `STdata` and is created using the function `createSTdata`. This function takes the data either as a space-wide matrix with the row names containing the date and the column names the station ID, or as a data frame in long form. Here we use the latter. This data frame needs to have the station ID as characters in the field `ID`, the data in the field `obs`, and the date in the field `date`. A new data frame of this form can be easily created using the function `transmute` from the package **dplyr**.

```
data("NOAA_df_1990", package = "STRbook")    # load NOAA data
NOAA_sub <- filter(NOAA_df_1990,          # filter data to only
                   year == 1993 &          # contain July 1993
                   month == 7 &
                   proc == "Tmax")        # and just max. temp.

NOAA_sub_for_STdata <- NOAA_sub %>%
                       transmute(ID = as.character(id),
                                 obs = z,
                                 date = date)
```

The covariates that will be used to model the spatially varying effects also need to be supplied as a data frame. In our case we only consider the station coordinates as covariates. The station coordinates are extracted from the maximum temperature data as follows.

```
covars <-  dplyr::select(NOAA_sub, id, lat, lon) %>%
           unique() %>%
           dplyr::rename(ID = id)       # createSTdata expects "ID"
```

Now we can construct the `STdata` object by calling the function `createSTdata`.

```
STdata <- createSTdata(NOAA_sub_for_STdata, covars = covars)
```

The model used in **SpatioTemporal** assumes that $\nu(\mathbf{s}; t)$ is temporally uncorrelated. Consequently, all temporal variability needs to be captured through the covariates or the basis functions. To check whether the data exhibit temporal autocorrelation (before adding any temporal basis functions), one can use the `plot` function. For example, we plot the estimated autocorrelation function for station 3812 in the left panel of Figure 4.11 (after the mean is removed from the data). The plot suggests that the data are correlated (the estimated lag-1 autocorrelation coefficient is larger than would be expected by chance at the 5% level of significance).

```
plot(STdata, "acf", ID = "3812")
```

The role of the temporal basis functions is to adequately capture temporal modes of variation. When modeling data over a time interval that spans years, one of these is typically a seasonal component. As another example, when modeling trace-gas emissions, one basis function to use would be one that captures weekday/weekend cycles typically found in gaseous pollutants (e.g., due to vehicular traffic). The package **SpatioTemporal** allows for user-defined basis functions (see the example at the end of this Lab) or data-driven basis functions (which we consider now). In both cases, the first temporal basis function, $\phi_1(t)$, is a constant; that is, $\phi_1(t) = 1$.

The basis functions extracted from the data are *smoothed, left singular vectors* (i.e., smoothed temporal EOFs) of the matrix $\widetilde{\mathbf{Z}}$, described in Technical Note 2.2. These make up the remaining $n_\alpha - 1$ basis functions, upon which smoothing is carried out using splines. In **SpatioTemporal**, these basis functions are found (or set) using the function `updateTrend`.

```
STdata <- updateTrend(STdata, n.basis = 2)
```

We can see that the lag-1 autocorrelation coefficient is no longer significant (at the 5% level) after adding in these basis functions; see the right panel of Figure 4.11. In practice, one should add basis functions until temporal autocorrelation in the data (at most stations) is considerably reduced. In this case study, it can be shown that 69% of stations record maximum temperature data that have lag-1 autocorrelation coefficients that are significant at the 5% level. On the other hand, with `n.basis = 2` (i.e., with two temporal basis functions for capturing temporal variation), the proportion of stations with residuals exhibiting a significant lag-1 autocorrelation coefficient is 26%.

```
plot(STdata, "acf", ID = "3812")
```

The basis functions, available in `STdata$trend`, are shown in the top panel of Figure 4.8.

In **SpatioTemporal**, the spatial quantities $\{\alpha_i(\mathbf{s})\}$ are themselves modeled as spatial fields. Once the $\{\phi_i(t)\}$ are declared, empirical estimates of $\{\alpha_i(\mathbf{s})\}$ can be found using the function `estimateBetaFields`. Note that we use the Greek letter "alpha" to denote these fields, which differs from the name "Beta" inside the command. The following and all subsequent references to "Beta" and "beta" should be interpreted as representing spatial fields $\{\alpha_i(\mathbf{s})\}$.

```
beta.lm <- estimateBetaFields(STdata)
```

The resulting object, `beta.lm`, contains two fields; `beta` (estimated coefficients) and `beta.sd` (standard error of the estimates) with row names equal to the station ID, and three columns corresponding to estimates of $\alpha_1(\mathbf{s})$, $\alpha_2(\mathbf{s})$, and $\alpha_3(\mathbf{s})$, respectively. We are interested in seeing whether the empirical estimates are correlated with our covariate, latitude. To this end, the authors of **SpatioTemporal** suggest using the package **plotrix**, and the function `plotCI`, to plot the estimates and covariance intervals against a covariate

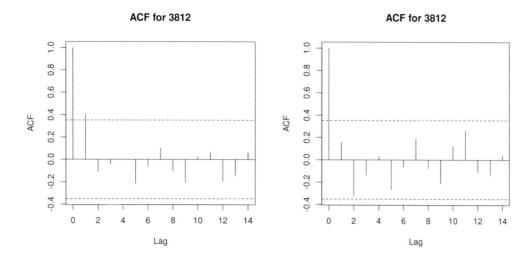

Figure 4.11: Left: Estimated autocorrelation function for the time series of maximum temperature `Tmax` at Station 3812. Right: Same as left panel, but with the data first detrended using an intercept and the two temporal basis functions shown in the top panel of Figure 4.8.

of choice. When plotting using `plotCI`, care should be taken that the ordering of the stations in `beta` and `beta.sd` is the same as that if the covariate data frame. For example, consider

```
head(row.names(beta.lm$beta))
```

```
## [1] "13865" "13866" "13871" "13873" "13874" "13876"
```

```
head(covars$ID)
```

```
## [1] 3804 3810 3811 3812 3813 3816
```

This illustrates a discrepancy, since the ordering of strings is not necessarily that of the ordered integers. For this reason we recommend employing best practice and always merging (e.g., using `left_join`) on a column variable; in this case, we choose the integer version of the field `ID`. In the following commands, we first convert the `beta` and `beta.sd` objects into data frames, add the column `ID`, join into a data frame `BETA`, and then combine with `covars` containing the latitude data.

```
beta.lm$beta <- data.frame(beta.lm$beta)
beta.lm$beta.sd <- data.frame(beta.lm$beta.sd)
beta.lm$beta$ID <- as.integer(row.names(beta.lm$beta))
```

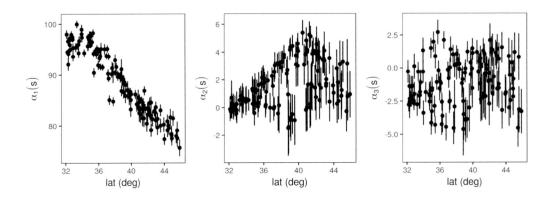

Figure 4.12: Empirical estimates of $\alpha_1(\mathbf{s})$, $\alpha_2(\mathbf{s})$, and $\alpha_3(\mathbf{s})$ at each station, with 95% confidence intervals, plotted as a function of latitude.

```
BETA <- cbind(beta.lm$beta, beta.lm$beta.sd)
colnames(BETA) <- c("alpha1", "alpha2", "alpha3", "ID",
                    "alpha1_CI", "alpha2_CI", "alpha3_CI")
BETA <- left_join(BETA, covars, by = "ID")
```

Once BETA is constructed, the empirical estimates can be plotted using ggplot, with geom_errorbar to also plot error bars, as follows.

```
ggplot(BETA) + geom_point(aes(x = lat, y = alpha1)) +
    geom_errorbar(aes(x = lat,
                      ymin = alpha1 - 1.96*alpha1_CI,
                      ymax = alpha1 + 1.96*alpha1_CI)) +
    ylab(expression(alpha[1](s))) +
    xlab("lat (deg)") + theme_bw()
```

The three empirical estimates, plotted as a function of latitude, are shown in Figure 4.12. The function $\alpha_1(\mathbf{s})$ exhibits a strong latitudinal trend, as expected; $\alpha_2(\mathbf{s})$ shows a weak latitudinal trend; and $\alpha_3(\mathbf{s})$ exhibits no trend. For this reason we model the expectations of these fields as in (4.32)–(4.34). Note that in this model we do not consider any spatio-temporal covariates, and hence the term $\mathbf{x}(\mathbf{s}; t)'\boldsymbol{\beta} = 0$ in (4.39). This does not mean that we do not have an intercept in our model: although it is random, the spatial field $\alpha_1(\mathbf{s})$ acts as a temporally invariant spatial covariate and includes a global space-time mean (α_{11} in (4.32)), which is estimated.

We let the covariance functions $\mathrm{cov}(\alpha_i(\mathbf{s}), \alpha_i(\mathbf{s} + \mathbf{h}))$, $i = 1, 2, 3$, be exponential covariance functions without a nugget-effect term. In **SpatioTemporal** these are declared as follows.

```
cov.beta <- list(covf = "exp", nugget = FALSE)
```

All that remains for constructing the spatio-temporal model is to define the spatial co-variance function of the zero-mean, temporally independent, residual process $\nu(\mathbf{s}; t)$; see (4.39). We choose this to be an exponential covariance function with a nugget effect to account for measurement error. The argument random.effect = FALSE is used to indicate that there is no random mean offset for the field at each time point.

```
cov.nu <- list(covf = "exp",
               nugget = ~1,
               random.effect = FALSE) # No random mean
                                      # for each nu
```

The function to create the spatio-temporal model is **createSTmodel**. This takes as data the object STdata, the covariates for the α-fields (an intercept and latitude for $\alpha_1(\mathbf{s})$ and $\alpha_2(\mathbf{s})$, and just an intercept for $\alpha_3(\mathbf{s})$; see (4.32)–(4.34)), the covariance functions of the α-fields and the ν-field, and a list containing the names of station coordinate fields (lon and lat).

```
locations <- list(coords = c("lon", "lat"))
LUR <- list(~lat, ~lat, ~1)   # lat trend for phi1 and phi2 only
STmodel <- createSTmodel(STdata,               # data
                         LUR = LUR,            # spatial covariates
                         cov.beta = cov.beta,  # cov. of alphas
                         cov.nu = cov.nu,      # cov. of nu
                         locations = locations) # coord. names
```

In order to fit the spatio-temporal model to the data, we need to provide initial values of the parameter estimates. The required parameter names can be extracted using the function **loglikeSTnames** and, for our model, are as follows.

```
parnames <- loglikeSTnames(STmodel, all = FALSE)
print(parnames)
```

```
## [1] "log.range.const.exp"        "log.sill.const.exp"
## [3] "log.range.V1.exp"           "log.sill.V1.exp"
## [5] "log.range.V2.exp"           "log.sill.V2.exp"
## [7] "nu.log.range.exp"           "nu.log.sill.exp"
## [9] "nu.log.nugget.(Intercept).exp"
```

Noting that all parameters are log-transforms of the quantities of interest, we let all of the initial values be equal to 3 (so that all initial ranges and sills are $e^3 \approx 20$). This seems reasonable when the temperature is varying on the order of several degrees Fahrenheit, and where the domain also spans several degrees (in latitude and longitude).

We use the function `estimate` below to fit the spatio-temporal model to the data. This may take several minutes on a standard desktop computer. In this instance, the resulting object `SpatioTemporalfit1` has been pre-computed and can be loaded directly from **STRbook** by typing `data("SpatioTemporalfit1", package = "STRbook")`.

```
x.init <- matrix(3, 9, 1)
rownames(x.init) <- loglikeSTnames(STmodel, all = FALSE)
SpatioTemporalfit1 <- estimate(STmodel, x.init)
```

The fitted coefficients for the parameters described by `parnames` above can be extracted from the fitted object using the function `coef`.

```
x.final <- coef(SpatioTemporalfit1, pars = "cov")$par
```

Having fitted the model, we now predict at unobserved locations. First, we establish the spatial and temporal grid upon which to predict; this proceeds by first initializing an `STdata` object on a grid. We construct the grid following a very similar approach to what was done in Lab 4.1.

```
## Define space-time grid
spat_pred_grid <- expand.grid(lon = seq(-100, -80, length = 20),
                   lat = seq(32, 46, length = 20))
spat_pred_grid$id <- 1:nrow(spat_pred_grid)
temp_pred_grid <- as.Date("1993-07-01") + seq(3, 28, length = 6)

## Initialize data matrix
obs_pred_wide <- matrix(0, nrow = 6, ncol = 400)

## Set row names and column names
rownames(obs_pred_wide) <- as.character(temp_pred_grid)
colnames(obs_pred_wide) <- spat_pred_grid$id

covars_pred <- spat_pred_grid                     # covariates
STdata_pred <- createSTdata(obs = obs_pred_wide,  # ST object
                   covars = covars_pred)
```

Now prediction proceeds using the function `predict`, which requires as arguments the model, the fitted model parameters, and the data matrix `STdata_pred`.

```
E <- predict(STmodel, x.final, STdata = STdata_pred)
```

The returned object E contains both the α-fields predictions as well as the Y-field prediction at the unobserved locations. For example, `E$beta$EX` contains the conditional expectations of $\alpha_1(\mathbf{s}), \alpha_2(\mathbf{s})$, and $\alpha_3(\mathbf{s})$ given the data. For conciseness, we do not illustrate the

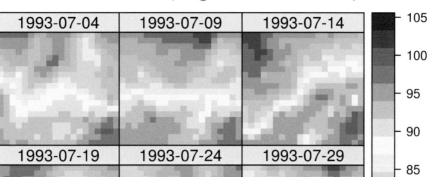

Figure 4.13: Predictions of `Tmax` in degrees Fahrenheit within a square lat-lon box defining the spatial domain of interest, for six days in July 1993, using temporal basis functions. Data for 14 July 1993 were deliberately omitted from the original data set.

plotting commands here. In the bottom panels of Figure 4.8, we show the conditional expectations, while in Figures 4.13 and 4.14 we show the predictions and prediction standard errors of maximum temperature over six days of interest in July 1993.

Using SpatioTemporal for Modeling Spatial Effects of Temporal Covariates

In the first part of this Lab, we extracted the temporal basis functions from the data. However, **SpatioTemporal** can also be used to model the spatially varying effect of exogenous temporal covariates. This can be done by manually setting the `STdata$trend` data frame. When modeling temperature, interesting covariates may include a periodic signal with period equal to one year, or an index such as the El Niño Southern Oscillation (ENSO) Index.

To use a pre-existing covariate, we need to use the `fnc` argument in **updateTrend** to define a function that takes a `Date` object as an input and returns the covariate at these dates. The easiest way to do this in this example is to specify a look-up table in the function containing the covariate for each date, but an interpolant can also be used when the covariate

Prediction errors (degrees Fahrenheit)

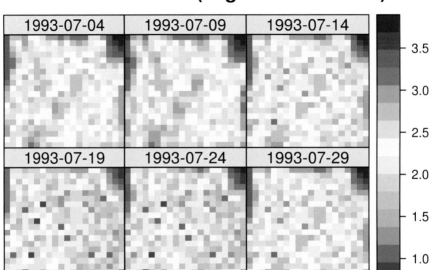

Figure 4.14: Prediction standard errors of `Tmax` in degrees Fahrenheit within a square lat-lon box enclosing the spatial domain of interest, for six days in July 1993, using temporal basis functions. Data for 14 July 1993 were deliberately omitted from the original data set.

has missing information for one or more dates.

As an exercise, repeat the Lab above, but this time use a single linear temporal trend as a temporal covariate. The look-up table we need is just a two-column data frame containing the `date` in the first column, and `V1` (first covariate) in the second column. This can be set up as follows.

```
all_dates <- NOAA_sub$date %>% unique()       # dates
lookup <- data.frame(date = all_dates,        # covariate (linear)
                     V1 = scale(as.numeric(all_dates)))
```

Type `plot(lookup)` to see the temporal covariate that we have just created. Now we need to create the function that takes a `Date` object as input and returns the required covariate values. This can be done using `left_join`.

```
## Function that returns the covariates in a data frame
## at the required dates
fnc <- function(dates) {
```

```
left_join(data.frame(date = dates),
          lookup, by = "date") %>%
select(-date)
}
```

Now we can call `updateTrend` with our covariate function as argument.

```
STdata <- updateTrend(STdata, fnc = fnc)
```

The rest of the code remains largely similar, except that now we are considering only two temporal basis functions and not three (the first basis function is constant in time, and the second one is linear in time). Changing the required parts of the code is left as an exercise.

Lab 4.4: Non-Gaussian Spatio-Temporal GAMs with mgcv

Generalized additive models (GAMs) and generalized additive mixed models (GAMMs) can be implemented quickly and efficiently with the package **mgcv** and the functions `gam` and `gamm`, respectively. For a comprehensive treatment of GAMs and GAMMs and their implementation through **mgcv**, see Wood (2017).

In this Lab we aim to predict the expected counts at arbitrary spatio-temporal locations, from the vector of observed counts **Z**. The data we use are the Carolina wren counts in the BBS data set described in Section 2.1. We require the package **mgcv** as well as **dplyr, tidyr, ggplot2** and **STRbook**.

```
library("dplyr")
library("ggplot2")
library("mgcv")
library("STRbook")
library("tidyr")
data("MOcarolinawren_long", package = "STRbook")
```

GAMs and GAMMs rely on constructing smooth functions of the covariates, and in a spatio-temporal context these will inevitably include space and time. In this Lab we consider the following simple GAM (see (4.36)):

$$g(Y(\mathbf{s};t)) = \beta + f(\mathbf{s};t) + \nu(\mathbf{s};t), \qquad (4.40)$$

where $g(\cdot)$ is a link function, β is an intercept, the function $f(\mathbf{s};t)$ is a random smooth function of space and time, and $\nu(\mathbf{s};t)$ is a spatio-temporal white-noise error process.

In **mgcv**, the random function $f(\mathbf{s};t)$ is generally decomposed using a separable *spline* basis. Now, there are several basis functions that can be used to reconstruct $f(\mathbf{s};t)$, some of which are knot-based (e.g., B-splines). For the purpose of this Lab, it is sufficient to know that splines, of whatever order, are decomposed into a set of basis functions. Thus, $f(\mathbf{s};t)$ is decomposed as $\sum_{i=1}^{r_1} \phi_{1i}(\mathbf{s};t)\alpha_{1i}$, where the $\{\alpha_{1i}\}$ are unknown random effects that need to be predicted, and the $\{\phi_{1i}\}$ are given below.

There are a number of basis functions that can be chosen. Those derived from thin-plate regression splines are convenient, as they are easily amenable to multiple covariates (e.g., functions of $(\mathbf{s};t) \equiv (s_1, s_2;t)$). Thin-plate splines are isotropic and invariant to rotation but not invariant to covariate scaling. Hence, the use of thin-plate splines for fitting a curve over space *and* time is not recommended, since units in time are different from those in space.

To combine interacting covariates with different units, such as space and time, **mgcv** implements a tensor-product structure, whereby the basis functions smoothing the individual covariates are combined productwise. That is,

$$f(\mathbf{s};t) = \sum_{i=1}^{r_1} \sum_{j=1}^{r_2} \phi_{1i}(\mathbf{s})\phi_{2j}(t)\alpha_{ij} \equiv \boldsymbol{\phi}(\mathbf{s};t)'\boldsymbol{\alpha}.$$

The function `te` forms the product from the marginals; for example, in our case this can achieved by using `te(lon,lat,t)`. Other arguments can be passed to `te` for added functionality; for example, the basis-function class is specifed through `bs`, the number of basis functions through `k`, and the dimension of each spline through `d`. In this case we employ a thin-plate spline basis over longitude and latitude (`"tp"`) and a cubic regression spline over time (`"cr"`). A GAM formula for (4.40) is implemented as follows

```
f <- cnt ~ te(lon, lat, t,            # inputs over which to smooth
              bs = c("tp", "cr"),     # types of bases
              k = c(50, 10),          # knot count in each dimension
              d = c(2, 1))            # (s,t) basis dimension
```

We chose $r_1 = 50$ basis functions for the spatial component and $r_2 = 10$ for the temporal component. These values were chosen after some trial and error. The number of knots could have been set using cross-validation; see Chapter 3. In general, the estimated degrees of freedom should be considerably lower than the total number of knots; if this is not the case, probably the number of knots should be increased.

In Lab 3.4 we saw that the Carolina wren counts are over-dispersed. To account for this, we use the negative-binomial distribution to model the response in (4.35) (a quasi-Poisson model would also be suitable). The `gam` function is called in the code below, where we specify the negative-binomial family and a log link (the function $g(\cdot)$ in (4.40)):

```
cnts <- gam(f, family = nb(link = "log"),
            data = MOcarolinawren_long)
```

The returned object is a `gam` object, which extends `glm` and `lm` objects (i.e., functions that can be applied to `glm` and `lm` objects, such as **residuals**, can also be applied to `gam` objects). The negative-binomial distribution handles over-dispersion in the data through a size parameter r, such that, for a fixed mean, the negative-binomial distribution approaches the Poisson distribution as $r \to \infty$. In this case the estimated value for r (named `Theta` in **mgcv**) is

```
cnts$family$getTheta(trans = 1)

## [1] 5.18
```

which is not large and suggestive of over-dispersion. Several graphical diagnostics relating to the fit can be explored using the **gam.check** function.

To predict the field at unobserved locations using the hierarchical model, we first construct a space-time grid upon which to predict.

```
MOlon <- MOcarolinawren_long$lon
MOlat <- MOcarolinawren_long$lat

## Construct space-time grid
grid_locs <- expand.grid(lon = seq(min(MOlon) - 0.2,
                                   max(MOlon) + 0.2,
                                   length.out = 80),
                         lat = seq(min(MOlat) - 0.2,
                                   max(MOlat) + 0.2,
                                   length.out = 80),
                         t = 1:max(MOcarolinawren_long$t))
```

Then we call the function **predict** which, when `se.fit = TRUE`, returns a list containing the predictions and their associated prediction standard errors.

```
X <- predict(cnts, grid_locs, se.fit = TRUE)
```

Specifically, the predictions and prediction standard errors are available in `X$fit` and `X$se.fit`, respectively. These can be plotted using **ggplot2** as follows.

```
## Put data to plot into data frame
grid_locs$pred <- X$fit
grid_locs$se <- X$se.fit

## Plot predictions and overlay observations
g1 <- ggplot() +
    geom_raster(data = grid_locs,
                aes(lon, lat, fill = pmin(pmax(pred, -1), 5))) +
```

```
    facet_wrap(~t, nrow = 3, ncol = 7) +
    geom_point(data = filter(MOcarolinawren_long, !is.na(cnt)),
               aes(lon, lat),
               colour = "black", size = 3) +
    geom_point(data=filter(MOcarolinawren_long, !is.na(cnt)),
               aes(lon, lat, colour = log(cnt)),
               size = 2) +
    fill_scale(limits = c(-1, 5),
               name = expression(log(Y[t]))) +
    col_scale(name = "log(cnt)", limits=c(-1, 5)) +
    theme_bw()

## Plot prediction standard errors
g2 <- ggplot() +
    geom_raster(data = grid_locs,
                aes(lon, lat, fill = pmin(se, 2.5))) +
    facet_wrap(~t, nrow = 3, ncol = 7) +
    fill_scale(palette = "BrBG",
               limits = c(0, 2.5),
               name = expression(s.e.)) +
    theme_bw()
```

The plots are shown in Figures 4.15 and 4.16, respectively. One may also use the `plot.gam` function on `cnts` to quickly generate plots of the tensor products.

Lab 4.5: Non-Gaussian Spatio-Temporal Models with INLA

Integrated Nested Laplace Approximation (INLA) is a Bayesian method that provides approximate marginal (posterior) distributions over all states and parameters. The package **INLA** allows for a variety of modeling approaches, and the reader is referred to the book by Blangiardo and Cameletti (2015) for an extensive treatment. Other useful resources are Lindgren and Rue (2015) and Krainski et al. (2019).

In this Lab we shall predict expected counts at arbitrary space-time locations from the vector of observed counts **Z**. The data we use are the Carolina wren counts in the BBS data set described in Section 2.1. For this Lab, we require the package **INLA** as well as **dplyr, tidyr, ggplot2** and **STRbook**.

```
library("INLA")
library("dplyr")
library("tidyr")
library("ggplot2")
library("STRbook")
data("MOcarolinawren_long", package = "STRbook")
```

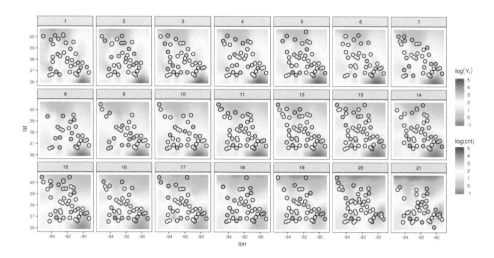

Figure 4.15: Posterior mean of $\log(Y(\cdot))$ on a grid for $t = 1$ (the year 1994) to $t = 21$ (the year 2014), based on a negative-binomial data model using the package **mgcv**. The log of the observed count is shown in circles using the same color scale.

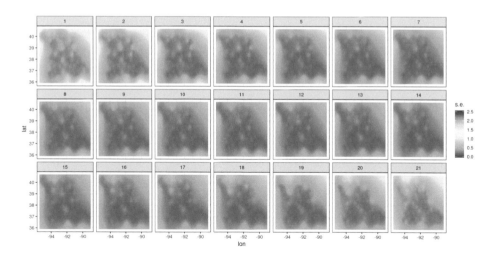

Figure 4.16: Posterior standard deviation (i.e., prediction standard error) of $\log(Y(\cdot))$ on a grid for $t = 1$ (the year 1994) to $t = 21$ (the year 2014), based on a negative-binomial data model using the package **mgcv**.

Consider the data model,

$$\mathbf{Z}_t | \mathbf{Y}_t \sim indep. \ NB(\mathbf{Y}_t, r), \tag{4.41}$$

and the process model,

$$\log(\mathbf{Y}_t) = \mathbf{X}_t \boldsymbol{\beta} + \boldsymbol{\Phi}_t \boldsymbol{\alpha}_t. \tag{4.42}$$

In (4.41) and (4.42), $\mathbf{Z}_t$ is an m_t-dimensional data vector of counts at m_t spatial locations, $E(\mathbf{Z}_t | \mathbf{Y}_t) = \mathbf{Y}_t$, $\mathbf{Y}_t$ represents the latent spatio-temporal mean process at m_t locations, $\boldsymbol{\Phi}_t$ is an $m_t \times n_\alpha$ matrix of spatial basis functions, r is the size parameter, and the associated random coefficients are modeled as $\boldsymbol{\alpha}_t \sim Gau(\mathbf{0}, \mathbf{C}_\alpha)$.

In order to fit this hierarchical model, we need to generate the basis functions with which to construct the matrices $\{ \boldsymbol{\Phi}_t : t = 1, \ldots, T \}$. In **INLA**, the basis functions used are typically "tent" (finite element) functions constructed over a triangulation of the domain. To establish a "boundary" for the domain, we can use the function `inla.nonconvex.hull`, as follows.

```
coords <- unique(MOcarolinawren_long[c("loc.ID", "lon", "lat")])
boundary <- inla.nonconvex.hull(as.matrix(coords[, 2:3]))
```

The triangulation of the domain is then carried out using the function `inla.mesh.2d`. This function takes several arguments (see its help file for details). Two of the most important arguments are `max.edge` and `cutoff`. When the former is supplied with a vector of length 2, the first element is the maximum edge length in the interior of the domain, and the second element is the maximum edge length in the exterior of the domain (obtained from a small buffer that is automatically created to reduce boundary effects). The second argument, `cutoff`, establishes the minimum edge length. Below we choose a maximum edge length of `0.8` in the domain interior. This is probably too large for the problem at hand, but reducing this considerably increases the computational burden when fitting the model.

```
MOmesh <- inla.mesh.2d(boundary = boundary,
                       max.edge = c(0.8, 1.2),  # max. edge length
                       cutoff = 0.1)            # min. edge length
```

The mesh and the data locations are plotted using the following commands.

```
plot(MOmesh, asp = 1, main = "")
lines(coords[c("lon", "lat")], col = "red", type = "p")
```

These are shown in Figure 4.17. Note that the triangulation is irregular and contains an extension with triangles that are larger than those in the interior of the domain.

As in the standard Gaussian case, the modeling effort lies in establishing the covariance matrix of $\boldsymbol{\alpha} \equiv (\boldsymbol{\alpha}_1', \ldots, \boldsymbol{\alpha}_T')'$. When using **INLA**, typically the covariance matrix of $\boldsymbol{\alpha}$ is

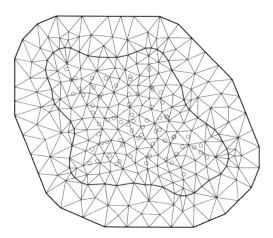

Figure 4.17: Triangulation for the Carolina wren data locations over which the "tent" functions are constructed (black), and the observation locations (red circles) are superimposed. The blue line denotes the interior non-convex domain of interest that includes all the data points.

chosen to be separable and of the form $\Sigma_t(\rho) \otimes \Sigma_s(\tau, \kappa, \nu)$ in such a way that its inverse (i.e., the precision matrix) is sparse. The matrix Σ_t is constructed assuming an AR(1) process, and thus it is parameterized using a single AR parameter, ρ. This parameter dictates the correlation of α across time; the closer ρ is to 1, the higher the temporal correlation. The matrix Σ_s is parameterized using three parameters, and it reflects the spatial covariance required such that the reconstructed field is, approximately, a solution to the stochastic partial differential equation (SPDE)

$$(\kappa^2 - \Delta)^{\alpha/2}(\tau Y(\cdot)) = \epsilon(\cdot),$$

where Δ is the Laplacian, $\epsilon(\cdot)$ is a white-noise process, and τ controls the variance. The resulting field has a Matérn covariance function. The parameter κ is a scaling parameter that translates to a "practical" spatial correlation length (i.e., the spatial separation at which the correlation is 0.1) $l = (\sqrt{8\nu})/\kappa$, while $\alpha = \nu + d/2$ is a smoothness parameter and d is the spatial dimension. Here we fix $\nu = 1$ ($\alpha = 2$); this parameter is notoriously difficult to estimate and frequently set using cross-validation. Note that there are other "practical" length scales used to characterize the range of a correlation function (e.g., "effective range" when the correlation is 0.05); our choice here is motivated by the **INLA** package that readily provides a marginal distribution over the parameter l as defined here.

    The SPDE can be constructed on the mesh using the function `inla.spde2.pcmatern`. The `pc` in `pcmatern` is short for "penalized complexity," and it is used to refer to prior distributions over the hyperparameters that are both interpretable and that have interesting theoretical properties (Simpson et al., 2017). We define prior distributions below over the range parameter l such that $P(l < 1) = 0.01$, and over the marginal standard deviation such that $P(\sigma > 4) = 0.01$. We elicited these distributions by looking at the count data – it is highly unlikely that the spatial correlation length is less than 1 degree and that the expected counts are of the order of 50 or more (we will use a log link, and $e^4 \approx 55$).

```
spde <- inla.spde2.pcmatern(mesh = MOmesh,
                            alpha = 2,
                            prior.range = c(1, 0.01),
                            prior.sigma = c(4, 0.01))
```

With the discretization shown in Figure 4.17, $\alpha_{t,i}$ can be viewed as the weight of the ith basis function at time t. The observation matrix $\mathbf{\Phi}_t$ then maps the observations to the finite-element space at time t; if the observation lies exactly on a vertex, then the associated row in $\mathbf{\Phi}_t$ will be 0 everywhere except for a 1 in the column corresponding to the vertex. Otherwise, the row has three non-zero elements, with each representing the proportion being assigned to each vertex. For point predictions or areal averages, all rows in $\mathbf{\Phi}_t$ sum to 1. Finally, for this example, we choose each element in $\mathbf{X}_t$ to be equal to 1. The coefficient β_0 is then the intercept.

The package **INLA** requires space and time to be "blocked up" with an ordering of the variables in which space runs faster than time (i.e., the first few variables are spatial nodes at the first time point, the next few are at the second time point, and so on). Hence we have the block-matrix structure

$$\log\left(\begin{bmatrix} \mathbf{Y}_1 \\ \vdots \\ \mathbf{Y}_T \end{bmatrix}\right) = \begin{bmatrix} \mathbf{X}_1 \\ \vdots \\ \mathbf{X}_T \end{bmatrix} \boldsymbol{\beta} + \begin{bmatrix} \mathbf{\Phi}_1 & \mathbf{0} & \cdots \\ \vdots & \ddots & \vdots \\ \mathbf{0} & \cdots & \mathbf{\Phi}_T \end{bmatrix} \begin{bmatrix} \boldsymbol{\alpha}_1 \\ \vdots \\ \boldsymbol{\alpha}_T \end{bmatrix}, \tag{4.43}$$

where $\log(\cdot)$ corresponds to a vector of elementwise logarithms. This can be further simplified to

$$\log(\mathbf{Y}) = \mathbf{X}\boldsymbol{\beta} + \mathbf{\Phi}\boldsymbol{\alpha}, \tag{4.44}$$

where $\mathbf{Y} = (\mathbf{Y}_1', \ldots, \mathbf{Y}_T')'$, $\mathbf{X} = (\mathbf{X}_1', \ldots, \mathbf{X}_T')'$, $\mathbf{\Phi} \equiv \text{bdiag}(\{\mathbf{\Phi}_t : t = 1, \ldots, T\})$, bdiag($\cdot$) constructs a block-diagonal matrix from its arguments, and $\boldsymbol{\alpha} \equiv (\boldsymbol{\alpha}_1', \ldots, \boldsymbol{\alpha}_T')'$.

A space-time index needs to be constructed for this representation. This index is a double index that identifies both the spatial location and the associated time point. In Lab 2.2 we saw how the function `expand.grid` can be used to generate such indices from a set of spatial locations and time points. In **INLA**, we instead use the function `inla.spde.make.index`. It takes as arguments the index name, the number of spatial points in the mesh, and the number of time points.

```
n_years <- length(unique(MOcarolinawren_long$t))
n_spatial <- MOmesh$n
s_index <- inla.spde.make.index(name = "spatial.field",
                                n.spde = n_spatial,
                                n.group = n_years)
```

The list `s_index` contains two important items, the `spatial.field` index, which runs from 1 to `n_spatial` for `n_years` times, and `spatial.field.group`, which runs from 1 to `n_years`, with each element replicated `n_spatial` times. Note how this is similar to what one would obtain from **`expand.grid`**.

The matrix $\mathbf{\Phi}$ in (4.44) is found using the **`inla.spde.make.A`** function. This function takes as arguments the mesh, the measurement locations `loc`, the measurement group (in our case the year of observation) and the number of groups.

```
coords.allyear <- MOcarolinawren_long[c("lon", "lat")] %>%
                   as.matrix()
PHI <- inla.spde.make.A(mesh = MOmesh,
                        loc = coords.allyear,
                        group = MOcarolinawren_long$t,
                        n.group = n_years)
```

Note that

```
dim(PHI)
```

```
## [1] 1575 5439
```

This is a matrix equal in dimension to (number of observations) × (number of indices) of our basis functions in space and time.

```
nrow(MOcarolinawren_long)
```

```
## [1] 1575
```

```
length(s_index$spatial.field)
```

```
## [1] 5439
```

The latent Gaussian model is constructed in **INLA** through a *stack*. Stacks are handy as they allow one to define data, effects, and observation matrices in groups (e.g., one accounting for the measurement locations and another accounting for the prediction locations), which can then be stacked together into one bigger stack. In order to build a stack we need to further block up (4.43) into a representation amenable to the `inla` function (called later on) as follows:

$$\log(\mathbf{Y}) = \mathbf{\Pi}\boldsymbol{\gamma},$$

where $\mathbf{\Pi} = (\mathbf{\Phi}, \mathbf{1})$ and $\boldsymbol{\gamma} = (\boldsymbol{\alpha}', \beta_0)'$.

A stack containing the data and covariates at the measurement locations is constructed by supplying the data (argument `data`), the matrix $\mathbf{\Pi}$ (argument `A`), and information on the vector $\boldsymbol{\gamma}$. The stack is then tagged with the label `"est"`.

```
## First stack: Estimation
n_data <- nrow(MOcarolinawren_long)
stack_est <- inla.stack(
                data = list(cnt = MOcarolinawren_long$cnt),
                A = list(PHI, 1),
                effects = list(s_index,
                               list(Intercept = rep(1, n_data))),
                tag = "est")
```

We next construct a stack containing the matrices and vectors defining the model at the prediction locations. In this case, we choose the triangulation vertices as the prediction locations; then $\mathbf{\Phi}$ is simply the identity matrix, and $\mathbf{X}$ is a vector of ones. We store the information on the prediction locations in `df_pred` and that for $\mathbf{\Phi}$ in `PHI_pred`.

```
df_pred <- data.frame(lon = rep(MOmesh$loc[,1], n_years),
                      lat = rep(MOmesh$loc[,2], n_years),
                      t = rep(1:n_years, each = MOmesh$n))
n_pred <- nrow(df_pred)
PHI_pred <- Diagonal(n = n_pred)
```

The prediction stack is constructed in a very similar way to the estimation stack, but this time we set the data values to `NA` to indicate that prediction should be carried out at these locations.

```
## Second stack: Prediction
stack_pred <- inla.stack(
                data = list(cnt = NA),
                A = list(PHI_pred, 1),
                effects = list(s_index,
                               list(Intercept = rep(1, n_pred))),
                tag = "pred")
```

The estimation stack and prediction stack are combined using the `inla.stack` function.

```
stack <- inla.stack(stack_est, stack_pred)
```

All `inla.stack` does is block-concatenate the matrices and vectors in the individual stacks. Denote the log-expected counts at the prediction locations as $\mathbf{Y}^*$, the covariates as $\mathbf{X}^*$, and the basis functions evaluated at the prediction locations as $\mathbf{\Phi}^*$. Then

$$\begin{bmatrix} \log(\mathbf{Y}) \\ \log(\mathbf{Y}^*) \end{bmatrix} = \begin{bmatrix} \mathbf{\Pi} \\ \mathbf{\Pi}^* \end{bmatrix} \boldsymbol{\gamma},$$

recalling that $\boldsymbol{\gamma} = (\boldsymbol{\alpha}', \beta_0)'$. Note that, internally, some columns of $\mathbf{\Pi}$ and $\mathbf{\Pi}^*$ corresponding to unobserved states are not stored. For example $\mathbf{\Phi}$, internally, has dimension

```
dim(stack_est$A)
```

```
## [1] 1575 1702
```

The number of rows corresponds to the number of data points, while the number of columns corresponds to the number of observed states (`sum(colSums(PHI) > 0)`) plus one for the intercept term.

All that remains before fitting the model is for us to define the formula, which is a combination of a standard R formula for the fixed effects and an **INLA** formula for the spatio-temporal residual component. For the latter, we need to specify the name of the index we created as the first argument (in this case `spatial.field`), the model (in this case `spde`), the name of the grouping/time index (in this case `spatial.field.group`) and, finally, the model to be constructed across groups (in this case an AR(1) model). The latter modeling choice implies that $E(\boldsymbol{\alpha}_{t+1} \mid \boldsymbol{\alpha}_t) = \rho\boldsymbol{\alpha}_t$, $t = 1, \ldots, T - 1$. Our choice for the prior on the AR(1) coefficient, ρ, is a penalized complexity prior, such that $P(\rho > 0) = 0.9$ to reflect the prior belief that we highly doubt a negative temporal correlation.

```
## PC prior on rho
rho_hyper <- list(theta=list(prior = 'pccor1',
                             param = c(0, 0.9)))

## Formula
formula <- cnt ~ -1 + Intercept +
                 f(spatial.field,
                   model = spde,
                   group = spatial.field.group,
                   control.group = list(model = "ar1",
                                        hyper = rho_hyper))
```

Now we have everything in place to run the main function for fitting the model, `inla`. This needs the data from the stack (extracted through `inla.stack.data`) and the exponential family (in this case negative-binomial). The remaining options indicate the desired outputs. In the command given below, we instruct `inla` to fit the model and also to compute the predictions at the required locations.

```
output <- inla(formula,
               data = inla.stack.data(stack, spde = spde),
               family = "nbinomial",
               control.predictor = list(A = inla.stack.A(stack),
                                        compute = TRUE))
```

This operation takes a long time. In **STRbook** we provide the important components of this object, which can be loaded through

```
data("INLA_output", package = "STRbook")
```

INLA provides approximate marginal posterior distributions for each α_t in α and $\{\beta, \rho, \tau\kappa\}$. The returned object, `output`, contains all the results as well as summaries of these results for quick analysis. From the posterior distributions over the precision parameter τ and scale parameter κ, we can readily obtain marginal posterior distributions over the more interpretable variance parameter σ^2 and practical range parameter l. Posterior distributions of some of the parameters are shown in Figure 4.18, where we can see that the AR(1) coefficient of the latent field, ρ, is large (most of the mass of the posterior distribution is close to 1), and the practical range parameter, l, is of the order of 2 degrees (≈ 200 km). The posterior distribution of the marginal variance of the latent field is largest between 2 and 4, These values suggest that there are strong spatial and temporal dependencies in the data. We give code below for plotting the posterior marginal distributions shown in Figure 4.18.

```
output.field <- inla.spde2.result(inla = output,
                                  name = "spatial.field",
                                  spde = spde,
                                  do.transf = TRUE)
## plot p(beta0 | Z)
plot(output$marginals.fix$Intercept,
     type = 'l',
     xlab = expression(beta[0]),
     ylab = expression(p(beta[0]*"|"*Z)))

## plot p(rho | Z)
plot(output$marginals.hyperpar$`GroupRho for spatial.field`,
     type = 'l',
     xlab = expression(rho),
     ylab = expression(p(rho*"|"*Z)))

## plot p(sigma^2 | Z)
plot(output.field$marginals.variance.nominal[[1]],
     type = 'l',
     xlab = expression(sigma^2),
```

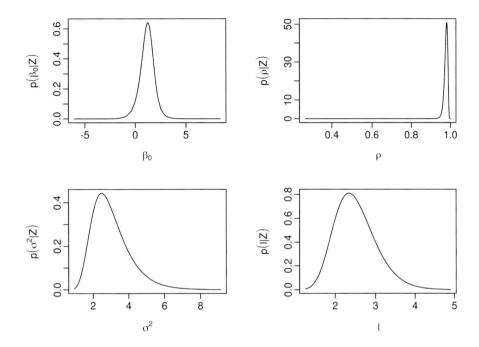

Figure 4.18: Marginal posterior distributions of β_0, the temporal correlation ρ, the variance σ^2, and the range parameter l.

```
    ylab = expression(p(sigma^2*"|"*Z)))

## plot p(range | Z)
plot(output.field$marginals.range.nominal[[1]],
     type = 'l',
     xlab = expression(l),
     ylab = expression(p(l*"|"*Z)))
```

We provide the prediction (posterior mean) and prediction standard error (posterior standard deviation) for $\log(Y(\cdot))$ in Figures 4.19 and 4.20, respectively. These figures were generated by linearly interpolating the posterior mean and posterior standard deviation of $\log(\mathbf{Y}^*)$ on a fine grid. Note how a high observed count at a certain location in one year affects the predictions at the same location in neighboring years, even if unobserved.

Plotting spatial fields, such as those shown in Figures 4.19 and 4.20, from the **INLA** output can be a bit involved since each prediction and prediction standard error of $\boldsymbol{\alpha}_t$ for each t needs to be projected spatially. First, we extract the predictions and prediction standard errors of $\boldsymbol{\alpha} = (\boldsymbol{\alpha}_1', \dots, \boldsymbol{\alpha}_T')'$ as follows.

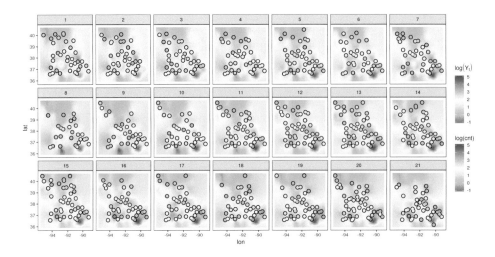

Figure 4.19: Posterior mean of $\log(Y(\cdot))$ on a grid for $t = 1$ (the year 1994) to $t = 21$ (the year 2014), based on a negative-binomial data model using the package **INLA**. The log of the observed count is shown in circles using the same color scale.

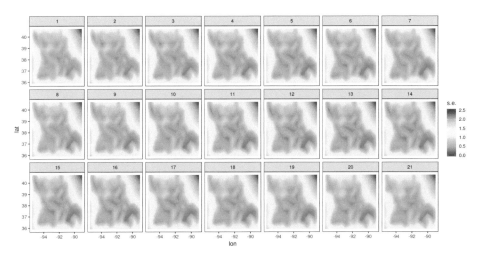

Figure 4.20: Posterior standard deviation (i.e., prediction standard error) of $\log(Y(\cdot))$ on a grid for $t = 1$ (the year 1994) to $t = 21$ (the year 2014), based on a negative-binomial data model using the package **INLA**.

```
index_pred <- inla.stack.index(stack, "pred")$data
lp_mean <- output$summary.fitted.values$mean[index_pred]
```

```
lp_sd <- output$summary.fitted.values$sd[index_pred]
```

Next, we need to create a spatial grid upon which we map the predictions and their associated prediction standard errors. This can be constructed using the function `expand.grid`. We construct an 80×80 grid below.

```
grid_locs <- expand.grid(
                lon = seq(min(MOcarolinawren_long$lon) - 0.2,
                          max(MOcarolinawren_long$lon) + 0.2,
                          length.out = 80),
                lat = seq(min(MOcarolinawren_long$lat) - 0.2,
                          max(MOcarolinawren_long$lat) + 0.2,
                          length.out = 80))
```

The function `inla.mesh.projector` provides all the information required, based on the created spatial grid, to carry out the mapping.

```
proj.grid <- inla.mesh.projector(MOmesh,
                xlim = c(min(MOcarolinawren_long$lon) - 0.2,
                         max(MOcarolinawren_long$lon) + 0.2),
                ylim = c(min(MOcarolinawren_long$lat) - 0.2,
                         max(MOcarolinawren_long$lat) + 0.2),
                dims = c(80, 80))
```

Now we have everything in place to map each α_t on our spatial grid. We iterate through t, and for each $t = 1, \ldots, T$ we map both the prediction and prediction standard errors of α_t on the spatial grid as follows.

```
pred <- sd <- NULL
for(i in 1:n_years) {
    ii <- (i-1)*MOmesh$n + 1
    jj <- i*MOmesh$n
    pred[[i]] <- cbind(grid_locs,
                    z = c(inla.mesh.project(proj.grid,
                                            lp_mean[ii:jj])),
                    t = i)
    sd[[i]] <- cbind(grid_locs,
                    z = c(inla.mesh.project(proj.grid,
                                            lp_sd[ii:jj])),
                    t = i)
}
```

The last thing we need to do is compile all the data (which are in lists) into one data frame for plotting with **ggplot2**. We concatenate all the list elements rowwise and remove those elements that are `NA` because they fall outside of the support of any basis function.

```
pred <- do.call("rbind", pred) %>% filter(!is.na(z))
sd <- do.call("rbind", sd) %>% filter(!is.na(z))
```

The data frames `pred` and `sd` now contain the spatio-temporal predictions and spatio-temporal prediction standard errors. Plotting of these fields using **ggplot2** is left as an exercise for the reader.

Chapter 5

Dynamic Spatio-Temporal Models

Chapter 4 presented the "descriptive" approach to incorporating spatio-temporal statistical dependence into models. This chapter discusses the "dynamic" approach, closer to that holy grail of *causation* that scientists talk and theorize about and that often drives their experiments. In contrast to descriptive models, which fit means and covariances to spatio-temporal data, dynamic models can more easily use scientific knowledge and probability distributions to capture the evolution of current and future spatial fields from past fields.

To convince yourself that this dynamic approach has merit, just look around. If you have ever been mesmerized by waves breaking on the beach, storm clouds building on the horizon, or huge flocks of birds flying collectively in formation, you have witnessed spatio-temporal dynamics in action. What these processes (and many others) have in common is the spatial arrangement of the objects or fields changing, or evolving, from one moment to the next. This is how nature works at the macro scale – the current state of nature evolves from past states. Why does this matter if you are interested in simply modeling data that are indexed in space and time? Don't the descriptive models presented in Chapter 4 represent nature as well?

The short answer to the second question is "yes," but less directly. As we discuss in this chapter, it is difficult to describe all the joint and marginal dependence structures that exist in nature and respect this natural dynamic evolution – which answers the first question. While there are important differences, common to both Chapter 4 and this chapter is a statistical modeling approach where we *always* attempt to account for uncertainty, both in our understanding of the process of interest and in the data we observe.

The primary focus of this chapter is on linear dynamic spatio-temporal models (DSTMs) in the univariate context. Although it is reasonable, and often quite useful, to consider such processes to be continuous in time, for the sake of brevity we focus here on the more practical case where time has been discretized. However, we note that many science-oriented mechanistic models are specified from a continuous-time perspective (e.g., stochastic differential equations), and these are used to motivate the dynamic portion of the

DSTM. It is beyond the scope of this book to take a continuous-time perspective, although it does fit into the DSTM framework.

For readers who have more of a time-series background, the DSTM could be thought of as a *time series of spatial processes.* We could consider an alternative perspective where the spatio-temporal process is a *spatial process of time series,* but the former perspective describes more naturally the dynamic evolutional aspect of the type of real-world phenomena discussed above. In particular, such a framework allows one not only to make predictions of spatial processes into the future, but also to make inference on parameters of models that correspond to mechanistic (e.g., physical or biological or economic ...) processes. This gives DSTMs a powerful insight into causative mechanisms.

We start the chapter with a general hierarchical DSTM formulation in Section 5.1, followed by a more complete discussion of the special case of the linear Gaussian DSTM in Section 5.2. This includes brief discussion of data models, process models, and parameter models. Section 5.3 considers approaches for dealing with the curse of dimensionality in spatial processes and parameter spaces that is often present in DSTM settings. Section 5.4 gives a brief discussion of nonlinear DSTMs. More details on the technical aspects are given in a number of appendices: we present some standard estimation and prediction algorithms in Appendix C and examples of parameter reduction and process motivation through mechanistic models in Appendix D. Finally, Appendices E and F present case studies on mechanistically motivated prediction of Mediterranean winds and a machine-learning-motivated nonlinear DSTM for forecasting tropical Pacific SSTs, respectively.

5.1   General Dynamic Spatio-Temporal Models

As discussed in Chapter 1, we like to consider statistical models from a hierarchical modeling (HM) perspective. In the context of DSTMs, this means that at a minimum we must specify: a "data model" that gives a model for the data, conditioned on the true process of interest and some parameters; a "process model" that specifies the dynamic evolution of the spatio-temporal process, given some parameters; and either models for the parameters from the previous two stages (Bayesian hierarchical model, BHM), or "plug-in" estimates of the parameters (empirical hierarchical model, EHM). In this section we give a general overview of hierarchical modeling in the context of a DSTM.

Recall from our preamble that we are considering discrete time here with temporal domain $D_t = \{0, 1, 2, \ldots\}$, where we assume a constant time increment $\Delta_t = 1$ (without loss of generality). We shall consider spatial locations for our observations and our latent process to be in some spatial domain D_s (which we may consider continuous or discrete and finite, depending on the context). The data can potentially come from anywhere and any time in the spatial and temporal domains; we denote data and potential data by $\{Z_t(\mathbf{s}) : \mathbf{s} \in D_s; \ t = 0, 1, \ldots\}$, although only a subset is actually observed. The latent process is denoted by $\{Y_t(\mathbf{s}) : \mathbf{s} \in D_s; \ t = 0, 1, \ldots\}$, and we may make inference on $Y_{t_0}(\mathbf{s}_0)$, even though there is no datum $Z_{t_0}(\mathbf{s}_0)$. Note that, unlike the models presented in Chapter 4, we change notation slightly and use a subscript t to represent time here, as is customary for discrete-time processes with $D_t = \{0, 1, 2, \ldots\}$.

5.1.1 Data Model

We begin with a data model that describes the relationship between the observations and the latent spatio-temporal process. Generally, we could write the data model in a DSTM as

$$Z_t(\cdot) = \mathcal{H}_t(Y_t(\cdot), \boldsymbol{\theta}_{d,t}, \epsilon_t(\cdot)), \quad t = 1, \ldots, T,$$

where $Z_t(\cdot)$ corresponds to the data at time t (and we use $(\cdot)$ to represent arbitrary spatial locations), $Y_t(\cdot)$ is the latent spatio-temporal process of interest, with a linear or nonlinear mapping, $\mathcal{H}_t$, that connects the data to the latent process. The data-model error, which is typically measurement error and sometimes small-scale spatio-temporal variability, is given by $\epsilon_t(\cdot)$. Finally, data-model parameters, which themselves may vary spatially and/or temporally, are represented by the vector $\boldsymbol{\theta}_{d,t}$. An important assumption here, and in many hierarchical representations of DSTMs, is that the data $Z_t(\cdot)$ are independent (in time) *when conditioned* on the true process $Y_t(\cdot)$ and parameters $\boldsymbol{\theta}_{d,t}$. Under this conditional-independence assumption, the joint distribution of the data conditioned on the true process and parameters can be represented in product form,

$$[\{Z_t(\cdot)\}_{t=1}^T \mid \{Y_t(\cdot)\}_{t=1}^T, \{\boldsymbol{\theta}_{d,t}\}_{t=1}^T] = \prod_{t=1}^T [Z_t(\cdot) \mid Y_t(\cdot), \boldsymbol{\theta}_{d,t}]. \quad (5.1)$$

This is one of two key independence/dependence assumptions in DSTMs (the other is discussed in Section 5.1.2 below). Most applications consider the component distributions on the right-hand side of (5.1) to be Gaussian, but it is not uncommon to consider other members of the exponential family of distributions (see Section 5.2.2 below). Indeed, a broader class of data models than the familiar Gaussian model is fairly easy to consider so long as the observations are conditionally independent given the true process. We consider specific examples of data models in Section 5.2.1.

5.1.2 Process Model

Perhaps the most important part of a DSTM is the decomposition of the joint distribution of the process in terms of conditional distributions that respect the time evolution of the spatial process. With $Y_t(\cdot)$ corresponding to the spatial process at time t, we can always factor the joint distribution using the chain rule of conditional probabilities:

$$
\begin{aligned}
[Y_0(\cdot), Y_1(\cdot), \ldots, Y_T(\cdot)] &= [Y_T(\cdot)|Y_{T-1}(\cdot), \ldots, Y_0(\cdot)] \\
&\times [Y_{T-1}(\cdot)|Y_{T-2}(\cdot), \ldots, Y_0(\cdot)] \times \cdots \\
&\times [Y_2(\cdot)|Y_1(\cdot), Y_0(\cdot)] \\
&\times [Y_1(\cdot)|Y_0(\cdot)] \\
&\times [Y_0(\cdot)],
\end{aligned}
$$

where, for notational simplicity, the dependence of these distributions on parameters has been suppressed. By itself, this decomposition is not all that useful because it requires a separate conditional model for $Y_t(\cdot)$ at each t. However, if we make a modeling *assumption* that utilizes *conditional independence*, then such a hierarchical decomposition can be quite useful. For example, we could make a *Markov assumption*; that is, conditioned on the past, only the recent past is important to explain the present. Under the first-order Markov assumption that the process at time t conditioned on all of the past is only dependent on the *most* recent past (and an additional modeling assumption that this process only depends on the current parameters), we get a very useful simplification,

$$
[Y_t(\cdot)|Y_{t-1}(\cdot), \ldots, Y_0(\cdot), \{\boldsymbol{\theta}_{p,t}\}_{t=0}^T] = [Y_t(\cdot)|Y_{t-1}(\cdot), \boldsymbol{\theta}_{p,t}], \tag{5.2}
$$

for $t = 1, 2, \ldots$, so that

$$
[Y_0(\cdot), Y_1(\cdot), \ldots, Y_T(\cdot)|\{\boldsymbol{\theta}_{p,t}\}_{t=0}^T] = \left(\prod_{t=1}^T [Y_t(\cdot)|Y_{t-1}(\cdot), \boldsymbol{\theta}_{p,t}]\right) [Y_0(\cdot)|\boldsymbol{\theta}_{p,0}]. \tag{5.3}
$$

This is the second of the key assumptions that is usually made for DSTMs (the first was discussed above in Section 5.1.1).

This first-order-Markov assumption, which is simple but powerful in spatio-temporal statistics, holds when $\{Y_t(\cdot)\}$ follows a *dynamic model* of the form

$$
Y_t(\cdot) = \mathcal{M}(Y_{t-1}(\cdot), \boldsymbol{\theta}_{p,t}, \eta_t(\cdot)), \quad t = 1, 2, \ldots, \tag{5.4}
$$

where $\boldsymbol{\theta}_{p,t}$ are parameters (possibly with spatial or temporal dependence) that control the process evolution described by the evolution operator $\mathcal{M}$, and $\eta_t(\cdot)$ is a spatial noise (error) process that is independent in time (i.e., $\eta_t(\cdot)$ and $\eta_r(\cdot)$ are independent for $r \neq t$). In general, this model can be linear or nonlinear and the associated conditional distribution, $[Y_t(\cdot)|Y_{t-1}(\cdot)]$, can be Gaussian or non-Gaussian. As in autoregressive modeling in time series, one can make higher-order Markov assumptions in this case as well, which requires additional time lags of the spatial process to be included on the right-hand side of the conditioning symbol " | " in the component distributions of (5.3). We focus primarily on the first-order-Markov case here, which is usually assumed; however, note that one can always reformulate a higher-order Markov model as a first-order model, albeit increasing the dimensionality of the process, so the first-order representation in terms of probability distributions is actually quite general. One also needs to specify a distribution for the initial state, $[Y_0(\cdot)|\boldsymbol{\theta}_{p,0}]$ or condition on it. We consider specific examples of DSTM process models in Section 5.2.1.

5.1.3   Parameters

A BHM requires distributions to be assigned to the parameters defined in the data model and the process model, namely $\{\boldsymbol{\theta}_{d,t}, \boldsymbol{\theta}_{p,t}\}$. Specific distributional forms for the parameters (e.g., spatially or temporally varying dependence on auxiliary covariate information) depend strongly on the problem of interest. Indeed, as mentioned in Chapter 1, one of the most important aspects of "deep" hierarchical modeling is the specification of these distributions, especially when one must deal with the curse of dimensionality. In that case, the primary modeling challenge in DSTMs is to come up with ways to effectively reduce the parameter space. This is illustrated in Section 5.2.1 with regard to linear DSTMs.

Despite the power of the BHM, in many cases it is possible and sufficient to simply estimate the parameters in an EHM context. This is commonly done in state-space models in time series and often utilizes the expectation-maximization (EM) algorithm or, as is done in the engineering literature, "state augmentation," where the process "vector" is augmented by the parameters. Again, the choice of the estimation approach is very problem-specific. We give a general EM algorithm for linear DSTMs in Appendix C.2.

5.2   Latent Linear Gaussian DSTMs

For illustration, we consider in this section the simplest (yet, most widely used) DSTM – where the process models in (5.2) are assumed to have additive Gaussian error distributions, and the evolution operator $\mathcal{M}$ in (5.4) is assumed to be linear. Let us suppose that we are interested in a latent process $\{Y_t(\mathbf{s}_i)\}$ at a set of locations given by $\{\mathbf{s}_i : i = 1, \ldots, n\}$, and that we have data at locations $\{\mathbf{r}_{jt} : j = 1, \ldots, m_t; \ t = 0, 1, \ldots, T\}$ (i.e., there could be a different number of data locations for each observation time, but we assume there is a finite set of m possible data locations to be considered; so $m_t \leq m$).

For simplicity, unless noted otherwise, we assume that the "locations" of interest can have either point or areal support (and, possibly different supports for the prediction locations and data locations).

5.2.1   Linear Data Model with Additive Gaussian Error

Consider the m_t-dimensional data vector, $\mathbf{Z}_t \equiv (Z_t(\mathbf{r}_{1t}), \ldots, Z_t(\mathbf{r}_{m_t t}))'$, and the n-dimensional latent-process vector, $\mathbf{Y}_t \equiv (Y_t(\mathbf{s}_1), \ldots, Y_t(\mathbf{s}_n))'$, that we wish to infer. For the jth observation at time t, the linear data model with additive Gaussian error is written as

$$Z_t(\mathbf{r}_{jt}) = b_t(\mathbf{r}_{jt}) + \sum_{i=1}^{n} h_{t,ji} Y_t(\mathbf{s}_i) + \epsilon_t(\mathbf{r}_{jt}), \qquad (5.5)$$

for $t = 1, \ldots, T$, where $b_t(\mathbf{r}_{jt})$ is an additive offset term for the jth observation at time t, $\{h_{t,ji}\}_{i=1}^{n} \equiv \mathbf{h}'_{t,j}$ are coefficients that map the latent process to the jth observation at time t, and the error term $\epsilon_t(\cdot)$ is independent of $Y_t(\cdot)$. Since $j = 1, \ldots, m_t$, the data model can be written in vector–matrix form as

$$\mathbf{Z}_t = \mathbf{b}_t + \mathbf{H}_t \mathbf{Y}_t + \boldsymbol{\varepsilon}_t, \quad \boldsymbol{\varepsilon}_t \sim Gau(\mathbf{0}, \mathbf{C}_{\epsilon,t}), \qquad (5.6)$$

where $\mathbf{b}_t$ is the m_t-dimensional offset term, $\mathbf{H}_t$ is the $m_t \times n$ mapping matrix (note that $\mathbf{h}'_{t,j}$ corresponds to the jth row of $\mathbf{H}_t$), and $\mathbf{C}_{\epsilon,t}$ is an $m_t \times m_t$ error covariance matrix, typically taken to be diagonal. Each of the data-model components is described briefly below.

Latent Spatio-Temporal Dynamic Process

The latent dynamic spatio-temporal process is represented by $\mathbf{Y}_t$. This is where most of the modeling effort is focused in the latent linear DSTM framework. It is convenient in many situations to assume that $\mathbf{Y}_t$ has mean zero; however, we present an alternative perspective in Section 5.3 below. As mentioned previously, we shall focus on first-order Markov models to describe the evolution of $\mathbf{Y}_t$.

Additive Offset Term

There are instances where there are potential biases between the observations and the process of interest, or where one would like to be able to model $\{Y_t(\cdot)\}$ as a mean-zero process. That is, the additive offset term, $\mathbf{b}_t$, accounts for non-dynamic spatio-temporal structure in the data vector, $\mathbf{Z}_t$, that allows us to consider $\mathbf{Y}_t$ to have mean zero. One might still be interested scientifically in predicting the sum $\mathbf{b}_t + \mathbf{H}_t \mathbf{Y}_t$ in (5.6). We may assume that the additive offset term $b_t(\mathbf{r}_{jt})$ is fixed through time, space, or constant across space and time (e.g., $b_t(\mathbf{r}_{jt}) \equiv b(\mathbf{r}_{jt})$, $b_t(\mathbf{r}_{jt}) \equiv b_t$, or $b_t(\mathbf{r}_{jt}) \equiv b$, respectively), or we may define it in terms of covariates (e.g., $b_t(\mathbf{r}_{jt}) \equiv \mathbf{x}'_{t,j}\boldsymbol{\beta}$, or $b_t(\mathbf{r}_{jt}) \equiv \mathbf{x}'_t\boldsymbol{\beta}_j$, where $\mathbf{x}_{t,j}$ and $\mathbf{x}_t$ are q-dimensional vectors of covariates and $\boldsymbol{\beta}$ and $\boldsymbol{\beta}_j$ are q-dimensional parameter vectors). Alternatively, we may consider the offset parameters to be either spatial or temporal random processes with distributions assigned at the next level of the model hierarchy (e.g., $\mathbf{b}_t \sim Gau(\mathbf{X}_t\boldsymbol{\beta}, \mathbf{C}_b)$, where $\mathbf{C}_b$ is a positive-definite matrix constructed using the methods described in Chapter 4).

Observation Mapping Function (or Matrix)

The observation mapping matrix $\mathbf{H}_t$ has elements $\{h_{t,ji}\}$ that are typically assumed known. They can be any linear basis that relates the process at the prediction locations to the observations. For example, it is often quite useful to let $\mathbf{h}_{t,j}$ correspond to a simple incidence vector (i.e., a vector of ones and zeros), so that each data location is associated with one or more of the process locations. The incidence vector can easily accommodate missing data or can serve as an "interpolator" such that each observation is related to some weighted combination of the process values.

In this simple illustration, consider the single observation equation (5.5) where $n = 3$. If $\mathbf{h}'_{t,j} = (0, 0, 1)$, it indicates that the observation $Z_t(\mathbf{r}_{jt})$ corresponds to the process value, $Y_t(\mathbf{s}_3)$, at time t. This is especially useful if the locations of the prediction grid are very close to (or a subset of) the observation locations and consequently are considered coincident. If $\mathbf{h}'_{t,j} = (0.2, 0.8, 0)$, then the observation at location $\mathbf{r}_{jt}$ corresponds to a weighted sum of the process at locations $\mathbf{s}_1$ and $\mathbf{s}_2$, with more weight being given to location $\mathbf{s}_2$. More generally, these weights can provide a simple way to deal with different spatial supports and orientations of the observations and the process. For example, the weights can correspond to the area of overlap between observation supports and process supports (see Chapter 7 of Cressie and Wikle, 2011, for details).

R tip: Recall from Section 2.2 that finding the intersections (areas or points of overlap) across spatial or spatio-temporal polygons, points, and grids can be done in a straight-forward manner using the function **over** from the packages **sp** and **spacetime**. This function can hence be used to construct the mapping matrices $\{\mathbf{H}_t\}$ in (5.6).

Finally, in the situation where the observation locations are a subset of the process locations and $m_t < n$, one has missing data, and this is easily accommodated via the mapping matrix. For example, if $m_t = 2$ and $n = 3$, then $\mathbf{Z}_t \equiv (Z_t(\mathbf{r}_{1t}), Z_t(\mathbf{r}_{2t}))'$, $\mathbf{Y}_t \equiv (Y_t(\mathbf{s}_1), Y_t(\mathbf{s}_2), Y_t(\mathbf{s}_3))'$, and $\boldsymbol{\varepsilon}_t \equiv (\epsilon_t(\mathbf{r}_{1t}), \epsilon_t(\mathbf{r}_{2t}))'$. If $\mathbf{r}_{1t} = \mathbf{s}_2$ and $\mathbf{r}_{2t} = \mathbf{s}_3$, the mapping matrix in (5.6) is given by the incidence matrix

$$\mathbf{H}_t = \begin{pmatrix} 0 & 1 & 0 \\ 0 & 0 & 1 \end{pmatrix}, \tag{5.7}$$

which indicates that observation $Z_t(\mathbf{r}_{1t})$ corresponds to process value $Y_t(\mathbf{s}_2)$, observation $Z_t(\mathbf{r}_{2t})$ corresponds to process value $Y_t(\mathbf{s}_3)$, and process value $Y_t(\mathbf{s}_1)$ does not have a corresponding observation at time t. This way to accommodate missing information is very useful for HMs, because it allows one to focus the modeling effort on the latent process $\{Y_t(\cdot)\}$, and the process is oblivious to which data are missing. Some would argue that a downside of this is the need to pre-specify the locations at which one is interested in modeling the process, but, with a sufficiently fine grid, this could effectively be everywhere in the spatial domain, D_s.

Although it is possible in principle to parameterize the mapping matrix and/or estimate it directly in some cases, we shall typically assume that it is known. Otherwise, one would have to be careful when specifying and estimating the process-model parameters to mitigate identifiability problems.

R tip: Recall from Section 4.4.1 that matrices such as that given in (5.7) tend to have many zeros and hence are *sparse*. Two R packages that cater to sparse matrices are **Matrix** and **spam**. Sparse matrices are stored and operated on differently than the standard dense R matrices, and they have favorable memory and computational properties. However, sparse matrix objects should only be used when the number of non-zero entries is small, generally on the order of 5% or less of the total number of matrix elements.

Error Covariance Matrix

In the linear Gaussian DSTM, the additive error process $\{\epsilon_t(\mathbf{r}_{jt})\}$ is assumed to have mean zero, is Gaussian, and can generally include dependence in space or time (although we will typically assume that the errors are independent in time, as is customary). So, when considering $\epsilon_t(\cdot)$ at a finite set of m_t observation locations, namely $\boldsymbol{\varepsilon}_t \equiv (\epsilon_t(\mathbf{r}_{1t}), \ldots, \epsilon_t(\mathbf{r}_{m_t t}))'$, we need to specify time-varying covariance matrices $\{\mathbf{C}_{\epsilon,t}\}$. In practice, given that most of the interesting dependence structure in the observations is contained in the process, and recalling that in the data model we are conditioning on that process, the structure of $\mathbf{C}_{\epsilon,t}$ should be pretty simple. Indeed, there is often an assumption that these data-model errors are independent with constant (in time and space) variance, so that $\mathbf{C}_{\epsilon,t} = \sigma_\epsilon^2 \mathbf{I}_{m_t}$, where σ_ϵ^2 represents the measurement-error variance. If this assumption is not reasonable, then in situations where the data-model error covariance matrix is assumed constant over time (e.g., $\mathbf{C}_{\epsilon,t} = \mathbf{C}_\epsilon$) and $m_t = m$, one might estimate the $m \times m$ covariance matrix $\mathbf{C}_\epsilon$ directly if m is not too large (see Appendix C.2), or one might parameterize $\mathbf{C}_\epsilon$. For example, in the case where the data-model error variances are heteroskedastic in space, one might model $\mathbf{C}_\epsilon = \mathrm{diag}(\mathbf{v}_\epsilon)$, and estimate the elements of $\mathbf{v}_\epsilon$. Alternatively, if there is spatial dependence one might parameterize $\mathbf{C}_\epsilon$ in terms of some valid spatial covariance functions (e.g., the Matérn class). The specific choice is very problem-dependent. It is important to recall the central principle of hierarchical modeling discussed in Chapter 1, in Section 3.5, and in greater detail in Section 4.3, which is that we attempt to place as much of the dependence structure as possible in the conditional mean, which simplifies the conditional-covariance specification dramatically.

5.2.2   Non-Gaussian and Nonlinear Data Model

Recall the general exponential family data model from Section 4.5, rewritten here to correspond to the discrete-time case. For $t = 1, 2, \ldots$, let

$$Z_t(\mathbf{s})|Y_t(\mathbf{s}), \gamma \sim EF(Y_t(\mathbf{s}), \gamma),$$

where EF corresponds to a distribution from the exponential family with scale parameter γ and mean $Y_t(\mathbf{s})$. Now, consider a transformation of the mean response $g(Y_t(\mathbf{s})) \equiv \tilde{Y}_t(\mathbf{s})$

using a specified monotonic link function $g(\cdot)$. Using a standard GLMM framework, we can model the transformed process $\tilde{Y}_t(\mathbf{s})$ as a latent Gaussian DSTM (Section 5.2.3) with or without the use of process/parameter reduction methods (Section 5.3). Note that we can also include a "mapping matrix" to this non-Gaussian data model as we did with the Gaussian data model in Section 5.2.1. That is, in a matrix formulation we could consider

$$\mathbf{Z}_t | \mathbf{Y}_t, \gamma \sim EF(\mathbf{H}_t \mathbf{Y}_t, \gamma),$$

where the distribution EF is taken elementwise, and $\mathbf{H}_t$ is an incidence matrix or change-of-support matrix as described in Section 5.2.1.

It is sometimes useful to consider a nonlinear transformation of the latent process $\{Y_t(\cdot)\}$ in a data model even if the error term is Gaussian. For example, analogous to equation (7.39) in Cressie and Wikle (2011), one can modify (5.6) above to accommodate a transformation of the elements of the process vector:

$$\mathbf{Z}_t = \mathbf{b}_t + \mathbf{H}_t \mathbf{Y}_t^a + \boldsymbol{\varepsilon}_t, \quad \boldsymbol{\varepsilon}_t \sim Gau(\mathbf{0}, \mathbf{C}_{\epsilon,t}), \tag{5.8}$$

where the coefficient $\{-\infty < a < \infty\}$ corresponds to a power transformation (applied to each element of $\mathbf{Y}_t$), which is one of the simplest ways to accommodate nonlinear or non-Gaussian processes in the data model. In general, $\{\mathbf{Y}_t^a\}$ may not generate a linear Gaussian model, but the additivity of the errors $\{\boldsymbol{\varepsilon}_t\}$ is an important part of (5.8). As an example, if $\{Y_t(\cdot)\}$ is positive valued, then this is analogous to the famed Box–Cox transformation. In some applications it is reasonable to assume that the transformation power a in (5.8) may vary with space or time, and may depend on covariates.

As with non-dynamic spatio-temporal models with non-Gaussian errors (Chapter 4), computation for estimation and prediction is more problematic when one considers non-Gaussian or nonlinear data models.

5.2.3 Process Model

Linear Markovian spatio-temporal process models generally assume that the value of the process at a given location at the present time is made up of a weighted combination (or is a "smoothed version") of the process throughout the spatial domain at previous times, plus an additive, Gaussian, spatially coherent "innovation" (see the schematic in Figure 5.1). This is perhaps best represented in a continuous-spatial context through an *integro-difference equation* (IDE). Specifically, a first-order spatio-temporal IDE process model is given by

$$Y_t(\mathbf{s}) = \int_{D_s} m(\mathbf{s}, \mathbf{x}; \boldsymbol{\theta}_p) Y_{t-1}(\mathbf{x}) \, d\mathbf{x} + \eta_t(\mathbf{s}), \quad \mathbf{s}, \mathbf{x} \in D_s, \tag{5.9}$$

for $t = 1, 2, \ldots$, where $m(\mathbf{s}, \mathbf{x}; \boldsymbol{\theta}_p)$ is a *transition kernel*, depending on parameters $\boldsymbol{\theta}_p$ that specify "redistribution weights" for the process at the previous time over the spatial domain, D_s, and $\eta_t(\cdot)$ is a time-varying (but statistically independent in time) continuous mean-zero

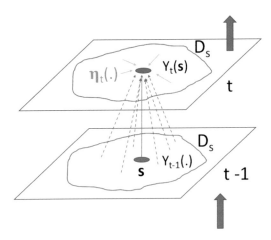

Figure 5.1: Cartoon illustration of a linear DSTM. The process at spatial location s and time t, $Y_t(\mathbf{s})$, is constructed from a linear combination of the process values at the previous time, $Y_{t-1}(\cdot)$, plus an "instantaneous" random spatial error process, $\eta_t(\cdot)$. The thick arrows indicate the passage from past to present to future.

Gaussian spatial process independent of $Y_{t-1}(\cdot)$. Generally, one of the parameters of $\boldsymbol{\theta}_p$ is just a multiplicative scalar that controls the temporal stability; see (5.23) in Lab 5.2. Note that we assume here, as one often does, that the parameter vector $\boldsymbol{\theta}_p$ does not vary with time, but it could do so in general. So, from (5.9), the process at location $\mathbf{s}$ and time t is given by the weighted average (integral) of the process throughout the domain at the past time, where the weights are given by the transition kernel, $m(\cdot,\cdot)$. The innovation given by $\eta_t(\cdot)$, which is independent of $Y_{t-1}(\cdot)$, has spatial dependence, is typically Gaussian, and accounts for spatial dependencies in $Y_t(\cdot)$ that are not captured by this weighted average. Another way to think about $\eta_t(\cdot)$ is that it adds back smaller-scale dependence that is removed in the inherent smoothing that occurs when $\{Y_{t-1}(\cdot)\}$ is averaged over space. In general, $\int_{D_s} m(\mathbf{s},\mathbf{x};\boldsymbol{\theta}_p)\,d\mathbf{x} < 1$ is needed for the process to be stable (non-explosive) in time. Note that the model in (5.9) implicitly assumes that the process $Y_t(\cdot)$ has mean zero. In some cases it may be appropriate to model a non-zero mean directly in the process, as is shown generally in (5.16) below and specifically for the IDE in Lab 5.2.

In the case where one has a finite set of prediction spatial locations (or regions) $D_s = \{\mathbf{s}_1, \mathbf{s}_2, \ldots, \mathbf{s}_n\}$ of interest (e.g., an irregular lattice or a regular grid), the first-order IDE evolution process model (5.9) can be discretized and written as a stochastic difference equation,

$$Y_t(\mathbf{s}_i) = \sum_{j=1}^{n} m_{ij}(\boldsymbol{\theta}_p)\, Y_{t-1}(\mathbf{s}_j) + \eta_t(\mathbf{s}_i), \tag{5.10}$$

for $t = 1, 2, \ldots$, with transition (redistribution) weights $m_{ij}(\boldsymbol{\theta}_p)$ that depend on parameters $\boldsymbol{\theta}_p$. In this case, the process at $Y_t(\mathbf{s}_i)$ considers a weighted combination of the values of the process at time $t - 1$ and at a discrete set of spatial locations.

Now, denoting the process vector $\mathbf{Y}_t \equiv (Y_t(\mathbf{s}_1), \ldots, Y_t(\mathbf{s}_n))'$, (5.10) can be written in vector–matrix form as a linear first-order vector autoregression DSTM,

$$\mathbf{Y}_t = \mathbf{M}\mathbf{Y}_{t-1} + \boldsymbol{\eta}_t, \tag{5.11}$$

where the $n \times n$ transition matrix is given by $\mathbf{M}$ with elements $\{m_{ij}\}$, and the additive spatial error process $\boldsymbol{\eta}_t \equiv (\eta_t(\mathbf{s}_1), \ldots, \eta_t(\mathbf{s}_n))'$ is independent of $\mathbf{Y}_{t-1}$ and is specified to be mean-zero and Gaussian with spatial covariance matrix $\mathbf{C}_\eta$. The stability (non-explosive) condition in this case requires that the maximum modulus of the eigenvalues of $\mathbf{M}$ (which may be complex-valued) be less than 1 (see Technical Note 5.1).

We have assumed in our discussion of the process model that the $\{Y_t(\mathbf{s}_i)\}$ have mean zero. Although it is possible to include an offset term in the Markovian process model at this stage, in this section we consider such an offset only in the data model as described above for (5.6). However, as we discuss below in Section 5.3, it is reasonable to consider the offset as part of this "process" decomposition, typically including covariate effects and/or seasonality.

Usually, $\mathbf{M}$ and $\mathbf{C}_\eta$ are assumed to depend on parameters $\boldsymbol{\theta}_p$ and $\boldsymbol{\theta}_\eta$, respectively, to mitigate the curse of dimensionality (here, the exponential increase in the number of parameters) that often occurs in spatio-temporal modeling. As discussed below in Section 5.3, the parameterization of these matrices (particularly $\mathbf{M}$) is one of the greatest challenges in DSTMs, and it is facilitated by using parameter models in a BHM. However, in relatively simple applications of fairly low dimensionality and large sample sizes (e.g., when n is small and $T \gg n$), one can estimate $n \times n$ matrices $\mathbf{M}$ and $\mathbf{C}_\eta$ directly in an EHM, as is commonly done in state-space models of time series (see Appendix C.2).

Technical Note 5.1: Eigenvalues of the Transition Matrix

Consider the first-order vector autoregressive model,

$$\mathbf{Y}_t = \mathbf{M}\mathbf{Y}_{t-1} + \boldsymbol{\eta}_t,$$

where $\mathbf{Y}_t$ is an n-dimensional vector, and $\mathbf{M}$ is an $n \times n$ real-valued transition matrix. The characteristic equation obtained from the determinant,

$$\det(\mathbf{M} - \lambda\mathbf{I}) = 0,$$

has n eigenvalues (latent roots), $\{\lambda_i : i = 1, \ldots, n\}$, some of which may be complex numbers. Each eigenvalue has a modulus and is associated with an eigenvector (taken together, an eigenvalue–eigenvector pair is sometimes referred to as an *eigenmode*) that describes the behavior associated with that eigenvalue. As discussed in Cressie and

Wikle (2011, Section 3.2.1), the eigenvalues and eigenvectors can tell us quite a bit about the dynamical properties of the model. First, assume in general that $\lambda_i = a_i + b_i\sqrt{-1}$ (where $b_i = 0$ if λ_i is real-valued), and define the complex magnitude (or "modulus") to be $|\lambda_i| = \sqrt{a_i^2 + b_i^2}$. We note that if $\max\{|\lambda_i| : i = 1, \ldots, n\} \geq 1$, then the eigenmode, and hence the model, is unstable, and $\mathbf{Y}_t$ will grow without bound as t increases. Conversely, if the maximum modulus of all the eigenvalues is less than 1, then the model is stable. Since $\mathbf{M}$ is real-valued, complex eigenvalues come in complex conjugate pairs, and their eigenmodes are associated with oscillatory behavior in the dynamics (either damped or exponentially growing sinusoids, depending on whether the modulus of the corresponding eigenvalue is less than 1 or greater than or equal to 1, respectively). In contrast, real-valued eigenvalues correspond to non-oscillatory dynamics.

Intuition for Linear Dynamics

Parameterizations of realistic dynamics should respect the fact that spatio-temporal interactions are crucial for dynamic propagation. For example, in the linear IDE model (5.9), the *asymmetry* and *rate of decay* of the transition kernel $m(\mathbf{s}, \mathbf{x}; \boldsymbol{\theta}_p)$, relative to a location (here, $\mathbf{s}$), control *propagation* (linear advection) and *spread* (diffusion), respectively. Figure 5.2 shows Hovmöller plots of four one-dimensional (in space) simulations of a spatio-temporal IDE process and their respective transition kernels evaluated at $s_0 = 0.5$. Panels (a) and (b) show the inherent diffusive nature of the process depending on kernel width; that is, spatially coherent disturbances tend to spread across space (diffuse) at a greater rate when the kernel is wider (i.e., has a larger aperture), which leads to more averaging from one time to the next. However, note that there is no "slanting" in the plot through time, indicating that there is no propagation through time (see Section 2.3.3 for an interpretation of Hovmöller plots). In contrast, panels (c) and (d) show clear evidence of propagation, to the left when the kernel is offset to the right, and to the right when the kernel is offset to the left. The intuition here is that the offset kernel pulls information from one particular direction, and redistributes it in the opposite direction, leading to propagation. More complex kernels (e.g., multimodal, or spatially varying) can lead to even more complex behavior. As we discuss in Section 5.3, these basic properties of the transition (redistribution) kernel can suggest efficient parameterizations of linear DTSM process models.

   As mentioned above, there are conditions on the transition kernel (or matrix in the discrete-space case) that correspond to unstable (explosive in time) behavior. From a dynamic perspective, a stable process implies that small perturbations to the spatial field will eventually decay to the equilibrium (mean) state. Because many real-world spatio-temporal processes are nonlinear, it can be the case that if one fits an unconstrained *linear* DSTM to data that come from such a *nonlinear* process, then the fitted model is unstable (explosive, with exponential "growth"). This is not necessarily a bad thing, as it provides immediate

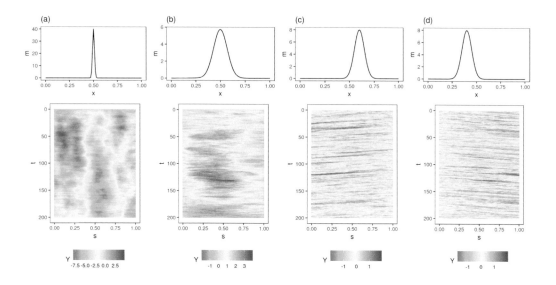

Figure 5.2: Transition kernels $m(0.5, x; \boldsymbol{\theta}_p)$ for different $\boldsymbol{\theta}_p$ and associated Hovmöller plots of spatio-temporal IDE process simulation (one-dimensional in space) with *iid* noise forcing. (a) Relatively narrow symmetric kernel. (b) Wider symmetric kernel. (c) Asymmetric kernel produced by shifting a symmetric kernel to the right. (d) Same as (c), but to the left. Note that the wider the kernel the greater the diffusion, and a shift implies propagation in the direction away from the shift.

feedback that the wrong model is being fitted or that the finite-time window for the observations suggests a transient period of growth (see Technical Note 5.2). In some cases, it can actually be helpful if the confidence (or credible) intervals of the transition-matrix parameters include the explosive boundary because the mean of the predictive distribution may show growth (a nonlinear feature) since it effectively averages over realizations that are both explosive and non-explosive. Of course, long-lead-time forecasts from such a model are problematic as exponential-growth models are only useful for very short-term predictions unless there is some nonlinear control mechanism (e.g., density-dependent carrying capacities in ecological applications).

> **Technical Note 5.2: Transient Growth**
> An interesting and less appreciated aspect of linear DSTMs is the fact that they can be stable yet still accommodate so-called "transient growth." That is, there are periods in time when the process does have brief excursions from its stable state. Essentially, if the transition operator is "non-normal" (i.e., in the discrete-space case, if $\mathbf{M}\mathbf{M}' \neq \mathbf{M}'\mathbf{M}$, in which case, the eigenvectors of $\mathbf{M}$ are non-orthogonal), but still stable (e.g.,

the maximum modulus of the eigenvalues of $\mathbf{M}$ is less than 1; see Technical Note 5.1), then the linear dynamic process can exhibit transient growth. This means that even though each eigenvector of the stable $\mathbf{M}$ is decaying asymptotically in time, there can be local-in-time (transient) periods where there is significant (even orders of magnitude) growth. This is due to the constructive interference of the non-orthogonal eigenvectors of the transition operator, $\mathbf{M}$. Since almost all real-world linear processes correspond to non-normal transition operators, this has important implications concerning how one might parameterize $\mathbf{M}$, as discussed in Section 5.3 below.

5.3   Process and Parameter Dimension Reduction

The latent linear Gaussian DSTM described in Section 5.2 above has unknown parameters associated with the data model $\mathbf{C}_\eta$, the transition operator $m(\mathbf{s}, \mathbf{x}; \boldsymbol{\theta}_p)$ or matrix $\mathbf{M}$, and the initial-condition distribution (e.g., $\boldsymbol{\mu}_0$ and $\mathbf{C}_0$). With the linear Gaussian data model, one typically considers a fairly simple parameterization of $\mathbf{C}_\epsilon$ (e.g., $\mathbf{C}_\epsilon = \sigma_\epsilon^2 \mathbf{I}$) or perhaps the covariance matrix implied by a simple spatial random process that has just a few parameters (e.g., a Matérn spatial covariance function or a spatial conditional autoregressive process). One of the greatest challenges when considering DSTMs in hierarchical statistical settings is the curse of dimensionality associated with the process-model level of the DSTM. For the fairly common situation where the number of spatial locations (n) is much larger than the number of time replicates (T), even the fairly simple linear DSTM process model (5.11) is problematic, as there are on the order of n^2 parameters to estimate. To proceed, one must reduce the number of free parameters to be inferred in the model and/or reduce the dimension of the spatio-temporal dynamic process. These two approaches are discussed briefly below.

5.3.1   Parameter Dimension Reduction

Consider the process-error spatial variance–covariance matrix, $\mathbf{C}_\eta$. In complex modeling situations, it is seldom the case that one would estimate this as a full positive-definite matrix in the DSTM. Rather, given that these are spatial covariance matrices, we would either use one of the common spatial covariance-function representations or a basis-function random-effects representation (as in Chapter 4 or in Section 5.3.2 below).

Generally, the transition-matrix parameters in the DSTM process model require the most care, as there there could be as many as n^2 of them and, as discussed above, the linear dynamics of the process are largely controlled by these parameters. In the case of the simple linear DSTM model (5.11), one could parameterize the transition matrix $\mathbf{M}$ simply as a random walk (i.e., $\mathbf{M} = \mathbf{I}$), a spatially homogeneous autoregressive process (i.e., $\mathbf{M} = \theta_p \mathbf{I}$), or a spatially varying autoregressive process ($\mathbf{M} = \mathrm{diag}(\boldsymbol{\theta}_p)$). The first two parameterizations are somewhat unrealistic for most real-world dynamic processes and are not recommended, but the last parameterization is more useful for real-world processes.

As an example of the last parameterization described above, consider the process model where $\mathbf{C}_\eta = \sigma_\eta^2 \mathbf{I}$, and $\mathbf{M} = \text{diag}(\boldsymbol{\theta}_p)$. We can decompose the first-order conditional distributions in this case as

$$[\mathbf{Y}_t|\mathbf{Y}_{t-1},\boldsymbol{\theta}_p,\sigma_\eta^2] = \prod_{i=1}^{n}[Y_t(\mathbf{s}_i)|Y_{t-1}(\mathbf{s}_i),\theta_p(i),\sigma_\eta^2], \quad t = 1,2,\ldots.$$

Thus, conditional on the parameters $\boldsymbol{\theta}_p = (\theta_p(1),\ldots,\theta_p(n))'$, we have spatially independent univariate AR(1) processes at each spatial location (i.e., only the Y-value at the previous time at the same spatial location influences the transition). However, if $\boldsymbol{\theta}_p$ is *random* and has spatial dependence, then if we integrate it out, the marginal conditional distribution, $[\mathbf{Y}_t|\mathbf{Y}_{t-1},\sigma_\eta^2]$, can imply that all of the elements of $\mathbf{Y}_{t-1}$ influence the transition to time t at all spatial locations (i.e., this is a non-separable spatio-temporal process). Recall from Section 4.3 that this building of dependence through marginalization is a fundamental principle of deep hierarchical modeling, and it provides a simple and often effective way to construct complex spatio-temporal models (see also Technical Note 4.3). But, although we can accommodate fairly complex non-separable spatio-temporal dependence in this marginalization, it is important to note that the conditional model does not account *directly* for interactions across space and time. This limits its utility in forecasting applications, where more realistic conditional dynamic specifications are required. Thus, we often seek parameterizations that directly include such interactions in the conditional model.

Recall from our discussion of the intuition behind linear dynamics in Section 5.2.3 that the transition kernel is very important. This suggests that we can model realistic linear behavior by parameterizing the kernel shape (particularly its decay in the spatial domain and its asymmetry) in terms of a relatively small number of parameters (e.g., in the transition kernel case, the kernel width, or variance, and shift, or mean, parameters). More importantly, if we allow these relatively few parameters to vary with space in a principled fashion, then we can accommodate a variety of quite complex dynamic behaviors. The strength of the HM approach is that one can fairly easily do this by endowing these kernel parameters with spatial structure at the parameter-model level of the hierarchy (e.g., allowing them to be a function of covariates and/or specifying them as spatial random processes).

As an example, consider the IDE process model given in (5.9), where we specify a Gaussian-shape transition kernel as a function of x relative to the location s (for simplicity, in a one-dimensional spatial domain):

$$m(s,x;\boldsymbol{\theta}_p) = \theta_{p,1} \exp\left(-\frac{1}{\theta_{p,2}}(x - \theta_{p,3} - s)^2\right), \tag{5.12}$$

where the kernel amplitude is given by $\theta_{p,1}$, the length-scale (variance) parameter $\theta_{p,2}$ corresponds to a kernel scale (aperture) parameter (i.e., the kernel width increases as $\theta_{p,2}$ increases), and the mean (shift) parameter $\theta_{p,3}$ corresponds to a shift of the kernel relative to

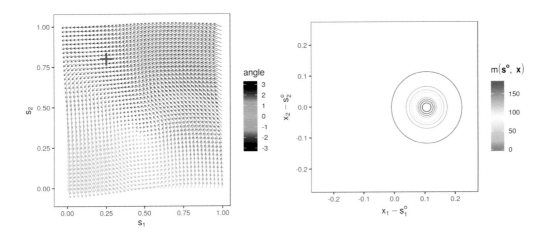

Figure 5.3: An example of a spatially varying kernel in an IDE spatio-temporal model. The left panel shows the direction (arrow orientation and color) and magnitude (arrow length) of flow induced by the kernel as a function of $\mathbf{x} = (x_1, x_2)'$ in two-dimensional space. The red cross indicates a specific location in space (s_1^o, s_2^o) around which the kernel is evaluated and plotted in the right panel. Note that the kernel, which is shifted to the right, induces a flow to the left.

location s. Notice that (5.12) is positive but need not integrate to 1 over x. Recall from Figure 5.2 the dynamical implication of changing the shift parameter. Specifically, if $\theta_{p,3}$ is positive (negative) it leads to leftward (rightward) movement because it induces asymmetry relative to location s. In addition, Figure 5.2 shows the dynamic implications when changing the kernel width/scale (e.g., wider kernels suggest faster decay). So, to obtain more complex dynamical behavior, we can allow these parameters to change with space. For example, suppose the mean (shift) parameter satisfies $\theta_{p,3}(s) = \mathbf{x}(s)'\boldsymbol{\beta} + \omega(s)$, where $\mathbf{x}(s)$ corresponds to covariates at spatial location s, $\boldsymbol{\beta}$ are the associated regression parameters, and $\omega(s)$ could correspond to a spatial Gaussian process (although, in some cases, it may be sufficient to omit the error term $\omega(s)$; see Lab 5.2 for an example). We can also allow the parameter $\theta_{p,2}$ to vary with space, but it is typically the case that $\theta_{p,3}$ is the more important of the two parameters. Figure 5.3 shows an example of a spatially varying kernel in two dimensions, and the kernel evaluated at one specific spatial location. Lab 5.1 implements the simple one-dimensional IDE process model and explores its simulation. Lab 5.2 shows how one can do spatio-temporal modeling and inference in R using the package **IDE**.

Although the IDE kernel representation suggests efficient parameterizations for linear dynamics in continuous space, there are many occasions where we seek efficient parameterizations in a discrete-space setting or in the context of random effects in basis-function expansions. In the case of the former, one of the most useful such parameterizations corresponds to transition operators that only consider local spatial neighborhoods. We describe these below and provide a mechanistically motivated example. We defer the discussion of dynamics for random effects in basis-function expansions to Section 5.3.2.

Lagged-Nearest-Neighbor Representations

The importance of the rate of decay and asymmetry in IDE transition-kernel representations suggests that for discrete space a very parsimonious, yet realistic, dynamic model can be specified in terms of a simple *lagged-nearest-neighbor* (LNN) parameterization, for example,

$$Y_t(\mathbf{s}_i) = \sum_{\mathbf{s}_j \in \mathcal{N}_i} m_{ij} Y_{t-1}(\mathbf{s}_j) + \eta_t(\mathbf{s}_i), \tag{5.13}$$

where $\mathcal{N}_i$ corresponds to a pre-specified neighborhood of the location $\mathbf{s}_i$ (including $\mathbf{s}_i$), for $i = 1, \ldots, n$, and where we specify $m_{ij} = 0$, for all $\mathbf{s}_j \notin \mathcal{N}_i$. Such a parameterization reduces the number of free parameters from the order of n^2 to the order of n. It is often reasonable to further parameterize the transition coefficients in (5.13) to account for decay (spread or diffusion) rate and asymmetry (propagation direction). In some cases, homogeneities of the transitions would result in a single parameter to control a particular type of neighbor (e.g., a parameter for the west neighbor and east neighbor transition coefficients), or, in other cases, it would be more appropriate to let these parameters vary in space as well (as with the IDE transition-kernel example above).

Motivation of an LNN with a Mechanistic Model

The LNN parameterization can be motivated by many mechanistic models, such as those suggested by standard discretization of integro-differential or partial differential equations (PDEs). In the latter case, the parameters m_{ij} in (5.13) can be parameterized in terms of other mechanistically motivated knowledge, such as spatially varying diffusion or advection coefficients. Again, in this framework the $\{m_{ij}\}$ are either estimated directly in an EHM or modeled at the next level of a BHM (typically, with some sort of spatial structure). As an example, consider the basic linear, non-random, advection–diffusion PDE,

$$\frac{\partial Y}{\partial t} = a \frac{\partial^2 Y}{\partial x^2} + b \frac{\partial^2 Y}{\partial y^2} + u \frac{\partial Y}{\partial x} + v \frac{\partial Y}{\partial y}, \tag{5.14}$$

conditional on a, b, u, and v, where a and b are diffusion coefficients that control the rate of spread, and u and v are advection parameters that account for the process "flow" (i.e., advection). Simple finite-difference discretization of such PDEs on a two-dimensional equally spaced finite grid (see Appendix D.1) can lead to LNN specifications of the form

$$\mathbf{Y}_t = \mathbf{M}(\boldsymbol{\theta}_p)\mathbf{Y}_{t-1} + \mathbf{M}_b(\boldsymbol{\theta}_p)\mathbf{Y}_{b,t} + \boldsymbol{\eta}_t,$$

where $\mathbf{Y}_t$ corresponds to a vectorization of the non-boundary grid points, with $\mathbf{M}(\boldsymbol{\theta}_p)$ a five-diagonal transition matrix with diagonals corresponding to functions of a, b, u, v and the discretization parameters (e.g., these five diagonals correspond to $\{\theta_{p,1}, \ldots, \theta_{p,5}\}$ in Appendix D.1). Such discretizations should account for boundary affects, and so we specify $\mathbf{Y}_{b,t}$ to be a boundary process (either fixed or assumed to be random) with $\mathbf{M}_b(\boldsymbol{\theta}_p)$ the associated transition operator based on the finite-difference discretization of the differential operator. The additive error process $\{\boldsymbol{\eta}_t\}$ is assumed to be Gaussian, mean-zero, and independent in time. In the more realistic case where the parameters a, b and/or u, v vary with space, the vector $\boldsymbol{\theta}_p$ varies with space as well, and we model it either in terms of covariates or as a spatial random process. The point is that we allow these mechanistic models to suggest or motivate LNN parameterizations rather than our specifying the structure directly. Appendix D presents detailed examples of DSTMs motivated by mechanistic models, and the case study in Appendix E presents an implementation of such a model for the Mediterranean winds data set described in Chapter 2.

5.3.2 Dimension Reduction in the Process Model

As discussed in Section 4.4, it is often the case that to reduce process dimensionality we could consider the spatio-temporal process of interest as a decomposition in terms of "fixed" effects and random effects in a basis-function expansion. This is particularly helpful for DSTM process models, as it is often the case that the important dynamics exist on a fairly low-dimensional space (i.e., manifold). Consider an extension to the spatial basis-function mixed-effects model (4.29) from Section 4.4.2,

$$Y_t(\mathbf{s}) = \mathbf{x}_t(\mathbf{s})'\boldsymbol{\beta} + \sum_{i=1}^{n_\alpha} \phi_i(\mathbf{s})\alpha_{i,t} + \sum_{j=1}^{n_\xi} \psi_j(\mathbf{s})\xi_{j,t} + \nu_t(\mathbf{s}), \tag{5.15}$$

where the term with covariates, $\mathbf{x}_t(\mathbf{s})'\boldsymbol{\beta}$, might be interpreted as a "fixed" or "deterministic" component with fixed effects $\boldsymbol{\beta}$; the first basis-expansion term, $\sum_{i=1}^{n_\alpha} \phi_i(\mathbf{s})\alpha_{i,t}$, contains known spatial basis functions $\{\phi_i(\cdot)\}$ and associated dynamically evolving random coefficients (i.e., random effects), $\{\alpha_{i,t}\}$; the residual basis-expansion term, $\sum_{j=1}^{n_\xi} \psi_j(\mathbf{s})\xi_{j,t}$, can account for non-dynamic spatio-temporal structure, where the basis functions, $\{\psi_j(\cdot)\}$, are again assumed known, and the random effects $\{\xi_{j,t}\}$ are typically non-dynamic or at least contain simple temporal behavior. The micro-scale term, $\nu_t(\cdot)$, is assumed to be a Gaussian process with mean zero and independent in time. The focus here is on the dynamically evolving random effects, $\{\alpha_{i,t}\}$.

As mentioned above, useful reductions in process dimension can be formulated with the understanding that the essential dynamics for spatio-temporal processes typically exist in a fairly low-dimensional space. This is helpful because, instead of having to model the evolution of, say, the n-dimensional vector $\mathbf{Y}_t$, one can model the evolution of a much lower-dimensional (of dimension n_α) process $\{\boldsymbol{\alpha}_t\}$, where $n_\alpha \ll n$. It is helpful to consider the vector form of (5.15):

$$\mathbf{Y}_t = \mathbf{X}_t\boldsymbol{\beta} + \boldsymbol{\Phi}\boldsymbol{\alpha}_t + \boldsymbol{\Psi}\boldsymbol{\xi}_t + \boldsymbol{\nu}_t, \tag{5.16}$$

where $\mathbf{X}_t$ is an $n \times (p + 1)$ matrix that could be time-varying and can be interpreted as a spatial offset corresponding to large-scale non-dynamical features and/or covariate effects, $\boldsymbol{\Phi}$ is an $n \times n_\alpha$ matrix of basis vectors corresponding to the latent dynamic coefficient process, $\{\boldsymbol{\alpha}_t\}$, and $\boldsymbol{\Psi}$ is an $n \times n_\xi$ matrix of basis vectors corresponding to the latent coefficient process, $\{\boldsymbol{\xi}_t\}$. Typically, $\{\boldsymbol{\xi}_t\}$ is assumed to have different dynamic characteristics than $\{\boldsymbol{\alpha}_t\}$, or this component might account for non-dynamic spatial variability. The error process $\{\boldsymbol{\nu}_t\}$ is Gaussian and assumed to have mean zero with relatively simple temporal dependence structure (usually independence).

The evolution of the latent process $\{\boldsymbol{\alpha}_t\}$ can proceed according to the linear equations involving a transition matrix, discussed earlier. For example, one could specify a first-order vector autoregressive model (VAR(1)),

$$\boldsymbol{\alpha}_t = \mathbf{M}_\alpha\boldsymbol{\alpha}_{t-1} + \boldsymbol{\eta}_t, \tag{5.17}$$

where $\mathbf{M}_\alpha$ is the $n_\alpha \times n_\alpha$ transition matrix, and $\boldsymbol{\eta}_t \sim Gau(\mathbf{0}, \mathbf{C}_\eta)$ (which are assumed to be independent of $\boldsymbol{\alpha}_{t-1}$ and independent in time). The matrices $\mathbf{M}_\alpha$ and $\mathbf{C}_\eta$ in (5.17) are often relatively simple in structure, depending on the nature of the real-world process and the type of basis functions considered. However, even in this low-dimensional context ($n_\alpha \ll n$), in many cases parameter-space reduction may still be necessary. One could consider the simple structures that were discussed in the context of linear DSTM process models (e.g., random walks, independent AR models, nearest-neighbor models). Typically, it is important, for the reasons discussed in Section 5.2.3 and Technical Note 5.2, that the transition operator be non-normal (i.e., $\mathbf{M}_\alpha'\mathbf{M}_\alpha \neq \mathbf{M}_\alpha\mathbf{M}_\alpha'$), so one should consider non-diagonal transition matrices in most cases. Also, the notion of "neighbors" is not always well defined in these formulations. If the basis functions given in $\boldsymbol{\Phi}$ are such that the elements of $\boldsymbol{\alpha}_t$ are not spatially indexed (e.g., in the case of global basis functions such as some types of splines, Fourier, EOFs, etc.), then a neighbor cannot be based on physical space (but perhaps it can be based on other characteristics, such as spatial scale). It is important to note that mechanistic knowledge can also be used in this case to motivate parameterizations for $\mathbf{M}_\alpha$. We illustrate a couple of such cases, one for a "spectral" representation of a PDE in Appendix D.2, and one for an IDE process in Appendix D.3. Lab 5.3 provides an example in which $\mathbf{M}_\alpha$ and $\mathbf{C}_\eta$ are estimated by the method of moments and by an EM algorithm (see Appendix C for more details about these algorithms).

> **R tip:** If one is able to write down the data model as $\mathbf{Z}_t = \mathbf{H}_t\mathbf{Y}_t + \boldsymbol{\varepsilon}_t$, $\boldsymbol{\varepsilon}_t \sim Gau(\mathbf{0}, \mathbf{C}_{\epsilon,t})$, and the process model as $\mathbf{Y}_t = \mathbf{M}\mathbf{Y}_{t-1} + \boldsymbol{\eta}_t$, $\boldsymbol{\eta}_t \sim Gau(\mathbf{0}, \mathbf{C}_\eta)$, where the $\{\mathbf{H}_t\}$ are known, then the problem of predicting $\{\boldsymbol{\alpha}_t\}$ and the estimation

of all the other parameters is the well-known dual state-parameter estimation problem for state-space models (see Appendix C). Several R packages are available for this, such as **KFAS**, **MARSS**, and **Stem**. Software for DSTMs is, however, less developed than that for descriptive models, and estimation/prediction with complex linear DSTMs and nonlinear DSTMs is likely to require customized R code.

Basis Functions

In the mechanistically motivated PDE and IDE cases presented in Appendices D.2 and D.3, the natural choice for basis functions are the Fourier modes (i.e., sines and cosines). This is typically not the case for DSTM process models. Indeed, there are many choices for the basis functions that could be used to define $\boldsymbol{\Phi}$ and $\boldsymbol{\Psi}$ in (5.16) (see, for example, Figure 4.7). In the context of DSTMs, it is usually important to specify basis functions such that interactions across spatial scales are allowed to accommodate transient growth. This can be more difficult to do in "knot-based" representations (e.g., splines, kernel convolutions, predictive processes), where the coefficients $\boldsymbol{\alpha}_t$ of $\boldsymbol{\Phi}$ are spatially referenced but not necessarily multi-resolutional. Most other basis-function representations are in some sense multi-scale, and the associated expansion coefficients $\{\boldsymbol{\alpha}_t\}$ are not indexed in space. In this case, the dynamical evolution in the DSTM can easily accommodate scale interactions. The example in Lab 5.3 uses EOFs as the basis functions in such a decomposition of SSTs. Recall that the coefficients $\{\boldsymbol{\xi}_t\}$ associated with the matrix $\boldsymbol{\Psi}$ are typically specified to have much simpler dynamic structure (if at all), since the controlling dynamics are assumed to be associated principally with $\{\boldsymbol{\alpha}_t\}$. Thus, one has more freedom in the choice of basis functions that define $\boldsymbol{\Psi}$.

5.4　Nonlinear DSTMs

The linear Gaussian DSTMs described in Sections 5.2 and 5.3 are widely used, but the state of the art for more complicated models is rapidly advancing. The purpose of this section is not to give a complete overview of these more advanced models but just a brief perspective on nonlinear DSTMs *without* the implementation details.

Many mechanistic processes are best modeled nonlinearly, at least at some spatial and temporal scales of variability. We might write this as a nonlinear spatio-temporal AR(1) process (of course, higher-order lags could be considered as well):

$$Y_t(\cdot) = \mathcal{M}(Y_{t-1}(\cdot), \eta_t(\cdot); \boldsymbol{\theta}_p), \quad t = 1, 2, \ldots, \tag{5.18}$$

where $\mathcal{M}$ is a nonlinear function that models the process transition from time $t-1$ to t, $\eta_t(\cdot)$ is an error process, and $\boldsymbol{\theta}_p$ are parameters. Unfortunately, although there is one basic linear model, there are an infinite number of nonlinear statistical models that could be considered. One could either take a nonparametric view of the problem and essentially learn the dynamics from the data, or one could propose specific model classes that can accommodate the type of behavior desired. In this section we briefly describe examples of these approaches to accommodate nonlinear spatio-temporal dynamics.

State-Dependent Models

The general nonlinear model (5.18) can be simplified by considering a *state-dependent model* (the terminology comes from the time-series literature, where these models were first developed), in which the transition matrix depends on the process (state) value at each time. For example, in the discrete spatial case, we can write

$$\mathbf{Y}_t = \mathbf{M}(\mathbf{Y}_{t-1}; \boldsymbol{\theta}_p)\, \mathbf{Y}_{t-1} + \boldsymbol{\eta}_t, \tag{5.19}$$

where the transition operator depends on $\mathbf{Y}_{t-1}$ and parameters $\boldsymbol{\theta}_p$ (which, more generally, may also vary with time and/or space). Models such as (5.19) are still too general for spatio-temporal applications and must be further specified. One type of state-dependent model is the *threshold vector autoregressive model*, given by

$$\mathbf{Y}_t = \begin{cases} \mathbf{M}_1 \mathbf{Y}_{t-1} + \boldsymbol{\eta}_{1,t}, & \text{if } f(\omega_t) \in d_1, \\ \vdots & \vdots \\ \mathbf{M}_K \mathbf{Y}_{t-1} + \boldsymbol{\eta}_{K,t}, & \text{if } f(\omega_t) \in d_K, \end{cases} \tag{5.20}$$

where $f(\omega_t)$ is a function of a time-varying parameter ω_t that can itself be a function of the process, $\mathbf{Y}_{t-1}$, in which case it is a state-dependent model. We implicitly assume that conditions on the right-hand side of (5.20) are mutually exclusive; that is, $d_1, \ldots, d_K$ are disjoint. A simpler threshold model results if the parameters $\{\omega_t\}$ do not depend on the process. Of course, the transition matrices $\{\mathbf{M}_1, \ldots, \mathbf{M}_K\}$ and error covariance matrices $\{\mathbf{C}_{\eta_1}, \ldots, \mathbf{C}_{\eta_K}\}$ depend on unknown parameters, and the big challenge in DSTM modeling is to reduce the dimensionality of this parameter space to facilitate estimation. Some of the approaches discussed above for the linear DSTM process model can also be applied in this setting.

> **R tip:** Threshold vector autoregressive time-series models can be implemented with the `TVAR` command in the package **tsDyn**.

General Quadratic Nonlinearity

A very large number of real-world processes in the physical and biological sciences exhibit quadratic interactions. For example, consider the following one-dimensional reaction–

diffusion PDE:

$$\frac{\partial Y}{\partial t} = \frac{\partial}{\partial x}\left(\delta\frac{\partial Y}{\partial x}\right) + Y\exp\left(\gamma_0\left(1 - \frac{Y}{\gamma_1}\right)\right), \tag{5.21}$$

where the first term corresponds to a diffusion (spread) term that depends on a parameter δ, and the second term corresponds to a density-dependent (Ricker) growth term with growth parameter γ_0 and carrying capacity parameter γ_1. More generally, each of these parameters could vary with space and/or time. Notice that the diffusion term is linear in Y but the density-dependent growth term is nonlinear in that it is a function of Y multiplied by a nonlinear transformation of Y. This can be considered a general case of a quadratic interaction.

A fairly general class of nonlinear statistical DSTM process models can be specified to accommodate such behavior. In discrete space and time, such a *general quadratic nonlinear* (GQN) DSTM can be written, for $i = 1, \ldots, n$, as

$$Y_t(\mathbf{s}_i) = \sum_{j=1}^{n} m_{ij}Y_{t-1}(\mathbf{s}_j) + \sum_{k=1}^{n}\sum_{\ell=1}^{n} b_{i,k\ell}\, g(Y_{t-1}(\mathbf{s}_\ell); \boldsymbol{\theta}_g)\, Y_{t-1}(\mathbf{s}_k) + \eta_t(\mathbf{s}_i), \tag{5.22}$$

where m_{ij} are the linear-transition coefficients seen previously, and the quadratic-interaction transition coefficients are denoted by $b_{i,k\ell}$. Importantly, a transformation of one of the components of the quadratic interaction is included through the function $g(\cdot)$, which can depend on parameters $\boldsymbol{\theta}_g$. This function $g(\cdot)$ is responsible for the term "general" in GQN, and such transformations are important for many processes such as density-dependent growth that one may see in an epidemic or invasive-species population processes (see, for example, (5.21) above), and they can keep forecasts from "blowing up" in time. The spatio-temporal error process $\{\eta_t(\cdot)\}$ is again typically assumed to be independent in time and Gaussian with mean zero and a spatial covariance matrix. Note that the conditional GQN model for $Y_t(\cdot)$ conditioned on $Y_{t-1}(\cdot)$ is Gaussian, but the marginal model for $Y_t(\cdot)$ will not in general be Gaussian because of the nonlinear interactions. The GQN model (5.22) can be shown to be a special case of the state-dependent model in (5.19).

There are multiple challenges when implementing models such as (5.22). Chief among these is the curse of dimensionality. There are $O(n^3)$ parameters and, unless one has an enormous number of time replicates ($T \gg n$), inference on them is problematic without some sort of regularization (shrinkage) and/or the incorporation of prior information. In addition to parameter estimation, depending on the specification of $g(\cdot)$ (which can act to control the growth of the process), these models can be explosive when used to forecast multiple time steps into the future. GQN models have been implemented on an application-specific basis in BHMs (see Chapter 7 of Cressie and Wikle, 2011, for more discussion).

Some Other Nonlinear Models

There are currently several promising approaches for nonlinear spatio-temporal modeling in addition to those mentioned above. For example, there are a wide variety of methods being developed in machine learning to predict and/or classify high volumes of dependent data, including spatio-temporal data (e.g., sequences of images). These methods often relate to variants of neural networks (e.g., convolutional and recurrent neural networks (RNNs)), and they have revolutionized many application areas such as image classification and natural-language processing. In their original formulations, these methods do not typically address uncertainty quantification. However, there is increasing interest in considering such models within broader uncertainty-based paradigms. As mentioned in Chapter 1, there is a connection between deep hierarchical statistical models (BHMs) and many of these so-called "deep learning" algorithms.

For example, the GQN model described above is flexible, interpretable, and can accommodate many different types of dynamic processes and uncertainty quantification strategies. Similarly, the typical RNN model is also flexible and can accommodate a wide variety of spatio-temporal dependence structures. However, both the GQN and RNN models can be difficult to implement computationally due to the high dimensionality of the hidden states and parameters, and it typically requires sophisticated regularization (and/or a large amount of data or prior information) to make them work. A computationally efficient alternative is the so-called *echo state network* (ESN) methodology that was developed as an alternative to RNNs in the engineering literature (for overviews, see Lukoševičius and Jaeger, 2009; Lukoševičius, 2012). Importantly, ESNs consider sparsely connected hidden layers that allow for sequential interactions yet assume most of the parameters ("weights") are randomly generated and then fixed, with the only parameters estimated being those that connect the hidden layer to the response. This induces a substantially more parsimonious structure in the model. Yet, these models traditionally do not explicitly include quadratic interactions or formal uncertainty quantification. McDermott and Wikle (2017) consider a quadratic spatio-temporal ESN model they call a quadratic ESN (QESN) and implement it in a bootstrap context to account for parameter uncertainty. Details concerning the QESN are given in Appendix F, and the associated case study provides an example of how to use an ensemble of QESNs to generate a long-lead forecast of the SST data.

Another type of nonlinear spatio-temporal model that is increasingly being considered in statistical applications is the *agent-based model* (or, in some literatures, the *individual-based model*). In this case, the process is built from local individual-scale interactions by means of fairly simple rules that lead to complex nonlinear behavior. Although these models are parsimonious in that they have relatively few parameters, they can be quite computationally expensive, and parameter inference can be challenging (although approximate likelihood methods and BHMs have shown recent promise). For examples, see Cressie and Wikle (2011, Section 7.3.4) and Wikle and Hooten (2016).

There is yet another parsimonious approach to nonlinear spatio-temporal modeling that is somewhat science-based and relies on so-called "analogs." Sometimes referred to as a "mechanism-free" approach, in its most basic form, analog forecasting seeks to find historical sequences of maps (analogs) that match a similar sequence culminating at the current time. Then it assumes that the forecast made at the current time will be what actually occurred with the best analog matches. (This is somewhat like the so-called "hot-deck imputation" in statistics.) Analog forecasting can be shown to be a type of spatio-temporal nearest-neighbor-regression methodology. There are many modifications to this procedure related to various conditions as to what "best" means when comparing analogs to the current state, distance metrics, how many analogs to use, and so forth. Traditionally, these methods have not been part of statistical methodology, and so uncertainty quantification and parameter estimation are not generally considered from a formal probabilistic perspective. Recent implementations have sought to consider uncertainty quantification and formal inference, including prediction, within a Bayesian inferential framework (McDermott and Wikle, 2016; McDermott et al., 2018).

5.5   Chapter 5 Wrap-Up

Recall that one of the big challenges with the descriptive spatio-temporal models described in Chapter 4 was the specification of realistic covariance structures. We showed that building such structures through conditioning on random effects could be quite useful. The present chapter considered spatio-temporal models from a conditional-in-time (dynamic) perspective that respected the belief that most real-world spatio-temporal processes are best described as spatial processes that evolve through time. Like the random-effects models of Chapter 4, this perspective relied very much on conditional-probability models. First, there was a strong assumption (which is also present in the descriptive models of Chapter 4) that the data, when conditioned on the true spatio-temporal process of interest, could be considered independent in time (and, typically, have fairly simple error structure as well). Second, a Markov assumption in time was made, so that the joint distribution of the process could be decomposed as the product of low-order Markov (in time) conditional-probability distributions. These conditional distributions corresponded to dynamic models that describe the transition of the spatial process from the previous time(s) to the current time. This dynamic model was further conditioned on parameters that control the transition and the innovation-error structure. We showed that the models can often benefit from these parameters being random processes (and/or dependent on covariates) as well.

We presented the most commonly used DSTMs with data models that have additive Gaussian error and process models that have linear transition structure with additive Gaussian error. In the simplest case, where time is discrete and interest is in a finite set of spatial locations, we showed that these models are essentially multivariate state-space time series models, and many of the sequential prediction and estimation algorithms from that literature (e.g., filters, smoothers plus estimation through EM, and Bayesian algorithms) can then be used. We also discussed that non-Gaussian data models are fairly easily accommodated if one can obtain conditional independence when conditioning on a latent Gaussian process model (e.g., a data model obtained from a generalized linear model). Additional details on such estimation methods can be found in Shumway and Stoffer (1982, 2006), Gamerman and Lopes (2006), Prado and West (2010), Cressie and Wikle (2011), and Douc et al. (2014).

We emphasized that the biggest challenge with these models is accommodating high dimensionality (either in data volume, number of prediction locations, or number of parameters to be estimated). Thus, one of the fundamental differences between DSTMs and multivariate time series models is that DSTMs require scalable parameterization of the evolution model. We showed that this modeling can be facilitated greatly by understanding some of the fundamental properties of linear dynamical systems and using this mechanistic knowledge to parameterize transition functions/matrices. Additional details on the mechanistic motivation for DSTMs can be found in Cressie and Wikle (2011).

We discussed that nonlinear DSTMs are an increasingly important area of spatio-temporal modeling. It is important that statistical models for such processes include realistic structural components (e.g., quadratic interactions) and account formally for uncertainty quantification. We mentioned that a significant challenge with these models in both the statistics and machine learning literature is to mitigate the curse of dimensionality in the parameter space (see, for example, Cressie and Wikle, 2011; Goodfellow et al., 2016). This often requires mechanistic-based parameterizations, informative prior distributions, and/or regularization approaches. This has led to increased interest in very parsimonious representations for nonlinear DSTMs, such as echo state networks, agent-based models, and analog models.

In general, DSTMs require many assumptions in order to build conditional models at each level of the hierarchy. They can also be difficult to implement in some cases due to complex dependence and deep levels, often requiring fully Bayesian implementations. This also makes it necessary to validate these assumptions carefully through model diagnostics and evaluation of their predictions. Some approaches to spatio-temporal model evaluation are discussed in Chapter 6.

Lab 5.1: Implementing an IDE Model in One-Dimensional Space

In this Lab we take a look at how one can implement a stochastic integro-difference equation (IDE) in one-dimensional space and time, from first principles. Specifically, we shall consider the dynamic model,

$$Y_t(s) = \int_{D_s} m(s, x; \boldsymbol{\theta}_p) Y_{t-1}(x) \mathrm{d}x + \eta_t(s), \quad s, x \in D_s,$$

where $Y_t(\cdot)$ is the spatio-temporal process at time t; $\boldsymbol{\theta}_p$ are parameters that we fix (in practice, these will be estimated from data; see Lab 5.2); and $\eta_t(\cdot)$ is a spatial process, independent of $Y_t(\cdot)$, with covariance function that we shall assume is known.

We only need the packages **dplyr**, **ggplot2**, and **STRbook** for this lab and, for reproducibility purposes, we fix the seed.

```
library("dplyr")
library("ggplot2")
library("STRbook")
set.seed(1)
```

Constructing the Process Grid and Kernel

We start off by constructing a discretization of the one-dimensional spatial domain $D_s = [0, 1]$. We shall use this discretization, containing cells of width Δ_s, for both approximate integrations as well as visualizations. We call this our spatial grid.

```
ds <- 0.01
s_grid <- seq(0, 1, by = ds)
N <- length(s_grid)
```

Our space-time grid is formed by calling **expand.grid** with s_grid and our temporal domain, which we define as the set of integers spanning 0 up to $T = 200$.

```
nT <- 201
t_grid <- 0:(nT-1)
st_grid <- expand.grid(s = s_grid, t = t_grid)
```

The transition kernel $m(s, x; \boldsymbol{\theta}_p)$ is a bivariate function on our spatial grid. It is defined below to be a Gaussian kernel, where the entries of $\boldsymbol{\theta}_p = (\theta_{p,1}, \theta_{p,2}, \theta_{p,3})'$ are the amplitude, the scale (aperture, twice the variance), and the shift (offset) of the kernel, respectively. Specifically,

$$m(s, x; \boldsymbol{\theta}_p) \equiv \theta_{p,1} \exp\left(-\frac{1}{\theta_{p,2}}(x - \theta_{p,3} - s)^2\right),$$

which can be implemented as an R function as follows.

```
m <- function(s, x, thetap) {
  gamma <- thetap[1]                    # amplitude
  l <- thetap[2]                        # length scale
  offset <- thetap[3]                   # offset
  D <- outer(s + offset, x, '-')        # displacements
  gamma * exp(-D^2/l)                   # kernel eval.
}
```

Note the use of the function **outer** with the subtraction operator. This function performs an "outer operation" (a generalization of the outer product) by computing an operation between every two elements of the first two arguments, in this case a subtraction.

We can now visualize some kernels by seeing how the process at $s = 0.5$ depends on x. Four such kernels are constructed below: the first is narrow and centered on 0.5; the second is slightly wider; the third is shifted to the right; and the fourth is shifted to the left. We store the parameters of the four different kernels in a list `thetap`.

```
thetap <- list()
thetap[[1]] <- c(40, 0.0002, 0)
thetap[[2]] <- c(5.75, 0.01, 0)
thetap[[3]] <- c(8, 0.005, 0.1)
thetap[[4]] <- c(8, 0.005, -0.1)
```

Plotting proceeds by first evaluating the kernel for all x at $s = 0.5$, and then plotting these evaluations against x. The first kernel is plotted below; plotting the other three is left as an exercise for the reader. The kernels are shown in the top panels of Figure 5.2.

```
m_x_0.5 <- m(s = 0.5, x = s_grid,          # construct kernel
             thetap = thetap[[1]]) %>%      # at s = 0.5
        as.numeric()                        # convert to numeric
df <- data.frame(x = s_grid, m = m_x_0.5)     # allocate to df
ggplot(df) + geom_line(aes(x, m)) + theme_bw() # plot
```

The last term we need to define is $\eta_t(\cdot)$. Here, we define it as a spatial process with an exponential covariance function with range parameter 0.1 and variance 0.1. The covariance matrix at each time point is then

```
Sigma_eta <- 0.1 * exp( -abs(outer(s_grid, s_grid, '-') / 0.1))
```

Simulating $\eta_t(s)$ over `s_grid` proceeds by generating a multivariate Gaussian vector with mean zero and covariance matrix `Sigma_eta`. To do this, one can use the function **mvrnorm** from the package **MASS**. Alternatively, one may use the lower Cholesky factor of `Sigma_eta` and multiply this by a vector of numbers generated from a mean-zero, variance-one, independent-elements Gaussian random vector (see Rue and Held, 2005, Algorithm 2.3).

```
L <- t(chol(Sigma_eta))   # chol() returns upper Cholesky factor
sim <- L %*% rnorm(nrow(Sigma_eta))   # simulate
```

Type `plot(s_grid, sim, 'l')` to plot this realization of $\eta_t(s)$ over `s_grid`.

Simulating from the IDE

Now we have everything in place to simulate from the IDE. Simulation is most easily carried out using a `for` loop as shown below. We shall carry out four simulations, one for each kernel constructed above, and store the simulations in a list of four data frames, one for each simulation. The following command initializes this list.

```
Y <- list()
```

For each simulation setting (which we iterate using the index `i`), we simulate the time points (which we iterate using `j`) to obtain the process. The "nested `for` loop" below accomplishes this. In the outer loop, the kernel is constructed and the process is initialized to zero. In the inner loop, the integration is approximated using a Riemann sum,

$$\int_{D_s} m(s, x; \boldsymbol{\theta}_p) Y_{t-1}(x) \mathrm{d}x \approx \sum_i m(s, x_i; \boldsymbol{\theta}_p) Y_{t-1}(x_i) \Delta_s,$$

where we recall that we have set $\Delta_s = 0.01$. Next, at every time point $\eta_t(s)$ is simulated on the grid and added to the sum (an approximation of the integral) above.

```
for(i in 1:4) {                          # for each kernel
  M <- m(s_grid, s_grid, thetap[[i]])    # construct kernel
  Y[[i]] <- data.frame(s = s_grid,       # init. data frame with s
                       t = 0,            # init. time point 0, and
                       Y = 0)            # init. proc. value = 0
  for(j in t_grid[-1]) {                 # for each time point
    prev_Y <- filter(Y[[i]],             # get Y at t - 1
                     t == j - 1)$Y
    eta <- L %*% rnorm(N)                # simulate eta
    new_Y <- (M %*% prev_Y * ds + eta) %>%
             as.numeric()               # Euler approximation

    Y[[i]] <- rbind(Y[[i]],              # update data frame
                    data.frame(s = s_grid,
                               t = j,
                               Y =  new_Y))
  }
}
```

Repeatedly appending data frames, as is done above, is computationally inefficient. For large systems it would be quicker to save a data frame for each time point in another list and then concatenate using `rbindlist` from the package **data.table**.

Since now `Y[[i]]`, for $i = 1, \ldots, 4$, contains a data frame in long format, it is straightforward to visualize. The code given below constructs the Hovmöller plot for the IDE process for $i = 1$. Plotting for $i = 2, 3, 4$ is left as an exercise for the reader. The resulting plots are shown in the bottom panels of Figure 5.2.

```
ggplot(Y[[1]]) + geom_tile(aes(s, t, fill = Y)) +
    scale_y_reverse() + theme_bw() +
    fill_scale(name = "Y")
```

Simulating Observations

Now assume that we want to simulate noisy observations from one of the process models that we have just simulated from. Why would we want to do this? Frequently, the only way to test whether algorithms for inference are working as they should is to mimic both the underlying true process *and* the measurement process. Working with simulated data is the first step in developing reliable algorithms that are then ready to be applied to real data.

To map the observations to the data we need an incidence matrix that picks out the process value that has been observed. This incidence matrix is simply composed of several rows, one for each observation, with zeros everywhere except for the entry corresponding to the process value that has been observed (recall Section 5.2.1). When the locations we are observing change over time, the incidence matrix correspondingly changes over time.

Suppose that at each time point we observe the process at 50 locations which, for convenience, are a subset of s_grid. (If this is not the case, some nearest-neighbor mapping or deterministic interpolation method can be used.)

```
nobs <- 50
sobs <- sample(s_grid, nobs)
```

Then the incidence matrix at time t, $\mathbf{H}_t$, can be constructed by matching the observation locations on the space-time grid using the function `which`.

```
Ht <- matrix(0, nobs, N)          # construct empty matrix
for(i in 1:nobs) {                # for each obs
  idx <- which(sobs[i] == s_grid) # find the element to set to 1
  Ht[i, idx] <- 1                 # set to 1
}
```

Note that `Ht` is sparse (contains many zeros), so sparse-matrix representations can be used to improve computational and memory efficiency; look up the packages **Matrix** or **spam** for more information on these representations.

We can repeat this procedure for every time point to simulate our data. At time t, the data are given by $\mathbf{Z}_t = \mathbf{H}_t \mathbf{Y}_t + \boldsymbol{\varepsilon}_t$, where $\mathbf{Y}_t$ is the latent process on the grid at time t, and $\boldsymbol{\varepsilon}_t$ is independent of $\mathbf{Y}_t$ and represents a Gaussian random vector whose entries are *iid* with mean zero and variance σ_ϵ^2. Assume $\sigma_\epsilon^2 = 1$ and that $\mathbf{H}_t$ is the same for each t. Then observations are simulated using the following `for` loop.

```
z_df <- data.frame()              # init data frame
for(j in 0:(nT-1)) {              # for each time point
  Yt <- filter(Y[[1]], t == j)$Y  # get the simulated process
  zt <- Ht %*% Yt + rnorm(nobs)   # map to obs and add noise
  z_df <- rbind(z_df,             # update data frame
            data.frame(s = sobs, t = j, z = zt))
}
```

Plotting of the simulated observations proceeds using **ggplot2** as follows.

```
ggplot(z_df) + geom_point(aes(s, t, colour = z))   +
  col_scale(name = "z") + scale_y_reverse() + theme_bw()
```

Note that the observations are noisy and reveal sizeable gaps. Filling in these gaps by first estimating all the parameters in the IDE from the data and then predicting at unobserved locations is the subject of Lab 5.2.

Lab 5.2: Spatio-Temporal Inference using the IDE Model

In this Lab we use the package **IDE** to fit spatio-temporal IDE models as well as predict and forecast from spatio-temporal data. We explore three cases. The first two cases consider simulated data where the true model is known, and the third considers the Sydney radar data set described in Chapter 2.

For this Lab, we need the package **IDE** and also the package **FRK**, which will be used to construct basis functions to model the spatially varying parameters of the kernel. In addition, we shall use the package **plyr** for binding data frames with unequal column number later on in the Lab.

```
library("plyr")
library("dplyr")
library("IDE")
library("FRK")
library("ggplot2")
library("sp")
library("spacetime")
library("STRbook")
```

The kernel $m(\mathbf{s}, \mathbf{x}; \boldsymbol{\theta}_p)$ used by the package **IDE** is given by

$$m(\mathbf{s}, \mathbf{x}; \boldsymbol{\theta}_p) = \theta_{p,1}(\mathbf{s}) \exp\left(-\frac{1}{\theta_{p,2}(\mathbf{s})}\left[(x_1 - \theta_{p,3}(\mathbf{s}) - s_1)^2 + (x_2 - \theta_{p,4}(\mathbf{s}) - s_2)^2\right]\right),$$
(5.23)

where $\theta_{p,1}(\mathbf{s})$ is the spatially varying amplitude, $\theta_{p,2}(\mathbf{s})$ is the spatially varying kernel aperture (or width), and the mean (shift) parameters $\theta_{p,3}(\mathbf{s})$ and $\theta_{p,4}(\mathbf{s})$ correspond to a spatially varying shift of the kernel relative to location $\mathbf{s}$. Spatially invariant kernels (i.e., where the elements of $\boldsymbol{\theta}_p$ are not functions of space) are also allowed.

The package **IDE** uses a bisquare spatial-basis-function decomposition for both the process $Y_t(\cdot)$ and the spatial process $\eta_t(\cdot)$, $t = 1, 2, \ldots$. The covariance matrix of the basis-function coefficients associated with $\eta_t(\cdot)$ is assumed to be proportional to the identity matrix, where the constant of proportionality is estimated. In **IDE**, the latent process $\widetilde{Y}_t(\mathbf{s})$

is the IDE dynamical process superimposed on some fixed effects, which can be expressed as a linear combination of known covariates $\mathbf{x}_t(\mathbf{s})$,

$$\widetilde{Y}_t(\mathbf{s}) = \mathbf{x}_t(\mathbf{s})'\boldsymbol{\beta} + Y_t(\mathbf{s}); \quad \mathbf{s} \in D_s, \tag{5.24}$$

for $t = 1, 2, \ldots,$ where $\boldsymbol{\beta}$ are regression coefficients. The data vector $\mathbf{Z}_t \equiv (Z_t(\mathbf{r}_{1t}), \ldots, Z_t(\mathbf{r}_{m_t t}))'$ is then the latent process observed with noise,

$$Z_t(\mathbf{r}_{jt}) = \widetilde{Y}_t(\mathbf{r}_{jt}) + \epsilon_t(\mathbf{r}_{jt}), \quad j = 1, \ldots, m_t,$$

for $t = 1, 2, \ldots,$ where $\epsilon_t(\mathbf{r}_{jt}) \sim iid\, Gau(0, \sigma_\epsilon^2)$.

Simulation Example with a Spatially Invariant Kernel

The package **IDE** contains a function `simIDE` that simulates the behavior of a typical dynamic system governed by linear transport. The function can simulate from a user-defined IDE model, or from a pre-defined one. In the latter case, the number of time points to simulate (`T`), the number of (spatially fixed) observations to use (`nobs`), and a flag indicating whether to use a spatially invariant kernel (`k_spat_invariant = 1`) or not (`k_spat_invariant = 0`), need to be provided. The pre-defined model includes a linear trend in s_1 and s_2.

```
SIM1 <- simIDE(T = 10, nobs = 100, k_spat_invariant = 1)
```

The returned list `SIM1` contains the simulated process in the data frame `s_df`, the observed data in the data frame `z_df`, and the observed data as an `STIDF` object `z_STIDF`. It also contains two **ggplot2** plots, `g_truth` and `g_obs`, which can be readily plotted as follows.

```
print(SIM1$g_truth)
print(SIM1$g_obs)
```

While the action of the transport is clearly noticeable in the evolution of the process, there is also a clear spatial trend. Covariates are included through the use of a standard R formula when calling the function `IDE`. Additional arguments to `IDE` include the data set, which needs to be of class `STIDF`, the temporal discretization to use (we will use 1 day) of class `difftime`, and the resolution of the grid upon which the integrations (as well as predictions) will be carried out. Other arguments include user-specified basis functions for the process and what transition kernel will be used, which for now we do not specify. By default, the IDE model will decompose the process using two resolutions of bisquare basis functions and will assume a spatially invariant Gaussian transition kernel.

```
IDEmodel <- IDE(f = z ~ s1 + s2,
                data = SIM1$z_STIDF,
                dt = as.difftime(1, units = "days"),
                grid_size = 41)
```

The returned object `IDEmodel` is of class `IDE` and contains initial parameter estimates, as well as predictions of α_t, for $t = 1, \ldots, T$, at these initial parameter estimates. The parameters in this case are the measurement-error variance, the variance of the random disturbance $\eta_t(\cdot)$ (whose covariance structure is fixed), the kernel parameters, and the regression coefficients β.

Estimating the parameters in the IDE model using maximum likelihood is a computationally intensive procedure. The default method currently implemented uses a differential evolution optimization algorithm from the package **DEoptim**, which is a global optimization algorithm that can be easily parallelized. Fitting takes only a few minutes on a 60-core high-performance computer, but can take an hour or two on a standard desktop computer. Fitting can be done by running the following code.

```
fit_results_sim1 <- fit.IDE(IDEmodel,
                            parallelType = 1)
```

Here `parallelType = 1` ensures that all available cores on the computer are used for fitting. Alternatively, the results can be loaded from cache using the following command.

```
data("IDE_Sim1_results", package = "STRbook")
```

The list `fit_results_sim1` contains two fields: `optim_results` that contains the output of the optimization algorithm, and `IDEmodel` that contains the fitted IDE model. The fitted kernel can be visualized by using the function **show_kernel**.

```
show_kernel(fit_results_sim1$IDEmodel)
```

Note how the fitted kernel is shifted to the left and upwards, correctly representing the southeasterly transport evident in the data. The estimated kernel parameters θ_p are given below.

```
fit_results_sim1$IDEmodel$get("k") %>% unlist()
```

```
##       par1       par2       par3       par4
## 152.83635    0.00198   -0.10160    0.10037
```

These estimates compare well to the true values `c(150, 0.002, -0.1, 0.1)` (see the help file for **simIDE**). The estimated regression coefficients are given below.

```
coef(fit_results_sim1$IDEmodel)
```

```
## Intercept           s1          s2
##     0.207        0.197       0.191
```

These also compare well to the true values `c(0.2, 0.2, 0.2)`. Also of interest are the moduli of the possibly complex eigenvalues of the evolution matrix **M**. These can be extracted as follows.

```
abs_ev <- eigen(fit_results_sim1$IDEmodel$get("M"))$values %>%
          abs()
summary(abs_ev)
```

```
##    Min. 1st Qu.  Median    Mean 3rd Qu.    Max.
##   0.003   0.290   0.366   0.331   0.399   0.464
```

Since the largest of these is less than 1, the IDE exhibits stable behavior.

For prediction, one may either specify a prediction grid or use the default one used for approximating the integrations set up by **IDE**. The latter is usually sufficient, so we use this without exception for the examples we consider. When a prediction grid is not supplied, the function **predict** returns a data frame with predictions spanning the temporal extent of the data (forecasts and hindcasts are explored later).

```
ST_grid_df <- predict(fit_results_sim1$IDEmodel)
```

The prediction map and prediction-standard-error map can now be plotted using standard **ggplot2** commands as follows.

```
gpred <- ggplot(ST_grid_df) +           # Plot the predictions
  geom_tile(aes(s1, s2, fill=Ypred)) +
  facet_wrap(~t) +
  fill_scale(name = "Ypred", limits = c(-0.1, 1.4)) +
  coord_fixed(xlim=c(0, 1), ylim = c(0, 1))

gpredse <- ggplot(ST_grid_df) +        # Plot the prediction s.es
  geom_tile(aes(s1, s2, fill = Ypredse)) +
  facet_wrap(~t) +
  fill_scale(name = "Ypredse") +
  coord_fixed(xlim=c(0, 1), ylim = c(0, 1))
```

In Figure 5.4, we show the observations, the true process, the predictions, and the prediction standard errors from the fitted model. Notice that the prediction standard errors are large in regions of sparse observations, as expected.

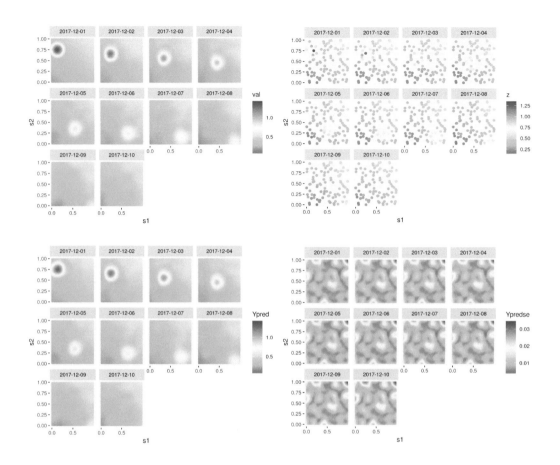

Figure 5.4: Simulated process (top left), simulated data (top right), predictions following the fitting of the IDE model (bottom left) and the respective prediction standard errors (bottom right).

Simulation Example with a Spatially Varying Kernel

In the previous example we considered the case of a spatially invariant kernel, that is, the case when the kernel $m(\mathbf{s}, \mathbf{x}; \boldsymbol{\theta}_p)$ is just a function of $\mathbf{x} - \mathbf{s}$. In this example, we consider the case when one or more of the $\boldsymbol{\theta}_p$ are allowed to be spatially referenced. Such models are needed when the spatio-temporal process exhibits, for example, considerable spatially varying drift (i.e., advection). Such a process can be simulated using the function `simIDE` by specifying `k_spat_invariant = 0`. To model data from a process of this sort, we need to have a large `nobs` and many time points; we set `T = 15`. This is important, as it is difficult to obtain reasonable estimates of spatially distributed parameters unless the data cover a large part of the spatial domain for a sustained amount of time.

```
SIM2 <- simIDE(T = 15, nobs = 1000, k_spat_invariant = 0)
```

As above, the process and the observed data can be plotted as two **ggplot2** plots.

```
print(SIM2$g_truth)
print(SIM2$g_obs)
```

Note how the process appears to rotate quickly counter-clockwise and come to a nearly complete standstill towards the lower part of the domain. The spatially varying advection that generated this field can be visualized using the following command.

```
show_kernel(SIM2$IDEmodel, scale = 0.2)
```

In this command, the argument `scale` is used to scale the arrow sizes by 0.2; that is, the shift per time point is five times the displacement indicated by the arrow.

Spatially varying kernels can be introduced by specifying the argument `kernel_basis` inside the call to **IDE**. The basis functions that **IDE** uses are of the same class as those used by **FRK**. We construct nine bisquare basis functions below that are equally spaced in the domain.

```
mbasis_1 <- auto_basis(manifold = plane(),   # fns on the plane
                       data = SIM2$z_STIDF,   # data
                       nres = 1,              # 1 resolution
                       type = 'bisquare')     # type of functions
```

To plot these basis functions, type `show_basis(mbasis_1)`.

Now, recall that $\theta_{p,1}$ (identified as `thetam1` in **IDE**) corresponds to the amplitude of the kernel, $\theta_{p,2}$ (`thetam2`) to the scale (width) or aperture, $\theta_{p,3}$ (`thetam3`) to the horizontal drift, and $\theta_{p,4}$ (`thetam4`) to the vertical drift. In what follows, suppose that $\theta_{p,1}$ and $\theta_{p,2}$ are spatially invariant (usually a reasonable assumption), and decompose $\theta_{p,3}$ and $\theta_{p,4}$ as sums of basis functions given in `mbasis_1`.

```
kernel_basis <- list(thetam1 = constant_basis(),
                     thetam2 = constant_basis(),
                     thetam3 = mbasis_1,
                     thetam4 = mbasis_1)
```

Modeling proceeds as before, except that now we specify the argument `kernel_basis` when calling **IDE**.

```
IDEmodel <- IDE(f = z ~ s1 + s2 + 1,
                data = SIM2$z_STIDF,
                dt = as.difftime(1, units = "days"),
                grid_size = 41,
                kernel_basis = kernel_basis)
```

Fitting also proceeds by calling the function `fit.IDE`. We use the argument `itermax = 400` below to specify the maximum number of iterations for the optimization routine to use.

```
fit_results_sim2 <- fit.IDE(IDEmodel,
                            parallelType = 1,
                            itermax = 400)
```

As above, since this is computationally intensive, we provide cached results that can be loaded using the following command.

```
data("IDE_Sim2_results", package = "STRbook")
```

The fitted spatially varying kernel can be visualized using the following command.

```
show_kernel(fit_results_sim2$IDEmodel)
```

The true and fitted spatially varying drift parameters are shown side by side in Figure 5.5. Note how the fitted drifts capture the broad directions and magnitudes of the true underlying process. Predictions and prediction standard errors can be obtained and mapped using `predict` as above. This is left as an exercise for the reader.

The Sydney Radar Data Set

Analysis of the Sydney radar data set proceeds in much the same way as in the simulation examples. In this case, we choose to have a spatially invariant kernel, since the data are not suggestive of spatially varying dynamics. We first load the Sydney radar data set as an `STIDF` object.

```
data("radar_STIDF", package = "STRbook")
```

As was seen in Chapter 2, the Sydney radar data set exhibits clear movement (drift), making the IDE a good modeling choice for these data. We now call the function `IDE` as before, with the added arguments `hindcast` and `forecast`, which indicate how many time intervals into the past, and how many into the future, we wish to predict for periods preceeding the training period (hindcast) and periods following the training period (forecast), respectively (see Section 6.1.3 for more information on hindcasts and forecasts). In this case the data are at 10-minute intervals (one period), and we forecast and hindcast for two periods each (i.e., 20 minutes).

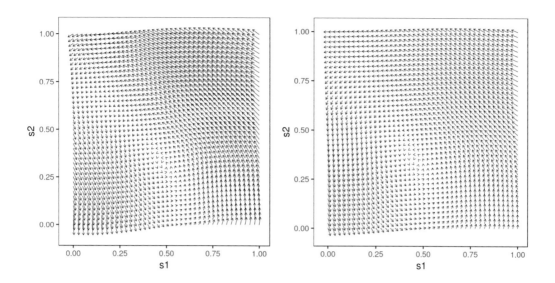

Figure 5.5: True drifts (left) and estimated drifts (right).

```
IDEmodel <- IDE(f = z ~ 1,
                data = radar_STIDF,
                dt = as.difftime(10, units = "mins"),
                grid_size = 41,
                forecast = 2,
                hindcast = 2)
```

Fitting proceeds by calling `fit.IDE`.

```
fit_results_radar <- fit.IDE(IDEmodel,
                             parallelType = 1)
```

Since this command will take a considerable amount of time on a standard machine, we load the results directly from cache.

```
data("IDE_Radar_results", package = "STRbook")
```

The fitted kernel can be visualized as it was above.

```
show_kernel(fit_results_radar$IDEmodel)
```

The kernel is again clearly shifted off-center and suggestive of transport in a predominantly easterly (and slightly northerly) direction. This is corroborated by visual inspection of the data. The estimated shift parameters are as follows.

```
shift_pars <- (fit_results_radar$IDEmodel$get("k") %>%
                      unlist())[3:4]
print(shift_pars)

## par3 par4
## -5.5 -1.9
```

The magnitude of the estimated shift vector is hence indicative of a transport of $\sqrt{(5.5)^2 + (1.9)^2} = 5.82$ km per 10-minute period, or 34.91 km per hour.

The modulus of the possibly complex eigenvalues of the evolution matrix $\mathbf{M}$ can be extracted as follows.

```
abs_ev <- eigen(fit_results_radar$IDEmodel$get("M"))$values %>%
          abs()
summary(abs_ev)

##    Min. 1st Qu.  Median    Mean 3rd Qu.    Max.
##    0.01    0.58    0.67    0.62    0.72    0.79
```

The largest absolute eigenvalue is considerably larger than that in the simulation study, suggesting more field persistence (although, since it is less than 1, the process is still stable). This persistence is expected, since the data clearly show patches of precipitation that are sustained and transported, rather than decaying, over time.

When calling the function `IDE`, we set up the object to be able to forecast 20 minutes into the future and hindcast 20 minutes into the past. These forecasts and hindcasts will be in the object returned from `predict`.

```
ST_grid_df <- predict(fit_results_radar$IDEmodel)
```

The data frame `ST_grid_df` contains the predictions in the field `Ypred` and the prediction standard errors in the field `Ypredse`. The field `t`, in both our data and predictions, contains the date as well as the time; we now create another field `time` that contains just the time of day.

```
radar_df$time <- format(radar_df$t, "%H:%M")
ST_grid_df$time <- format(ST_grid_df$t, "%H:%M")
```

The code given below plots the data as well as the smoothed fields containing the hindcasts, the predictions, and the forecasts. So that we match the data plots with the prediction plots, timewise, we create empty fields corresponding to hindcast and forecast periods in the data frame containing the observations. This can be achieved easily using `rbind.fill` from the package **plyr**.

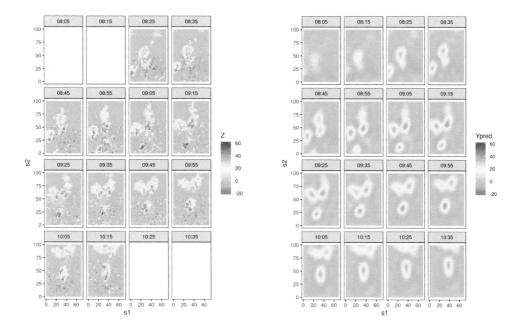

Figure 5.6: Observed data (left), and hindcasts, predictions, and forecasts using the IDE model (right).

```
## Add time records with missing data
radar_df <- rbind.fill(radar_df,
                    data.frame(time = c("08:05", "08:15",
                                        "10:25", "10:35")))

## Plot of data, with color scale capped to (-20, 60)
gobs <- ggplot(radar_df) +
  geom_tile(aes(s1, s2, fill = pmin(pmax(z, -20), 60))) +
  fill_scale(limits = c(-20, 60), name = "Z") +
  facet_wrap(~time) + coord_fixed() + theme_bw()

## Plot of predictions with color scale forced to (-20, 60)
gpred <- ggplot(ST_grid_df) +
  geom_tile(aes(s1, s2, fill = Ypred)) +
  facet_wrap(~time) + coord_fixed() + theme_bw() +
  fill_scale(limits = c(-20, 60), name = "Ypred")
```

The plots are shown in Figure 5.6. Notice how both the forecasts and the hindcasts incorporate the information on transport that is evident in the data. We did not plot prediction standard errors in this case, which is left as an exercise for the reader.

Lab 5.3: Spatio-Temporal Inference with Unknown Evolution Operator

If we have no prior knowledge to guide us on how to parameterize $\mathbf{M}$, then $\mathbf{M}$ can be estimated in full in the context of a standard state-space modeling framework. When taking this approach, it is important that a very low-dimensional representation of the spatio-temporal process is adopted – the dimension of the parameter space increases quadratically with the dimension of the process, and thus the model can easily become over-parameterized.

Empirical orthogonal functions (EOFs) are ideal basis functions to use in this case, since they capture most of the variability in the observed signal, by design. In this Lab we look at the SST data set, take the EOFs that we generated in Lab 2.3, and estimate all unknown parameters, first within a classical time-series framework based on a vector autoregression and using the method of moments (see Appendix C.1), and then in a state-space framework using the EM algorithm (see Appendix C.2).

Time-Series Framework

The aim of this first part of the Lab is to show how even simple methods can be used in a dynamical setting to provide prediction and prediction standard errors on a variable of interest. These methods work particularly well when we have complete spatial coverage and a high signal-to-noise ratio; this is the case with the SST data.

First, we load the usual packages.

```
library("ggplot2")
library("STRbook")
```

Then we load **expm** for raising matrices to a specified power and **Matrix**, which here we only use for plotting purposes.

```
library("expm")
library("Matrix")
```

We now load the SST data, but this time we truncate it at April 1997 in order to forecast the SSTs 6 months ahead, in October 1997.

```
data("SSTlandmask", package = "STRbook")
data("SSTlonlat", package = "STRbook")
data("SSTdata", package = "STRbook")
delete_rows <- which(SSTlandmask == 1)    # remove land values
SST_Oct97 <- SSTdata[-delete_rows, 334]    # save Oct 1997 SSTs
SSTdata <- SSTdata[-delete_rows, 1:328]    # until April 1997
SSTlonlat$mask <- SSTlandmask              # assign mask to df
```

Next, we construct the EOFs using only data up to April 1997. The following code follows closely what was done in Lab 2.3, where the entire data set was used.

```
Z <- t(SSTdata)                     # data matrix
spat_mean <- apply(SSTdata, 1, mean) # spatial mean
nT <- ncol(SSTdata)                 # no. of time points
Zspat_detrend <- Z - outer(rep(1, nT), # detrend data
                      spat_mean)
Zt <- 1/sqrt(nT-1)*Zspat_detrend    # normalize
E <- svd(Zt)                        # SVD
```

The number of EOFs we use here to model the SST data is $n = 10$. These 10 leading EOFs capture 74% of the variability in the data.

```
n <- 10
```

Recall that the object `E` contains the SVD, that is, the matrices $\mathbf{U}$ and $\mathbf{V}$ and the singular values. The dimension-reduced time series of coefficients are given by the EOFs multiplied by the spatially detrended data, that is, $\boldsymbol{\alpha}_t = \boldsymbol{\Phi}'(\mathbf{Z}_t - \hat{\boldsymbol{\mu}})$, $t = 1, \ldots, T$, where $\hat{\boldsymbol{\mu}} = (1/T) \sum_{t=1}^{T} \mathbf{Z}_t$ is the estimated spatial mean.

```
TS <- Zspat_detrend %*% E$v[, 1:n]
summary(colMeans(TS))
```

```
##       Min.   1st Qu.    Median       Mean   3rd Qu.       Max.
## -2.09e-16 -8.00e-17  3.20e-16  3.11e-16  6.92e-16  8.89e-16
```

In the last line above, we have verified that the time series have mean zero, which is needed to compute covariances by taking outer products. Next, we estimate the matrices $\mathbf{M}$ and $\mathbf{C}_\eta$ using the method of moments. First we create two sets of time series that are shifted by τ time points with respect to each other; in this case we let $\tau = 6$, so that we analyze dynamics on a six-month scale. The ith column in `TStplustau` below corresponds to the time series at the $(i + 6)$th time point, while that in `TSt` corresponds to the time series at ith time point.

```
tau <- 6
nT <- nrow(TS)
TStplustau <- TS[-(1:tau), ] # TS with first tau time pts removed
TSt <- TS[-((nT-5):nT), ]    # TS with last tau time pts removed
```

The lag-0 empirical covariance matrix and the lag-τ empirical cross-covariance matrices are now computed by taking their matrix cross-product and dividing by the appropriate number of time points; see (2.4).

```
Cov0 <- crossprod(TS)/nT
Covtau <- crossprod(TStplustau,TSt)/(nT - tau)
```

The estimates for $\mathbf{M}$ and $\mathbf{C}_\eta$ can now be estimated from these empirical covariance matrices. As discussed in Appendix C.1, this can be done using the following code.

```
C0inv <- solve(Cov0)
Mest <- Covtau %*% C0inv
Ceta <- Cov0 - Covtau %*% C0inv %*% t(Covtau)
```

There are more efficient ways to compute the quantities above that ensure symmetry and positive-definiteness of the results. In particular, the inverse rarely needs to be found explicitly. For further information, the interested reader is referred to standard books on linear algebra (see, for example, Schott, 2017).

The matrices can be visualized using the function `image`.

```
image(Mest)
image(Ceta)
```

From visual inspection, the estimate of the propagator matrix, `Mest`, is by no means diagonally dominant, implying that there is benefit in assuming interactions between the EOFs across time steps (see Section 5.3.2). Further, the estimated variances along the diagonal of the covariance matrix of the additive disturbance in the IDE model, $\mathbf{C}_\eta$, decrease with the EOF index; this is expected as EOFs with higher indices tend to have higher-frequency components.

Forecasting using this EOF-reduced model is straightforward as we take the coefficients at the final time point, α_t, propagate those forward, and re-project onto the original space. For example, $\hat{\mu} + \Phi \mathbf{M}^2 \alpha_t$ gives a one-year forecast. Matrix powers (which represent multiple matrix multiplications and do not come from elementwise multiplications) can be implemented using the operator `%^%` from the package **expm**; this will be used when we implement the state-space model below. Here we consider six-month-ahead forecasts; in the code below we project ahead the EOF coefficients of the time series at the 328th time point (which corresponds to April 1997) six months into the future.

```
SSTlonlat$pred <- NA
alpha_forecast <- Mest %*% TS[328, ]
```

The projection onto the original space is done by pre-multiplying by the EOFs and adding back the estimated spatial mean (see Section C.1).

```
idx <- which(SSTlonlat$mask == 0)
SSTlonlat$curr[idx]  <- as.numeric(E$v[, 1:n] %*% TS[328, ] +
```

```
                                        spat_mean)
SSTlonlat$pred[idx]  <- as.numeric(E$v[, 1:n] %*% alpha_forecast +
                                        spat_mean)
```

Now we add the data to the data frame for plotting purposes.

```
SSTlonlat$obs1[idx]  <- SSTdata[, 328]
SSTlonlat$obs2[idx]  <- SST_Oct97
```

The six-month-ahead prediction variances can also be computed (see Appendix C.1).

```
C <- Mest %*% Cov0 %*% t(Mest) + Ceta
```

The prediction variances are found by projecting the covariance matrix C onto the original space and extracting the diagonal elements. The prediction standard errors are the square root of the prediction variances and hence obtained as follows.

```
SSTlonlat$predse[idx] <-
    sqrt(diag(E$v[, 1:n] %*% C %*% t(E$v[, 1:n])))
```

Plotting proceeds in a straightforward fashion using **ggplot2**. In Figure 5.7 we show the April 1997 data and the EOF projection for that month, as well as the October 1997 data and the forecast for that month. From visual inspection, the El Niño pattern of high SSTs is captured but the predicted anomaly is too low. This result is qualitatively similar to what we obtained in Lab 3.3 using linear regression models.

State-Space Framework

The function **DSTM_EM**, provided with the package **STRbook**, runs the EM algorithm that carries out maximum likelihood estimation in a state-space model. The function takes the data Z, the initial covariance $\mathbf{C}_0$ in Cov0, the initial state μ_0 in muinit, the evolution operator $\mathbf{M}$ in M, the covariance matrix $\mathbf{C}_\eta$ in Ceta, the measurement-error variance σ_ϵ^2 in sigma2_eps, the matrix $\mathbf{H}$ in H, the maximum number of EM iterations in itermax, and the tolerance in tol (the tolerance is the smallest change in the log-likelihood, multiplied by 2, required across two consecutive iterations of the EM algorithm, before terminating). All parameters supplied to the function need to be initial guesses (usually those from the method of moments suffice); these will be updated using the EM algorithm.

```
DSTM_Results <- DSTM_EM(Z = SSTdata,
                        Cov0 = Cov0,
                        muinit = matrix(0, n, 1),
                        M = Mest,
                        Ceta = Ceta,
```

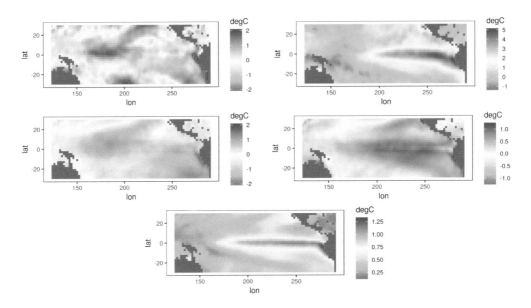

Figure 5.7: Top: SST data for April 1997 (left) and October 1997 (right). Middle: the EOF projection for April 1997 (left), and the forecast for October 1997 (right). Note the different color scales for the predictions (up to 1°C) and for the observations (up to 5°C). Bottom: Prediction standard errors for the forecast.

```
sigma2_eps = 0.1,
H = H <- E$v[, 1:n],
itermax = 10,
tol = 1)
```

The returned object `DSTM_Results` contains the estimated parameters, the smoothed states and their covariances, and the complete-data negative log-likelihood. In this case, estimates of $\{\boldsymbol{\alpha}_t\}$ using the state-space framework are practically identical to those obtained using the time-series framework presented in the first part of the Lab. We plot estimates of $\alpha_{1,t}$, $\alpha_{2,t}$, and $\alpha_{3,t}$ below for the two methods; see Figure 5.8.

```
par(mfrow = c(1,3))
for(i in 1:3) {
  plot(DSTM_Results$alpha_smooth[i, ], type = 'l',
       xlab = "t", ylab = bquote(alpha[.(i)]))
  lines(TS[, i], lty = 'dashed', col = 'red')
}
```

Let us turn now to inference on the parameters. From Appendix C.2 note that the EM algorithm utilizes a Kalman filter that processes the data one time period (e.g., month) at

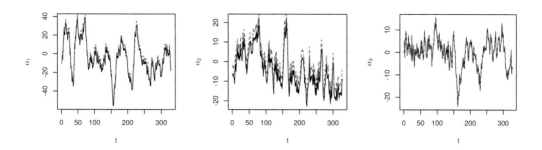

Figure 5.8: Estimates of $\alpha_{1,t}$ (left), $\alpha_{2,t}$ (center), and $\alpha_{3,t}$ (right) using the method of moments (red dashed line) and the EM algorithm (black solid line).

a time. (Recall that with the method of moments we let $\tau = 6$ months, so we estimated directly the state transitions over six months.) Therefore, inferences on the parameters and their interpretations differ considerably. For example, the left and right panels of Figure 5.9 show the estimates of the evolution matrix for the two methods. At first sight, it appears that the matrix estimated using the EM algorithm is indicating a random-walk behavior. However, if we multiply the matrix $\mathbf{M}$ by itself six times (which then describes the evolution over six months), we obtain something that is relatively similar to what was estimated using the method of moments using a time lag of $\tau = 6$ months.

To make the plots in Figure 5.9, we first cast the matrices into objects of class `Matrix`. Note that using the function **`image`** on objects of class `matrix` generates similar plots that are, however, less informative. On the other hand, plots of **Matrix** objects are done using the function **`levelplot`** in the **lattice** package.

```
image(as(DSTM_Results$Mest, "Matrix"))
image(as(DSTM_Results$Mest %^% 6, "Matrix"))
image(as(Mest, "Matrix"))
```

Forecasting proceeds the same way as in the method of moments. Specifically, we take the last smoothed time point (which corresponds to April 1997) and use the EM-estimated one-month propagator matrix to forecast the SSTs six months ahead. This is implemented easily using a `for` loop.

```
alpha <- DSTM_Results$alpha_smooth[, nT]
P <- DSTM_Results$Cov0
for(t in 1:6) {
    alpha <- DSTM_Results$Mest %*% alpha
    P <- DSTM_Results$Mest %*% P %*% t(DSTM_Results$Mest) +
```

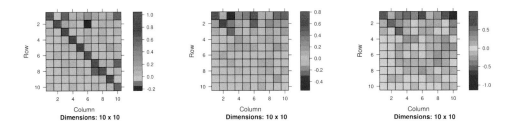

Figure 5.9: Estimate of a one-step-ahead evolution operator using the EM algorithm (left); EM estimate raised to the sixth power (center); and estimate of the six-steps-ahead evolution operator using the method of moments (right).

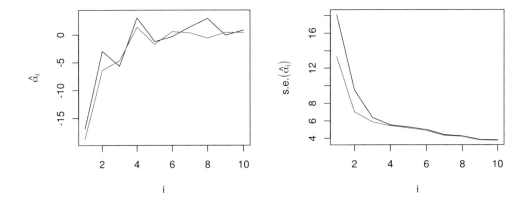

Figure 5.10: Forecasts (left) and prediction standard errors (right) for the EOF coefficients in October 1997 using a lag-6 time-series model estimated using the method of moments (black) and a lag-1 state-space model estimated using the EM algorithm (red).

```
DSTM_Results$Ceta
}
```

It is instructive to compare the predictions and the prediction standard errors of the forecasted EOF coefficients using the two models; see Figure 5.10. While the prediction standard errors for the state-space model are slightly lower (which is expected since a measurement-error component of variability has been filtered out), it is remarkable that forecasts from the lag-6 time-series model are quite similar to those of the lag-1 state-space model. These two models and their respective inferences can be expected to differ when there is more nonlinearity in the process and/or the data are less complete in space and time.

In our concluding remarks, we remind the reader that in this Lab we considered a *linear* DSTM for modeling SSTs. Recent research has suggested that *nonlinear* DSTMs may provide superior prediction performance; see the case study in Appendix F.

Chapter 6

Evaluating Spatio-Temporal Statistical Models

How do you know that the model you fitted actually fits well? At the core of our approach to the analysis of spatio-temporal data is a more or less detailed model containing statistical components that are designed to capture the spatio-temporal variability in the data. This chapter is about evaluating the spatio-temporal model that you fitted to describe (or to some extent explain) the variability in your data.

Model building is an iterative process. We have data and/or a scientific hypothesis and we build the model around them (e.g., using the methods of Chapters 3–5). Then we must evaluate whether that model is a reasonable representation of the real world, and we should modify it accordingly if it is not. Sometimes this process is called *model criticism* because we are *critiquing* the strengths and weaknesses of our model, analogously to a movie critic summing up a film in terms of the things that work and the things that do not. In our case, we already know our model is wrong (recall Box's aphorism), but we do not know just *how wrong* it is. Just as there is no correct model, there is no correct way to do model evaluation either. Rather, think of it as an *investigation*, using evidence from a variety of sources, into whether the model is reasonable or not. In this sense, we are "detectives" searching for evidence that our model can represent what we hope it represents in our particular application, or we are like medical doctors running tests on their patients. With that in mind, this chapter is about providing helpful suggestions on how to evaluate models for spatio-temporal data.

We split our model-evaluation suggestions into three primary components: *model checking*, *model validation*, and *model selection*. From our perspective, *model checking* consists of evaluating our model diagnostically to check its assumptions and its sensitivity to these assumptions and/or model choices. *Model validation* consists of evaluating how well our model actually reproduces the real-world quantities that we care about. *Model selection* is a framework in which to compare several plausible models. We consider each of these in some detail in this chapter.

It is important to note that the boundaries between these three components of model evaluation are fairly "fluid," and the reader may well notice that there is a great deal of overlap in the sense that approaches discussed in one of these sections could be applied in other sections. Such is the nature of the topic, especially in the context of spatio-temporal modeling where, we must say, it is not all that well developed.

In this chapter we focus less on how to implement the methods in R and more on the methods themselves. The reasons for this are twofold. First, several diagnostics are straight-forward to calculate once predictive distributions are available. Second, there are only a few packages that have a comprehensive suite of diagnostic tools. We also note that quite a few more primary literature citations are included in this chapter than are given in the other chapters of the book, because there has not been extensive discussion of the topic in the spatio-temporal modeling literature.

In the next section, we digress slightly to discuss how model-based predictions can be compared to observations appropriately, since the two have different statistical properties.

6.1   Comparing Model Output to Data: What Do We Compare?

Before we can talk about model evaluation, we have to decide what we will compare our model to. Hierarchical spatio-temporal modeling procedures give us the predictive distribution of the latent spatio-temporal process, Y, given a (training) set of observations, $\mathbf{Z}$, which we represent as $[Y|\mathbf{Z}]$. Because we are most often interested in this latent process, we would like to evaluate our model based on its ability to provide reasonable representations of Y. But by definition this process is *hidden* or *latent* – meaning that it is not observed directly – and thus we cannot directly evaluate our *modeled* process against the *true* process unless we do it through simulation (see Section 6.1.1 below). Alternatively, we can evaluate our model using *predictive distributions of data*, where we compare predictions of data (not of Y), based on our model, against the actual observed data. In particular, there are four types of predictive distributions of the data that we might use: the *prior predictive distribution*, the *posterior predictive distribution*, what we might call the *empirical predictive distribution*, and the *empirical marginal distribution*. Note that considering predictions of the data Z instead of Y involves the additional uncertainty associated with the measurement process. This is similar to standard regression modeling where the uncertainty of the prediction of an unobserved response is higher than the uncertainty of inferring the corresponding mean response. The four types of predictive distributions are defined in Section 6.1.2. Finally, given that we have simulated predictive distributions of either the data or the latent process, there is still the issue of which samples to compare. We touch on this in Section 6.1.3, with a brief discussion of various types of validation and cross-validation samples that we might use to evaluate our model.

6.1.1 Comparison to a Simulated "True" Process

Although we do not have access to the latent process, Y, for evaluating our model, there is a well-established simulation-based alternative for complex processes known as an *observation system simulation experiment* (OSSE; see Technical Note 6.1). The basic idea of an OSSE is that one uses a complex simulation model to generate the true underlying process, say Y_{osse}, and then, one generates simulated data, say $\mathbf{Z}_{osse}$, by applying an observation/sampling scheme to this true process that mimics the real-world sampling design and measurement technology. One can then use these OSSE-simulated observations in the statistical model and compare Y obtained from the predictive distribution based on the statistical model (i.e., $[Y|\mathbf{Z}_{osse}]$) against the simulated Y_{osse}. The metrics used for such a comparison could be any of the metrics that are described in the following sections of this chapter. Not surprisingly, OSSEs are very useful when exploring different sampling schemes and, in the geophysical sciences, they are important for studying complex earth observing systems *before* expensive observing-system hardware is deployed. They are also very useful for comparing competing methodologies that infer Y or scientifically meaningful functions of Y.

Technical Note 6.1: Observation System Simulation Experiment, OSSE

Observation system simulation experiments are model-based simulation experiments that are designed to consider the effect of potential observing systems on the ability to recover the true underlying process of interest, especially when real-world observations are not available. For example, these are used extensively in the geophysical sciences to evaluate new remote sensing observation systems and new data assimilation forecast systems. However, they can also be used to evaluate the effectiveness of process modeling for complex real-world processes in the presence of incomplete observations, or when observations come at different levels of spatial and temporal support (see, for example, Berliner et al., 2003). The typical OSSE consists of the following steps. Steps 1 and 2 correspond to simulation, and steps 3–5 are concerned with the subsequent statistical analysis.

1. Simulate the spatio-temporal process of interest with a well-established (usually mechanistic) model. This simulation corresponds to the "true process." Note that this is usually *not* a simulation from the statistical model of interest, since as much real-world complexity as possible is put into the simulation; call it Y_{osse}.

2. Apply an observation-sampling protocol to the simulated true process to obtain synthetic observations. This sampling protocol introduces realistic observation error (bias, uncertainty, and change of support) and typically considers various missing-data scenarios; call the observations $\mathbf{Z}_{osse}$.

3. Use $\mathbf{Z}_{osse}$ from step 2 in the spatio-temporal statistical model of interest, and ob-

tain the predictive distribution $[Y|\mathbf{Z}_{\text{osse}}]$ of the true process given the synthetic observations.

4. Compare features of the predictive distribution of the true process from step 3 to Y_{osse} simulated in step 1.

5. Use the results of step 4 to either (a) refine the statistical model that was used to obtain $[Y|\mathbf{Z}_{\text{osse}}]$, or (b) refine the observation process, or both.

6.1.2 Predictive Distributions of the Data

The *posterior predictive distribution* (ppd) is best thought of in the context of a Bayesian hierarchical model (BHM) and is given by (e.g., Gelman et al., 2014)

$$[\mathbf{Z}_{\text{ppd}}|\mathbf{Z}] = \iint [\mathbf{Z}_{\text{ppd}}|\mathbf{Y}, \boldsymbol{\theta}][\mathbf{Y}, \boldsymbol{\theta}|\mathbf{Z}]d\mathbf{Y}d\boldsymbol{\theta}, \qquad (6.1)$$

where $\mathbf{Z}_{\text{ppd}}$ is a vector of predictions at some chosen spatio-temporal locations. We have assumed that if we are given the true process, $\mathbf{Y}$, and parameters, $\boldsymbol{\theta}$, then $\mathbf{Z}_{\text{ppd}}$ is independent of the observations $\mathbf{Z}$. (Note that we are using the vector $\mathbf{Y}$ to represent the process here to emphasize the fact that we are dealing with high-dimensional spatio-temporal processes.) In the models considered in this book, one can easily generate samples of $\mathbf{Z}_{\text{ppd}}$ through *composition sampling*. For example, generating posterior samples of $\mathbf{Y}$ and $\boldsymbol{\theta}$ in the BHM context comes naturally with Markov chain Monte Carlo (MCMC) implementations, and these samples are just "plugged into" the data model $[\mathbf{Z}_{\text{ppd}}|\mathbf{Y}, \boldsymbol{\theta}]$ to generate the random draws of $\mathbf{Z}_{\text{ppd}}$.

The *prior predictive distribution* (pri) corresponds to the marginal distribution of the data and is given by

$$[\mathbf{Z}_{\text{pri}}] = \iint [\mathbf{Z}_{\text{pri}}|\mathbf{Y}, \boldsymbol{\theta}][\mathbf{Y}|\boldsymbol{\theta}][\boldsymbol{\theta}]d\mathbf{Y}d\boldsymbol{\theta}, \qquad (6.2)$$

where $\mathbf{Z}_{\text{pri}}$ is a vector of predictions at selected spatio-temporal locations. As with the ppd, realizations from this distribution can be easily generated through composition sampling, where in this case we simply generate samples of $\boldsymbol{\theta}$ from its prior distribution, use those to generate samples of the process $\mathbf{Y}$ from the process model, and then use these samples in the data model to generate realizations of the data, $\mathbf{Z}_{\text{pri}}$. In contrast to the ppd, no MCMC posterior samples need to be generated for this distribution.

Finally, in the empirical hierarchical model (EHM) context we define the *empirical predictive distribution* (epd) as

$$[\mathbf{Z}_{\text{epd}}|\mathbf{Z}] = \int [\mathbf{Z}_{\text{epd}}|\mathbf{Y}, \widehat{\boldsymbol{\theta}}][\mathbf{Y}|\mathbf{Z}, \widehat{\boldsymbol{\theta}}]d\mathbf{Y}, \qquad (6.3)$$

and the *empirical marginal distribution* (emp) as

$$[\mathbf{Z}_{\text{emp}}] = \int [\mathbf{Z}_{\text{emp}}|\mathbf{Y}, \widehat{\boldsymbol{\theta}}][\mathbf{Y}|\widehat{\boldsymbol{\theta}}]d\mathbf{Y}, \tag{6.4}$$

where $\mathbf{Z}_{\text{epd}}$ and $\mathbf{Z}_{\text{emp}}$ are vectors of predictions at selected spatio-temporal locations. The difference between (6.3) and (6.1), and between (6.4) and (6.2), is that instead of integrating over $\boldsymbol{\theta}$ (which is assumed to be random in the BHM framework), we substitute an estimate $\widehat{\boldsymbol{\theta}}$ (e.g., a ML or REML estimate). Again, it is easy to sample from (6.3) and (6.4) by composition sampling since, once $\widehat{\boldsymbol{\theta}}$ is obtained, we can generate samples of $\mathbf{Y}$ easily from $[\mathbf{Y}|\mathbf{Z}, \widehat{\boldsymbol{\theta}}]$ and from $[\mathbf{Y}|\widehat{\boldsymbol{\theta}}]$ with an MCMC. In the spatio-temporal Gaussian case, these are known multivariate normal distributions. Then $\widehat{\boldsymbol{\theta}}$ and the samples of $\mathbf{Y}$ are "plugged into" the data model to obtain samples of $\mathbf{Z}_{\text{epd}}$ and $\mathbf{Z}_{\text{emp}}$, respectively.

For illustration, consider the IDE model fitted to the Sydney radar data set in Lab 5.2. The top panels of Figure 6.1 show two samples from the epd for the time points 08:45, 08:55, and 09:05, while the bottom panels show two samples from the emp at the same time points. We shall discuss model validation using the predictive distributions of the data in Section 6.3.1, but simply "eyeballing" the plots may also reveal interesting features of the fitted model. First, the two samples from the epd are qualitatively quite similar, and this is usually an indication that the data have considerable influence on our predictive distributions. These epd samples are also very different from the emp samples, adding weight to the argument that the predictions in the top panels are predominantly data driven. Second, the samples from emp are very useful in revealing potential flaws and strengths of the model. For example, in this case the samples reveal that negative values for dBZ (green and blue) are just as likely as positive values for dBZ (orange and red), while we know that this is not a true reflection of the underlying science. On the other hand, the spatial length scales, and the persistence of the spatial features in time are similar to what one would expect just by looking at the data (see Section 2.1). These qualitative impressions, which will be made rigorous in the following sections, play a big role in selecting and tuning spatio-temporal models to improve their predictive ability.

R tip: Several R packages contain built-in functionality for sampling from one or more of the predictive distributions listed in (6.1)–(6.4). For example, the function `krige` in the package **gstat** can be used to generate simulations from both epd and emp, while the function `simIDE` in the package **IDE** can be used to generate simulations from emp after fitting an IDE model.

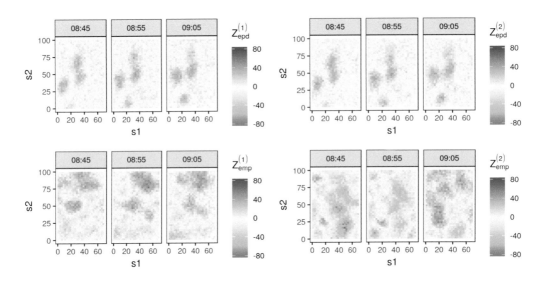

Figure 6.1: Two samples from the empirical predictive distribution (top) and the empirical marginal distribution (bottom), respectively, using the IDE model fitted to the Sydney radar data set in Lab 5.2.

6.1.3 Validation and Cross-Validation

Most often we will have to compare real-world validation observations, say $\mathbf{Z}_v$, to observations predicted from our model, say $\mathbf{Z}_p$, from one (or all) of the four possibilities (ppd, pri, epd, emp) given in the previous section. The question here is, to which observations do we compare $\mathbf{Z}_p$? The *generalization* ability of a model is a property that says how well it can predict a test data set (also referred to as a validation data set) that is different from the data used to train the model. (Note that the words "test" and "validation" are often used interchangeably in this context; we prefer to use "validation.") So, assume that we have used a sample of data, $\mathbf{Z}$, to train our model. Before we describe the different possibilities for selecting validation data $\mathbf{Z}_v$, note that spatio-temporal processes have certain properties that should be considered when comparing model predictions to real-world observations. In particular, as with time series, spatio-temporal processes have a unidirectional time dependence and, like spatial processes, they have various degrees of spatial dependence. These dependencies should be considered whenever possible when evaluating a spatio-temporal model.

In general, the choice for validation observations $\mathbf{Z}_v$ can then be one of the following.

(a) *Training-data validation.* It can be informative to use predicted observations of the training data set ($\mathbf{Z}_v = \mathbf{Z}$) to evaluate our model, particularly when evaluating the model's ability to fit the data and for checking model assumptions via diagnostics.

However, in the context of prediction, it is not typically recommended to use the training data for validating the model's predictive ability, as the model's *training error* is typically *optimistic* in the sense that it underestimates the predictive error that would be observed in an independent sample. Perhaps not surprisingly, the amount of this optimism is related to how strongly a predicted value from the training data set affects its own prediction (see Hastie et al., 2009, Chapter 7, for a comprehensive overview).

(b) *Within-sample validation.* It is often useful to consider validation samples in which one leaves out a collection of spatial observations at time(s) within the spatio-temporal window defined by the extent of the training data set. Although one can leave out data at random in such settings, a more appropriate evaluation of spatio-temporal models results from leaving out "chunks" of data. This is because the spatio-temporal dependence structure must be very well characterized to adequately fill in large gaps for spatio-temporal processes (particularly dynamic processes). We saw such an example in Chapter 4, where we left out one period of the NOAA maximum temperature data but had observations both before and after that period.

(c) *Forecast validation.* One of the most-used validation methods for time-dependent data is to leave out validation data beyond the last time period of the training period, and then to use the model to forecast at these future time periods. To predict the evolution of spatial features through time, the spatio-temporal model must adequately account for (typically non-separable) spatio-temporal dependence. Hence, forecast validation provides a gold standard for such evaluations.

(d) *Hindcast validation. Hindcasting* (sometimes known as *backtesting*) refers to using the model to predict validation data at time periods *before* the first time period in the training sample. Of course, this presumes that we have access to data that pre-dates our training sample! This type of out-of-sample validation has similar advantages to forecast validations.

(e) *Cross-validation.* There are many modeling situations where one needs all of the available observations to train the model, especially at the beginning and end of the data record. Or perhaps one is not certain that the periods in the forecast or hindcast validation sample are representative of the entire period (e.g., when the process is non-stationary in time). This is a situation where cross-validation can be quite helpful. Recall that we described cross-validation in Technical Note 3.1. In the context of spatio-temporal models with complex dependence, one has to be careful that the cross-validation scheme chosen respects the dependence structure. In addition, many implementations of spatio-temporal models are computationally demanding, which can make traditional cross-validation very expensive.

In Lab 6.1 we provide an example of within-sample validation, where a 20-minute interval from the Sydney radar data set is treated as validation data, and a model using spatio-temporal basis functions is compared to an IDE model through their prediction performances in this 20-minute interval.

Spatio-Temporal Support of Validation Data and Model Predictions

So far we have assumed that the validation data set, $\mathbf{Z}_v$, and the model-predicted observations, $\mathbf{Z}_{\text{ppd}}$ (say), are available at the same spatial and temporal support. In many applications this is not the case. For example, our model may produce spatial fields (defined over a grid) at daily time increments, but observations may be station data observed every hour. In some sense, if our data model is realistic, then we may have already accounted for these types of change of support. In other cases, one may perform *ad hoc* interpolation or aggregation to bring the validation and model support into agreement. This is a standard approach in many meteorological forecasting studies (see, for example, Brown et al., 2012). The hierarchical modeling paradigm discussed here does provide the flexibility for incorporating formal change of support, but this is beyond the scope of this book (for more details, see Cressie and Wikle, 2011, Chapter 7, and the references therein). In the remainder of this chapter we shall assume that the validation sample and the associated model predictions are at the same spatio-temporal support.

6.2   Model Checking

Now that we know what to compare to what, consider the first of our three types of model evaluation: *model checking*. From our perspective, this corresponds to checking model assumptions and the sensitivity of the model output to these assumptions and/or model choices. That is, we evaluate our spatio-temporal model using statistical *diagnostics*. We begin with a brief description of possible extensions of standard regression diagnostics, followed by some simple graphical diagnostics, and then we give a brief description of robustness checks.

6.2.1   Extensions of Regression Diagnostics

As in any statistical-modeling problem, one should evaluate spatio-temporal modeling assumptions by employing various diagnostic tools. In regression models and GLMs, one often begins such an analysis by evaluating residuals, usually obtained by subtracting the estimated or predicted response from the data. Looking at residuals may bring our attention to certain aspects of the data that we have missed in our model.

As discussed in Chapter 3, for additive Gaussian measurement error, we can certainly do this in the spatio-temporal case by evaluating the spatio-temporal residuals,

$$\widehat{e}(\mathbf{s}_i; t_j) \equiv Z(\mathbf{s}_i; t_j) - \widehat{Z}_p(\mathbf{s}_i; t_j), \tag{6.5}$$

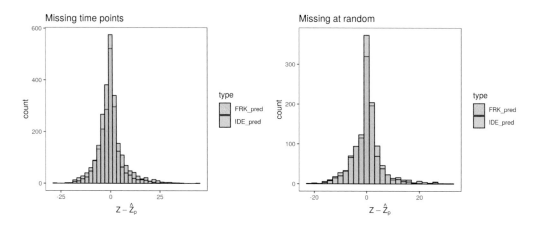

Figure 6.2: Histograms of errors at validation locations for the fitted IDE (blue) and FRK (red) models for the time points that are omitted from the data (left) and for space-time locations that are missing at random (right).

for $i = 1, \ldots, m$ and $j = 1, \ldots, T$, where $\widehat{Z}_p(\mathbf{s}_i; t_j)$ is the mean of the ppd or epd as discussed in Section 6.1.2. Note that for notational simplicity we assume in this chapter that we have the same number of observations (m) at the same spatial locations for each time point. This need not be the case, and the equations can easily be modified to represent the more general setting of a different number of observations at different locations for each time point.

In Figure 6.2 we show the histograms of the spatio-temporal residuals obtained for the two models evaluated in Lab 6.1 using validation data for an entire 20-minute block (left panel) and at random locations (right panel). It is clear from these histograms that for both types of missingness, the variance of the residuals based on the IDE model is slightly lower than that based on the model used by the FRK model.

In addition to the classical residuals given in (6.5), we can consider deviance or Pearson chi-squared residuals for non-Gaussian data models (as discussed in Chapter 3). Given spatio-temporal residuals, it is usually helpful to visualize them using the various tools discussed in Chapter 2 (see Lab 6.1). In addition, as discussed in Chapter 3, one can consider quantitative summaries to evaluate residual temporal, spatial, or spatio-temporal dependence, such as with the PACF, Moran's I, and S-T covariogram summaries. In the case of the latter, one may also consider more localized summaries, known as *local indicators of spatial association (LISAs)* or their spatio-temporal equivalents (*ST-LISAs*) where the component pieces of a summary statistic are indexed by their location and evaluated individually (see Cressie and Wikle, 2011, Section 5.1).

Diagnostics have also been developed specifically for models with spatial dependence that are easily extended to spatio-temporal models. For example, building on the ground-breaking work of Cook (1977), Haslett (1999) considered a simple approach for "deletion diagnostics" in models with correlated errors. For example, if one has a model such as $\mathbf{Z} \sim Gau(\mathbf{X}\boldsymbol{\beta}, \mathbf{C}_z)$, then interest is in the effect of leaving out elements of $\mathbf{Z}$ on the estimation of $\boldsymbol{\beta}$. Analogously to K-fold cross-validation discussed in Chapter 3, assume we split our observations into two groups, $\mathbf{Z} = \{\mathbf{Z}_b, \mathbf{Z}_v\}$, and then we predict $\mathbf{Z}_v$ based only on training data $\mathbf{Z}_b$, which we denote by $\widehat{\mathbf{Z}}^{(-v)}$. Then, as with standard (independent and identically distributed (iid) errors) regression, one can form diagnostics in the correlated-error context, analogous to the well-known *DFBETAS* and *Cook's distance* diagnostics. These compare the regression coefficients estimated under the hold-out scenario (say, $\widehat{\boldsymbol{\beta}}^{(-v)}$) to the parameters estimated using all of the data ($\widehat{\boldsymbol{\beta}}$), and Haslett (1999) provides some efficient approaches to obtain $\widehat{\boldsymbol{\beta}}^{(-v)}$. It is important to note that these diagnostics are based on the cross-validated residuals,

$$\widehat{\mathbf{e}}_v \equiv \mathbf{Z}_v - \widehat{\mathbf{Z}}^{(-v)}, \tag{6.6}$$

rather than the within-sample residuals given by (6.5).

6.2.2 Graphical Diagnostics

Several diagnostic plots have proven useful for evaluating predictive models, and these largely depend on the observation type. Recall, from our discussions in Chapters 3–5, that it is fairly straightforward to model spatio-temporal binary or count data using the techniques we described within a GLM framework. Our discussion below on graphical diagnostics covers the most common types of data encountered in practice.

When considering binary outcomes, which are common when observing processes such as occupancy (presence–absence) in ecology, and precipitation (rain or no rain) in meteorology, there is a long tradition in statistics and engineering of considering a *receiver operating characteristic* (ROC) curve. For binary data, a statistical model (say, a Bernoulli data model with a logit link function) provides an estimate of the probability that the outcome is a 1 (versus a 0). Then, for predictions, a threshold probability is typically set, and the predicted outcome is put equal to 1 if the estimated probability is larger than the threshold, and put equal to 0 if not. Clearly, the performance of the predictions will depend on the threshold. The ROC plot presents the true positive rate (i.e., sensitivity, namely the percentage of 1s that were correctly predicted) on the y-axis versus the false positive rate (i.e., 1 minus the specificity, namely the percentage of 0s that were incorrectly predicted to be 1s) on the x-axis as the value of the threshold probability changes (from 0 to 1). Since we prefer a model that gives a high true positive rate and low false positive rate, we like to see ROC curves that are well above the 45-degree line. One often summarizes an ROC curve by the *area under the ROC* curve (sometimes abbreviated as "area under the curve" (AUC)), with the best possible area being 1.0 and with a value of 0.5 corresponding to a "no information" (i.e., a coin-flipping) model. Figure 6.3 shows two ROC curves for a data set based on 100 simulated Bernoulli responses from a logistic regression model with simulated covariates.

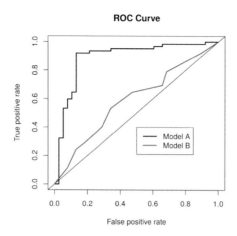

Figure 6.3: ROC curves for two models fitted to a simulated Bernoulli (binary) data set with 100 observations that was generated from a logistic regression model with simulated covariates. The black line corresponds to the ROC curve for a simpler logistic regression model ("Model A") with a corresponding AUC = 0.89, and the red line is the ROC curve for a model ("Model B") based just on random guessing with AUC = 0.59. This figure was obtained using the `roc.plot` function in the **verification** R package.

The black ROC curve corresponds to a simple model (a logistic regression model with fewer covariates than used for the simluation) and the red ROC curve corresponds to flipping a coin (random guessing). The AUCs for the two models are 0.89 and 0.59, respectively. Although useful for evaluating prediction, the ROC curve is limited in that it is generally insensitive to prediction biases (see Wilks, 2011, Chapter 8).

> **R tip:** ROC curves can be easily generated in R using the functions `prediction` and `performance` from the package **ROCR** or the function `roc.plot` from the package **verification**.

There are several diagnostic plots that are used for meteorological forecast validation but are less commonly used in statistics (see Wilks, 2011, Chapter 8). Some of these plots attempt to show elements of the joint distribution of the prediction and the corresponding validation observation. As an example, *conditional quantile plots* are used for continuous responses (e.g., temperature). In particular, these plots consider predicted values on the x-axis and the associated quantiles from the *empirical predictive distribution of the observations associated with the predictions* on the y-axis. This allows one to observe potential problems with the predictive model (e.g., biases). This is better seen in an example. The left panel of Figure 6.4 shows a conditional quantile plot for simulated data in a situation where the predictive model is, on average, biased high relative to the observations by about 3 units. This can easily be seen in this plot since the conditional distribution of the observations given the predictions is shifted below the 45-degree line. In the right panel of Figure 6.4 we show the conditional quantile plot for the IDE model predictions in Lab 6.1 for the missing 20-minute interval. The predictions appear to be unbiased except when the observed reflectivity is close to the zero.

Similar decomposition-based plots can be used for probabilistic predictions of discrete events (e.g., the *reliability diagram* and the *discrimination diagram*; see Wilks, 2011, Chapter 8, and the R package **verification**) and have an advantage over the ROC plot since they display the joint distribution of prediction and corresponding observations and thus can reveal forecast biases.

> **R tip:** Conditional quantile plots can be generated in R using the function `conditional.quantile` from the package **verification**.

When one has samples from a predictive distribution (as described in Section 6.1.2) or an ensemble forecasting model (such as described in Appendix F), there are additional graphical assessments that can be informative to evaluate a model's predictive performance. Consider the so-called *verification ranked histogram*. Suppose we have n_f different predictive situations, each with an observation (an element of $\mathbf{Z}_v$, say Z_v^i, for $i = 1, \ldots, n_f$) to be used in verification, and for each of these predictions we have n_s samples from the predictive distribution, say $[\mathbf{Z}_{epd}^i | \mathbf{Z}_b]$, $i = 1, \ldots, n_f$, where $\mathbf{Z}_{epd}$ is a sample of size n_s (note that we could just as easily consider the ppd here). For each of the n_f predictive situations we calculate the rank of the observation relative to the ordered n_s samples; for example, if the observation is less than the smallest sample member, then it gets a rank of 1, if it is larger than the largest sample member, it gets a rank of $n_s + 1$, and so on. If the observation and the samples are from the same distribution, then the rank of the observation should be uniformly distributed (since it is equally likely to fall anywhere in the sample). Thus, we plot the n_f ranks in a histogram and look for deviations from uniformity. As shown in Wilks (2011, Chapter 8), deviations from uniformity can suggest problems such as bias or over-/under-dispersion.

As an example, Figure 6.5 shows verification histograms for three cases of (simulated) observations using the `Rankhist` and `PlotRankhist` functions in the package **SpecsVerification**. Each example is based on $n_f = 2000$ verification observations (i.e., $\{Z_v^i, i = 1, \ldots, 2000\}$) and $n_s = 20$ samples from the associated predictive distribution $[\mathbf{Z}_{epd}^i | \mathbf{Z}_b]$ for each of these verification observations. The left panel shows a case where the predictive distribution is under-dispersed relative to the observations and the right panel shows a case where the predictions are biased low relative to the observations. The center

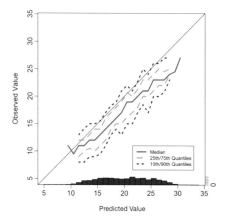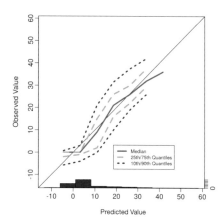

Figure 6.4: Left: Conditional quantile plot for 1000 simulated observations and predictions in which the model produces predictions that are biased approximately 3 units high relative to the observations. Right: Conditional quantile plot for the IDE model predictions in the missing 20-minute gap in Lab 6.1. These figures were obtained using the `conditional.quantile` function in the **verification** R package. Note that the x-axis gives the histogram associated with the verification observations $\{Z_v^i, \ i = 1, \ldots, n_f\}$ and the colored lines in the plot correspond to smooth quantiles from the conditional distribution of predicted values for each of these verification observations.

panel shows a case where the observations and predictions are from the same distribution, which implies *rank uniformity*. Note that there is a reasonable amount of sampling variability in these rank histograms. It is fairly straightforward to use a chi-squared test to test a null hypothesis that the histogram corresponds to a uniform distribution (see Weigel, 2012). The **SpecsVerification** package will implement this test in the context of the rank histogram. For the simulated example, the p-values for the left-panel and right-panel cases in Figure 6.5 are very close to 0, resulting in rejection of the null hypothesis of rank uniformity, whereas the case represented by the center panel has a p-value close to 0.8, so that rank uniformity is not rejected.

The graphical methods described here are not really designed for spatio-temporal data. One might be able to consider predictions at different time periods and spatial locations as different cases for comparison, but spatio-temporal dependence is not explicitly accounted for in such comparisons. This could be problematic as predictions in close proximity in space and time are spatio-temporally correlated, and it is therefore relatively easy to select a subset of points that indicate that predictions are biased, when in reality they are not.

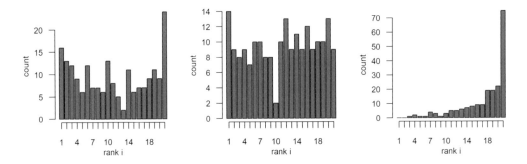

Figure 6.5: Verification ranked histograms corresponding to $n_f = 2000$ simulated observations and $n_s = 20$ samples from the associated predictive distribution for these 2000 observations. Left: the predictive distribution is under-dispersed relative to the observations; Center: the predictive distribution and the observation distribution appear the same (rank uniformity); Right: the predictive distribution is biased low relative to the observations. This figure was obtained using the `Rankhist` function in the **SpecsVerification** R package.

Any apparent bias could be a fortuitous outcome of the space-time locations chosen for validation. One way to get around this issue is to consider predictions at different (well-separated) time points at the same location in space (so as to break the spatio-temporal dependence). Then one could look at several such plots for different locations in space to gain an appreciation for the geographical influence on model performance. In the context of the rank histogram, there have been some attempts to consider multivariate predictands (e.g., multiple locations in space and/or multivariate responses), but the challenge then is to develop ranks in this multivariate setting. Perhaps the most useful such approach is based on the so-called *minimum spanning tree* histograms; see the summary in Wilks (2011, Chapter 8). The development of graphical diagnostics for spatio-temporal data is very much a research topic at the time of writing.

6.2.3   Sensitivity Analysis

An important part of model evaluation is the notion of *robustness*. Informally, we might say that model robustness is an evaluation of whether certain model assumptions have too much influence on model predictions. (This is a bit different from the more classical topic of robust estimation of model parameters.) Here we focus on the relatively simple notion of *sensitivity analysis* in the context of spatio-temporal modeling. In a sensitivity analysis, we evaluate how much our predictions change as we vary some aspect of our model (e.g., the number of basis functions or the degree of spatial dependence in an error distribution). We briefly describe some heuristic approaches to sensitivity analysis in this section, but we note that the validation statistics described below in Section 6.3 could also be used as metrics to evaluate model sensitivity.

In the case where we fix certain parameters at their estimates (e.g., covariance parameters in S-T kriging), we should evaluate the sensitivity of the model predictions to the estimated values. Note that a common criticism of such empirical plug-in approaches (used in an EHM implementation) is that they do not capture sufficient variability (e.g., relative to a BHM implementation) because they do not take the uncertainty of the parameter estimates directly into account. Nonparametric bootstraping could be used, but it can be challenging to take bootstrap samples that adequately represent the dependence structure in the spatio-temporal data. So, to evaluate the sensitivity of model predictions to fixing parameters at their data-based estimates, one might consider how sensitive the prediction errors are to the fixed parameters, $\boldsymbol{\theta}$, being estimated by two different methods, say using MLE and REML. Then, as in the spatial setting of Kang et al. (2009), we can consider heuristic measures such as the ratio of predictive standard deviations. In the spatio-temporal setting, this can be written as

$$\left[\frac{\text{var}(Y(\mathbf{s};t)|\mathbf{Z},\widehat{\boldsymbol{\theta}}_a)}{\text{var}(Y(\mathbf{s};t)|\mathbf{Z},\widehat{\boldsymbol{\theta}}_b)} \right]^{1/2}, \tag{6.7}$$

where $\mathbf{Z}$ represents the data, $\widehat{\boldsymbol{\theta}}_a$ and $\widehat{\boldsymbol{\theta}}_b$ are two parameter estimates (e.g., ML and REML estimates), and $\text{var}(Y(\mathbf{s};t)|\mathbf{Z},\boldsymbol{\theta})$ represents the process' predictive variance at $(\mathbf{s};t)$ for fixed $\boldsymbol{\theta}$. Clearly, if the ratio in (6.7) is close to 1, then it suggests that there is little sensitivity in the predictive standard deviations relative to differences in the parameter estimates $\widehat{\boldsymbol{\theta}}_a$ and $\widehat{\boldsymbol{\theta}}_b$.

Similarly, we might compare the standardized differences in predictive means,

$$\frac{E(Y(\mathbf{s};t)|\mathbf{Z},\widehat{\boldsymbol{\theta}}_a) - E(Y(\mathbf{s};t)|\mathbf{Z},\widehat{\boldsymbol{\theta}}_b)}{\{\text{var}(Y(\mathbf{s};t)|\mathbf{Z},\widehat{\boldsymbol{\theta}}_b)\}^{1/2}}, \tag{6.8}$$

where $E(Y(\mathbf{s};t)|\mathbf{Z},\boldsymbol{\theta})$ is the predictive mean for fixed $\boldsymbol{\theta}$. In this case, if (6.8) is close to 0, it suggests that the predictive means are not overly sensitive to these parameter-estimate differences. We also note that (6.7) and (6.8) are given for an individual location $(\mathbf{s};t)$ in space and time, but one could do additional averaging over regions in space and/or time periods and/or produce plots in space and time.

For illustration, consider the maximum temperature in the NOAA data set fitted using a Gaussian process, as in Lab 4.1. One can fit the theoretical semivariogram to the data using either least squares or weighted least squares. What, then, is the sensitivity of our predictions to the choice of fitting method? With **gstat**, one can call `fit.StVariogram` with `fit.method = 6` (default) for least squares, or `fit.method = 2` for weights based on the number of data pairs in the spatio-temporal bins used to construct the empirical semivariogram (see Cressie, 1993, Chapter 2). For a grid cell at (100°W, 34.9°N) on 14 July 1993, the ratio of the predictive standard deviations is 1.03, while the standardized difference in the predictive means is 0.00529. When can a spatio-temporal model be considered *robust* in terms of its predictions? The answer to this question largely depends on the reason why the model was fitted in the first place and is application-dependent, but, in the context of these maximum-temperature data, it is reasonable to say that the estimation method chosen does not seem to impact the predictions at the chosen space-time location in a substantial way.

In Bayesian implementations of spatio-temporal models, we may still be interested in the sensitivity of our posterior distributions to certain parameters or model assumptions. In this case we could make different model assumptions and compare samples of Y from the posterior distribution or samples of $\mathbf{Z}_{\mathrm{ppd}}$ from the ppd. Comparisons could be made using measures analogous to (6.7) and (6.8) or more general measures of distributional comparisons discussed below in Section 6.3. In the context of MCMC algorithms that generate posterior samples, this can be costly in complex models as it requires that one fit the full model with possibly many different data-model, process-model, and parameter-model distributions.

6.3   Model Validation

Recall that *model validation* is simply an attempt to determine how closely our model represents the real-world process of interest, as manifested by the data we observe. Specifically, after checking our model assumptions through diagnostics and sensitivity analysis, we can validate it against the real world. Although by no means exhaustive, this section presents some of the more common model-validation approaches that are used in practice.

6.3.1   Predictive Model Validation

One of the simplest ideas in model validation is to assess whether the data that are generated from our fitted model "look" like data that we have observed. That is, we can consider samples of $\mathbf{Z}_{\mathrm{ppd}}$ or $\mathbf{Z}_{\mathrm{epd}}$ from the ppd or the epd, respectively, as described in Section 6.1.2. Given that we have samples of $\mathbf{Z}_{\mathrm{ppd}}$ or of $\mathbf{Z}_{\mathrm{epd}}$, what do we do with them?

As in Section 6.2, we refer to these samples simply as $\mathbf{Z}_p$. We can look at any diagnostics we like to help us discern how similar these draws from the ppd or the epd are to the observed data – remember, we are trying to answer the question as to whether the observed data look reasonable based on the predictive distribution obtained from our model. These diagnostics are sometimes called *predictive diagnostics*. Here, discussion focuses on *posterior predictive diagnostics* based on the ppd, but there is an obvious analog of empirical predictive diagnostics where one considers the epd rather than the ppd.

As outlined in Gelman et al. (2014), a formalization of this notion is to consider a *discrepancy measure*, $T(\mathbf{Z}; \mathbf{Y}, \boldsymbol{\theta})$. The discrepancy $T(\cdot)$ is specified by the modeler and may be a measure of overall fit (e.g., a scoring rule such as described in Section 6.3.4) or any other feature of the data, the process, and the parameters. So, one calculates $T(\cdot)$ for each of L replicates of the simulated data, and also for the observed data.

We now change notation slightly to show in detail how posterior predictive diagnostics can be constructed. Specifically, for the simulated observations, we calculate $\{T(\mathbf{Z}_p^{(\ell)}; \mathbf{Y}^{(\ell)}, \boldsymbol{\theta}^{(\ell)}) : \ell = 1, \ldots, L\}$ for the L replicates $\{\mathbf{Z}_p^{(\ell)}\}$ sampled from $[\mathbf{Z}_p|\mathbf{Z}]$ based on the samples $\{\mathbf{Y}^{(\ell)}, \boldsymbol{\theta}^{(\ell)}\}$ from $[\mathbf{Y}, \boldsymbol{\theta}|\mathbf{Z}]$. Simple scatter plots of the discrepancy measures from the replicated data samples, $T(\mathbf{Z}_p^{(\ell)}; \mathbf{Y}^{(\ell)}, \boldsymbol{\theta}^{(\ell)})$, versus the discrepancy measure from the observed data, $T(\mathbf{Z}_p; \mathbf{Y}^{(\ell)}, \boldsymbol{\theta}^{(\ell)})$, can be informative. For example, if the points are scattered far from a 45-degree line, then we can assume that for this choice of T the model is not generating data that behave like the observations (e.g., see Gelman et al., 2014, Section 6.3).

We can make this procedure less subjective by considering *posterior predictive p-values*, which are given by

$$p_B = \Pr(T(\mathbf{Z}_p; \mathbf{Y}, \boldsymbol{\theta}) \geq T(\mathbf{Z}; \mathbf{Y}, \boldsymbol{\theta})|\mathbf{Z}),$$

where the probability is calculated based on the samples $\{T(\mathbf{Z}_p^{(\ell)}; \mathbf{Y}^{(\ell)}, \boldsymbol{\theta}^{(\ell)}) : \ell = 1, \ldots, L\}$. In general, values of p_B close to 0 or 1 cast doubt on whether the model produces data similar to the observed $\mathbf{Z}$ (relative to the chosen discrepancy measure), in which case one may need to reconsider the model formulation. It is important to reiterate that this "p-value" is best used as a diagnostic procedure, not for formal statistical testing. As mentioned, one can also construct analogous predictive diagnostics based on the prior predictive distribution (i.e., *prior predictive p-values*), the empirical predictive distribution (i.e., *empirical predictive p-values*), and the empirical marginal distribution (i.e., *empirical marginal p-values*). In the epd context, this has been formulated as a Monte Carlo test for validation (e.g., Kornak et al., 2006).

For illustration, consider the example of Section 6.1.2 (Sydney radar data set and the IDE model). We chose discrepancy measures to be the minimum ($T_{\min}$) and maximum ($T_{\max}$) radar reflectivity across the grid boxes with centroid at $s_1 = 26.25$ (i.e., a vertical transect) over the three time points shown in Figure 6.1. In Figure 6.6 we plot the empirical marginal distributions and empirical predictive distributions for these two discrepancy measures as obtained from $L = 500$ replications, together with the observed minimum and maximum. In both of these cases, and for both distributions, the p-values are greater than 0.05, suggesting a reasonable fit. Specifically, the empirical marginal p-value and empirical predictive p-value for $T_{\min}$ were 0.09 and 0.442, respectively, while the p-values for $T_{\max}$ were 0.364 and 0.33, respectively (note that the p-values we report are $\min(p_B, 1 - p_B)$).

6.3.2 Spatio-Temporal Validation Statistics

Perhaps the most common scalar validation statistic for continuous-valued spatio-temporal processes is the mean squared prediction error (MSPE), which for spatio-temporal validation sample $\{Z_v(\mathbf{s}_i; t_j) : j = 1, \ldots, T; \ i = 1, \ldots, m\}$, and corresponding predictions

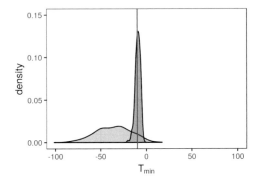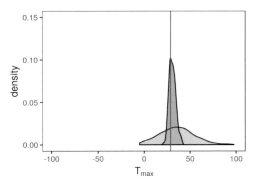

Figure 6.6: Empirical marginal distribution (green) and empirical predictive distribution (blue) densities for the minimum ($T_{\min}$, left) and maximum ($T_{\max}$, right) radar reflectivities across all grid boxes with centroid at $s_1 = 26.25$ (i.e., a vertical transect) for the times shown in Figure 6.1. In both panels, the red line denotes the observed statistic.

$\{\widehat{Z}_v(\mathbf{s}_i; t_j)\}$, is given by

$$MSPE = \frac{1}{Tm} \sum_{j=1}^{T} \sum_{i=1}^{m} \{Z_v(\mathbf{s}_i; t_j) - \widehat{Z}_v(\mathbf{s}_i; t_j)\}^2,$$

where again, for convenience, we have assumed the same number of spatial observations for each time period (which simplifies the notation, but different numbers of spatial locations for each time are easily accommodated). In this section we assume that $\{\widehat{Z}_v(\mathbf{s}_i; t_j)\}$ are predictions based on *all* of the data, $\mathbf{Z}$ (we relax that assumption in Section 6.3.3). Sometimes one might be interested in looking at MSPE for a particular time point, averaged across space, or for a particular spatial location (or region), averaged across time. The MSPE summary is so popular because it is an empirical measure of *expected squared error loss* which, when minimized, results in the S-T kriging predictor. In addition, the MSPE can be decomposed into a term corresponding to the bias (squared) of the predictor plus a term corresponding to the variance of the predictor. This is important because a large part of model-building consists of exploring the trade-offs between bias and variance. It is equally common to consider the *root mean squared prediction error* (RMSPE), which is simply the square root of the MSPE. This is sometimes favored because the units of the RMSPE are the same as those of the observations.

In cases where one wishes to protect against the influence of outliers, it is common to consider the *mean absolute prediction error* (MAPE), which can be computed from

$$MAPE = \frac{1}{Tm} \sum_{j=1}^{T} \sum_{i=1}^{m} |Z_v(\mathbf{s}_i; t_j) - \widehat{Z}_v(\mathbf{s}_i; t_j)|.$$

Although a useful summary for validation, the MAPE does not have the natural decomposition into bias and variance components that the MSPE does. But we note that for errors that do not exhibit bias, the MAPE can be interpreted as a robust version of the RMSPE.

Another common scalar validation statistic for spatio-temporal data is the so-called *anomaly correlation coefficient* (ACC). This is the usual *Pearson product moment* formula for correlation (i.e., the empirical correlation) applied to *anomalies* of the observations and predictions. Anomalies (which is a term that comes from the atmospheric sciences) are just deviations with respect to a long-term average of the observations (e.g., climatology in atmospheric applications). That is, let $Z_v'(\mathbf{s}_i; t_j) \equiv Z_v(\mathbf{s}_i; t_j) - Z_a(\mathbf{s}_i)$ and $\widehat{Z}_v'(\mathbf{s}_i; t_j) \equiv \widehat{Z}_v(\mathbf{s}_i; t_j) - Z_a(\mathbf{s}_i)$ be the anomalies of the validation observations and corresponding predictions relative to the time-averaged observation, $Z_a(\mathbf{s}_i)$, at location $\mathbf{s}_i$, for $i = 1, \ldots, m$. Then the ACC is just the empirical correlation between $\{Z_v'(\mathbf{s}_i; t_j)\}$ and $\{\widehat{Z}_v'(\mathbf{s}_i; t_j)\}$. This can be calculated across all time periods and spatial locations, or across time for each spatial location separately (and plotted on a map), or across space for each time period separately (and plotted as a time series). As with any correlation measure, the ACC does not account for bias in predictions relative to the observations, but it is still useful for spatial-field validation as it does detect phase differences (shifts) between fields. In contrast, the MSPE captures bias and variance and is not invariant to linear association.

The statistics literature has considered several simple heuristic validation metrics for spatio-temporal data. For example, in the context of within-sample validation, for spatio-temporal validation data $\{Z_v(\mathbf{s}_i; t_j)\}$ and corresponding mean predictions $\{\widehat{Z}_v(\mathbf{s}_i; t_j)\}$, one can consider the following spatial validation statistics based on residuals and predictive variances as outlined in Carroll and Cressie (1996):

$$V_1(\mathbf{s}_i) = \frac{(1/T) \sum_{j=1}^{T} \{Z_v(\mathbf{s}_i; t_j) - \widehat{Z}_v(\mathbf{s}_i; t_j)\}}{(1/T)\{\sum_{j=1}^{T} \mathrm{var}(Z_v(\mathbf{s}_i; t_j)|\mathbf{Z})\}^{1/2}}, \tag{6.9}$$

$$V_2(\mathbf{s}_i) = \left[\frac{(1/T) \sum_{j=1}^{T} \{Z_v(\mathbf{s}_i; t_j) - \widehat{Z}_v(\mathbf{s}_i; t_j)\}^2}{(1/T) \sum_{j=1}^{T} \mathrm{var}(Z_v(\mathbf{s}_i; t_j)|\mathbf{Z})} \right]^{1/2}, \tag{6.10}$$

$$V_3(\mathbf{s}_i) = \left[\frac{1}{T} \sum_{j=1}^{T} \{Z_v(\mathbf{s}_i; t_j) - \widehat{Z}_v(\mathbf{s}_i; t_j)\}^2 \right]^{1/2}, \tag{6.11}$$

where $\mathrm{var}(Z_v(\mathbf{s}_i; t_j)|\mathbf{Z})$ is the predictive variance. The summary $V_1(\mathbf{s}_i)$ provides a sense of the bias of the predictors in space (i.e., we expect this value to be close to 0 if there is no predictive bias). Similarly, $V_2(\mathbf{s}_i)$ provides a measure of the accuracy of the MSPEs and should be close to 1 if the model estimate of prediction error is reasonable. Finally, $V_3(\mathbf{s}_i)$ is a measure of goodness of prediction, with smaller values being better – this is more useful when our model is compared to some baseline model or when there is a comparison of several models. It is often helpful to plot these summary measures as a function of space to identify if certain regions in space show better predictive performance. Note that equivalent temporal validation statistics, in obvious notation $V_1(t), V_2(t), V_3(t)$, can be obtained by replacing the averages over the time points with averages over the spatial locations. These can then be evaluated analogously to the spatial versions, and plotted as time series to see if certain time periods show better performance than others.

R tip: Several R packages contain functionality for computing these simple validation statistics. However, these can be implemented directly by the user with a few lines of code using functions that take three arguments (the data, the predictions, and the prediction standard errors) as input. For example,

```
V1 <- function(z, p, pse) sum(z - p) / sqrt(sum(pse^2))
```

implements (6.9). Our suggestion is to implement them once and keep them handy!

6.3.3 Spatio-Temporal Cross-Validation Measures

The validation measures presented in Section 6.3.2 above are often used for within-sample validation, and thus they are naturally optimistic measures in the sense that the data are being used twice (once to train the model and once again to validate the model). As we have discussed in Section 6.1.3, it is much better to use a hold-out validation sample if possible, but such validation may be difficult to come by (or, in the case of spatio-temporal dependence, difficult to select). In that case, it is common to use cross-validation methods (recall Technical Note 3.1) with your favorite validation measures (e.g., $MSPE$, $MAPE$, (6.9)–(6.11) above or the scoring rules presented in Section 6.3.4). There have been a few examples in the literature of specific cross-validation statistics for spatio-temporal data, which we briefly describe here.

As a direct example in the case of leave-one-out-cross-validation (LOOCV), one might extend the notion of cross-validation residuals given in (6.6) (e.g., Kang et al., 2009) to

$$\left\{ \frac{Z(\mathbf{s}_i; t_j) - E(Z(\mathbf{s}_i; t_j)|\mathbf{Z}^{(-i,-t_j)})}{\{\text{var}(Z(\mathbf{s}_i; t_j)|\mathbf{Z}^{(-i,-t_j)})\}^{1/2}} \right\},$$

where $\mathbf{Z}^{(-i,-t_j)}$ corresponds to the data with observation $Z(\mathbf{s}_i; t_j)$ removed. These residuals can be explored for outliers and potential spatio-temporal dependence (as described in Section 6.2.1 above). Similarly, we can consider *predictive cross-validation* (PCV) and *standardized cross-validation* (SCV) measures (e.g., Kang et al., 2009),

$$PCV \equiv \left(\frac{1}{mT} \right) \sum_{j=1}^{T} \sum_{i=1}^{m} \{Z(\mathbf{s}_i; t_j) - E(Z(\mathbf{s}_i; t_j)|\mathbf{Z}^{(-i,-t_j)})\}^2 \qquad (6.12)$$

and

$$SCV \equiv \left(\frac{1}{mT} \right) \sum_{j=1}^{T} \sum_{i=1}^{m} \frac{\{Z(\mathbf{s}_i; t_j) - E(Z(\mathbf{s}_i; t_j)|\mathbf{Z}^{(-i,-t_j)})\}^2}{\text{var}(Z(\mathbf{s}_i; t_j)|\mathbf{Z}^{(-i,-t_j)})}. \tag{6.13}$$

Note the similarity between (6.11) and (6.12), and between (6.10) and (6.13). If our model is performing well, we would like to see values of PCV near 0 and values of SCV close to 1. Of course, these evaluation criteria can be considered from a K-fold cross-validation perspective as well.

6.3.4 Scoring Rules

One of the benefits of the statistical methods presented in Chapters 4 and 5 is that they give probabilistic predictions – that is, we do not just get a single prediction but, rather, a predictive distribution. This is a good thing as it allows us to account for various sources of uncertainty in our predictions. However, it presents a bit of a problem in that ideally we want to verify a distributional prediction but we have just one set of observations. We need to find a way to compare a distribution of predictions to a single realized (validation) observation. Formally, this can be done through the notion of a *score*, where the predictive distribution, say $p(z)$, is compared to the validation value, say Z, with the *score function* $S(p(z), Z)$. There is a long history in probabilistic forecast "verification," originating in the meteorology community, of favoring scoring functions that are *proper*; see Technical Note 6.2 for a description of proper scoring rules and Gneiting and Raftery (2007) for technical details.

Intuitively, proper scoring rules are expressed in such a way that a forecaster receives the best score (on average) if their forecast distribution aligns with their true beliefs. This relates to the notion of "forecast consistency" discussed in Murphy (1993), which concerns how closely the forecaster's prediction matches up with their judgement. The point here is that there may be incentives for a forecaster to *hedge* their forecast away from their true beliefs, and this should be discouraged. For example, Carvalho (2016) and Nakazono (2013) describe a situation where an expert with an established reputation might tend to report a forecast closer to the consensus of a particular group, whereas a forecaster who is just starting out might seek to increase her reputation by overstating the probabilities of particular outcomes that she thinks might be understated in the consensus. Proper scoring rules are designed such that there is no reward for this type of hedging.

Three common, and related, (strictly) proper scoring rules used in spatial and spatio-temporal prediction are the *Brier score* (BRS), the *ranked probability score* (RPS), and the *continuous ranked probability score* (CRPS). The BRS can be used to compare probability predictions for categorical variables. It is most often used when the outcomes are binary, $\{0, 1\}$, events. Assuming Z is a binary observation and $p = \text{Pr}(Z = 1|\text{data})$ comes from the model, the BRS is defined as

$$BRS(p, Z) = (Z - p)^2, \tag{6.14}$$

where, as in golf, small scores are good. (Note that in this section, where possible, we omit the space and time labels for notational simplicity and just present the rules in terms of arbitrary predictive distributions and observations.) In practice, we calculate the average BRS for a number of predictions and associated observations in the validation data set. The BRS can be decomposed into components associated with prediction "reliability, resolution, and uncertainty" (see, for example, Wilks, 2011, Chapter 8). Note that there are several other skill scores that could also be used for binary responses (e.g., the Heidke skill score, Peirce skill score, Clayton skill score, and Gilbert skill score) that are based on comparing components of a 2×2 contingency table (see Wilks, 2011, Chapter 8).

Some of the scoring rules used for binary data can be extended to multi-category predictions, although in the case of ordinal data one should take into account the relative "distance" (spread or dispersion) between categories (see, for example, the Gandin–Murphy skill score and Gerrity skill score described in Wilks, 2011, Chapter 8). The ranked probability score is a multi-category extension to the BRS given by

$$RPS(p, Z) = \frac{1}{J-1} \sum_{i=1}^{J} \left(\sum_{j=1}^{i} Z_j - \sum_{j=1}^{i} p_j \right)^2, \tag{6.15}$$

where J is the number of outcome categories, p_j is the predicted probability of the jth category, and $Z_j = 1$ if the category occurred, and $Z_j = 0$ otherwise. Note that (6.15) depends on an ordering of the categories and, when $J = 2$, we recover the Brier score (6.14). A perfect prediction leads to the case where $RPS = 0$ and the worst possible score is $RPS = 1$. RPS is strictly proper and accounts for the distance between groups, which is important for ordinal data. As with the BRS, in practice we typically calculate the average RPS for a number of predictions at different spatio-temporal locations in the validation data set.

A natural extension of the RPS to the case of a continuous response occurs if we imagine that we bin the continuous response into J ordered categories and let $J \to \infty$. The continuous ranked probability score has become one of the more popular proper scoring rules in spatio-temporal statistics. It is formulated in terms of the predictive cumulative distribution function (cdf), say $F(z)$, and is given by

$$CRPS(F, Z) = \int (\mathbb{1}\{Z \leq x\} - F(x))^2 dx,$$

where $\mathbb{1}\{Z \leq x\}$ is an indicator variable that takes the value 1 if $Z \leq x$, and the value 0 otherwise. An illustration of the procedure by which the CRPS is evaluated is shown in Figure 6.7, for an observation $Z = 6$ and $F(x)$ the normal distribution function with mean 6.5 and standard deviation 1. In this example, $CRPS = 0.331$.

In the case where the cdf F has a finite first moment, the CRPS can be written as

$$CRPS(F, Z) = E_F |z - Z| - \frac{1}{2} E_F |z - z'|, \tag{6.16}$$

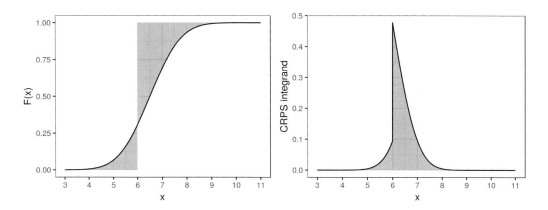

Figure 6.7: Left: The cumulative distribution function, $F(x)$, of a prediction with mean 6.5 and prediction standard error 1. The observation is $Z = 6$, and the shaded area denotes the difference between the cumulative distribution function of the observation (a step function) and the predictive distribution. Right: The integrand used to compute the CRPS, the area under the curve.

where z and z' are independent random variables with distribution function F (e.g., Gneiting and Raftery, 2007). Thus, analytical forms for the CRPS can be derived for many standard predictive cumulative distribution functions, and hence for these functions it can be computed efficiently (see, for example, the **scoringRules** R package). However, the CRPS can be difficult to compute for complex predictive distributions such as one might get from a BHM. In such situations, one can approximate the CRPS by using an empirical predictive cdf.

For example, given samples of predictions, $Z_1, \ldots, Z_m$, from F, one can show that (e.g., Jordan et al., 2017b)

$$CRPS(\widehat{F}_m, Z) = \frac{1}{m} \sum_{i=1}^{m} |Z_i - Z| - \frac{1}{2m^2} \sum_{i=1}^{m} \sum_{j=1}^{m} |Z_i - Z_j|, \qquad (6.17)$$

where the empirical cdf,

$$\widehat{F}_m(x) = \frac{1}{m} \sum_{i=1}^{m} \mathbb{1}\{Z_i \leq x\}, \qquad (6.18)$$

is substituted into (6.16). More efficient computational approaches can be used to estimate (6.16), as discussed in Jordan et al. (2017b). Note that (6.18) implicitly assumes that the $\{Z_i\}$ are iid, which is a reasonable assumption when one has multiple predictions (widely separated in time) for a given location. However, the iid assumption is not typically realistic for spatio-temporal validation data sets with multiple observations (see the discussion below on multivariate scoring rules for an alternative).

In the common case where one is only interested in evaluating the predictive distribution through its first two central moments, say μ_F and σ_F^2, Gneiting and Katzfuss (2014) suggest considering the *Dawid–Sebastiani score* (DSS),

$$DSS(F, Z) = \frac{(Z - \mu_F)^2}{\sigma_F^2} + 2 \log \sigma_F, \tag{6.19}$$

which is a proper scoring rule and is simple to compute. In the case of a Gaussian predictive density function $f(z)$, it can be shown that the DSS in (6.19) is equivalent to the so-called *logarithmic score* (LS),

$$LS(F, Z) = -\log f(Z), \tag{6.20}$$

where f is the density function associated with F. This is one of the most-used proper scoring rules in machine learning. Note that sometimes the LS is defined without the negative sign (i.e., $\log f(Z)$), in which case a larger score is better. We prefer to define it as in (6.20) so that a smaller score is better, and as we show below in Section 6.4, this form of the LS is often used when comparing models.

It can be quite useful to consider the *skill (S)* of a predictive model, which we define here as the average of the scoring rule over a range of prediction cases. For pairs $\{(F_i, Z_i) : i = 1, \ldots, N\}$, the skill is given by

$$\mathcal{S} = \frac{1}{N} \sum_{i=1}^{N} S(F_i, Z_i), \tag{6.21}$$

where S is a generic score function. We can use a *skill score (SS)* to compare predictions from models to some reference prediction method. For example,

$$SS_{\mathcal{M}} = \frac{\mathcal{S}_{\mathcal{M}} - \mathcal{S}_{\text{ref}}}{\mathcal{S}_{\text{opt}} - \mathcal{S}_{\text{ref}}}, \tag{6.22}$$

where $\mathcal{S}_{\mathcal{M}}$, $\mathcal{S}_{\text{ref}}$, and $\mathcal{S}_{\text{opt}}$ represent the skill of the model $\mathcal{M}$, the reference method, and a hypothetical optimal predictor, respectively. The skill score (6.22) takes a maximum value of 1 when the model $\mathcal{M}$ prediction is optimal, a value of 0 when the model $\mathcal{M}$ has skill equivalent to the reference method, and a value less than 0 when the model $\mathcal{M}$ has lower skill than the reference method. As noted by Gneiting and Raftery (2007), $SS_{\mathcal{M}}$ is not proper in general, even if the scoring rule used in its construction is proper.

> **R tip:** Functions to compute the Brier score, the ranked probability score, the continuous ranked probability score, and the logarithmic score can be found in the R package **verification**.

Multivariate Scoring Rules

The scoring rules given above are univariate quantities that can be averaged or more generally summarized across time and space in our setting. Although less common, there are scoring rules that explicitly account for the multivariate nature of a multivariate prediction, which can be important when there are dependencies in the process model (between variables in space or time). This addresses the *iid* caveat we put on the CRPS calculation in (6.17) and (6.18), and it applies also to the skill defined by (6.21). For example, the **scoringRules** R package implements the *energy score (ES)* discussed in Gneiting and Raftery (2007), which is given by

$$ES(F, \mathbf{Z}) = E_F ||\mathbf{z} - \mathbf{Z}|| - \frac{1}{2} E_F ||\mathbf{z} - \mathbf{z}'||, \qquad (6.23)$$

where, say, $\mathbf{Z} = (Z(\mathbf{s}_i; t_j) : i = 1, \ldots, m; j = 1, \ldots, T)$, $|| \cdot ||$ represents the Euclidean norm, and $\mathbf{z}$ and $\mathbf{z}'$ are independent random vectors with multivariate cdf F. Notice from comparison to (6.16) that (6.23) is a multivariate extension of the CRPS. Scheuerer and Hamill (2015) state that numerous studies have shown that a good performance of this score function requires a correct specification of the dependence structure in the model. When only the first and second moments are of interest, an alternative is to consider the multivariate version of the DSS given by (6.19), which we define as

$$DSS_{mv}(F, \mathbf{Z}) = \log |\mathbf{C}_F| + (\mathbf{Z} - \boldsymbol{\mu}_F)' \mathbf{C}_F^{-1} (\mathbf{Z} - \boldsymbol{\mu}_F), \qquad (6.24)$$

where $\boldsymbol{\mu}_F = E(\mathbf{Z}|\text{data})$ and $\mathbf{C}_F = \text{var}(\mathbf{Z}|\text{data})$ are the mean vector and covariance matrix of the multivariate predictive cdf F.

Scheuerer and Hamill (2015) note that variograms (which, as we discuss in Chapter 4, account for spatial and spatio-temporal dependence) consider the expected squared difference between observations, and they generalized this to define a multivariate score that they call the *variogram score of order p (VS_p)*. This can be written as

$$VS_p(F, \mathbf{Z}) = \sum_{i=1}^{mT} \sum_{j=1}^{mT} w_{ij} (|Z_i - Z_j|^p - E_F |z_i - z_j|^p)^2,$$

where w_{ij} are non-negative weights, and z_i and z_j are the ith and jth elements of a random vector, $\mathbf{z}$, from the multivariate cdf, F, and for ease of notation we write the data vector as $\mathbf{Z} = (Z_1, \ldots, Z_{mT})'$. The weights can be used to de-emphasize certain difference pairs (e.g., those that are farther apart) and $p = 2$ corresponds to the variogram defined in Chapter 4. In Lab 6.1, we illustrate the use of the ES and VS_p.

Technical Note 6.2: Proper Scoring Rules

This note follows the very intuitive description found in Bröcker and Smith (2007). Let $p(z)$ be a probability distribution of predictions of Z, which we wish to compare to an observation Z with cdf F (i.e., we wish to validate our predictive model). Let a *score* be some comparison measure between the predictive distribution and the observed value, denoted $S(p, Z)$. Typically, scores are defined so that smaller scores indicate better predictions. The score S is said to be *proper* if

$$E_F\{S(p, Z)\} \geq E_F\{S(q, Z)\} \tag{6.25}$$

for any two predictive distributions, $p(z)$ and $q(z)$, where $q(z)$ is the "true" predictive distribution. That is, (6.25) says that the expected score is minimized when the predictive distribution coincides exactly with the true predictive distribution. The scoring rule is *strictly proper* if this minimum in the expected score occurs only when $p(z) = q(z)$ for all z, that is, when the predictive distribution is the same as the true distribution. The concept of propriety is very intuitive in that it formalizes the notion that if our predictive distribution coincided with the true distribution, $q(z)$, then it should be at least as good as some other forecast distribution, $p(z)$, not equal to $q(z)$.

6.3.5   Field Comparison

A special case of validation concerns comparing spatial or spatio-temporal "fields." The idea of *field comparison* is to compare two or more spatial or spatio-temporal fields (typically gridded observations and/or model output, but note that they do not need to be gridded), in some sense, to decide if they are "different." This has been of interest for quite some time in the geophysical sciences such as meteorology, where data and processes are naturally dependent in space and time. As an example, assume we have a model that provides short-term predictions (i.e., nowcasts) of precipitation, and we wish to validate our model's predictions with weather radar data by comparing the two fields. Field comparison can also be used for inference where we would like to formally test whether two spatial fields are significantly different. Many of the validation summaries and scoring rules discussed above can be used in this context, although rigorous statistical inference has proved challenging. For example, the MSPE, MAPE, RMSPE, and ACC measures are often used for field comparison. Further, some specialized summaries have been designed to compare spatial (and, in principle, spatio-temporal) features of the process and data in these comparisons, and we discuss a few of these below.

Field-Matching Methods

One of the biggest challenges in comparing spatial fields is to decide how well features match up. For example, in the context of the aforementioned radar-nowcasting problem, the goal might be to predict a feature (say, a storm cell) that is present in the observed radar data, but the prediction might be shifted in space relative to the observations. Is such a prediction better than if the prediction of the feature is not shifted, but covers an overly broad area compared to the observed feature? Another issue is that the two fields may agree at some spatial scales of resolution, but not at others. One of the primary challenges in field comparison is to account for differences in feature location, orientation, and scale.

When comparing two spatial fields of discrete outcomes, particularly in the context of validating a predictive model, we can adapt many of the score functions to the spatial case, beyond the simple averaging in a score function, where we try to account for the different ways that spatial fields may match up. One of the most famous is the *threat score* (TS) (also known as the *critical success index*). The TS is a simple summary that was originally designed for 2×2 contingency tables. That is, it is the ratio of the number of successful predictions of an event divided by the number of situations where that event was predicted or observed, so notice that the number of correct predictions of the non-event is not considered. In the context of field comparison, consider

$$TS = \frac{A_{11}}{A_{11} + A_{10} + A_{01}}, \tag{6.26}$$

where A_{11} is the area associated with the intersection of the region where the predicted event was expected to occur with the region where it did occur, A_{10} is the area where the event was predicted to occur but did not occur, and A_{01} is the area where the event occurred but was not predicted to occur.

For illustration, we consider the example in Lab 6.1 (Sydney radar data set), where we leave out data in the 10-minute periods at 09:35 and 09:45, and then we predict the reflectivities at these time points using both an IDE model and an FRK model with spatio-temporal basis functions. When using the TS, we first need to identify the presence, or otherwise, of an event, and we do this by setting a *threshold* parameter: an observation or prediction greater than this threshold is classified as an event, while an observation or prediction less than this threshold is classified as a non-event (in practice, we often compare across multiple threshold values). Figure 6.8 shows the events and non-events in the data and in the predictions from the two models at 09:35, for a threshold of 25 dBZ. Clearly, the IDE model has been more successful in capturing "events" in this instance. The TSs for both models for thresholds varying between 15 dBZ and 25 dBZ are given in Table 6.1: we see that the IDE model outperforms the FRK models for all thresholds using this field-matching diagnostic. Of course, kriging is not designed to predict events above a threshold (e.g., Zhang et al., 2008), but neither is IDE prediction. Incorporating the dynamics appears to carry extra advantages!

R tip: Check out the **SpatialVx** package for a comprehensive suite of field-matching methods. In this example, we used the `vxstats` function from **SpatialVx** to obtain the threat scores; this function also returns other useful diagnostics, such as the probability of event detection and the false-alarm rate.

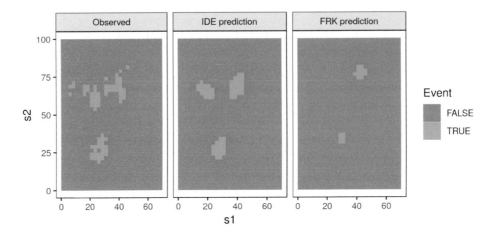

Figure 6.8: Plots showing the presence or absence of events at 09:35, obtained by thresholding the observations (left) or the IDE/FRK predictions (center and right) at 25 dBZ.

Table 6.1: Threat scores (TS) calculated using (6.26) for both the IDE predictions and the FRK predictions at 09:35 for different thresholds.

Threshold (dBZ)	TS for IDE	TS for FRK
15.00	0.73	0.32
20.00	0.58	0.21
25.00	0.37	0.11

Field-matching approaches have attempted to deal with questions of scale decompositions and feature properties (location, orientation, phase, amplitude), and a summary of such methods from a geophysical perspective can be found in Brown et al. (2012) and Gilleland et al. (2010). A brief summary of field matching from a statistical perspective can be found in Cressie and Wikle (2011, Section 5.7). In addition to using the MSPE, ACC, and score functions, methods based on scale decomposition such as EOF-based diagnostics (Branstator et al., 1993) and wavelet decompositions (Briggs and Levine, 1997) have been used successfully for field matching. In these cases, the usual measures are applied to the various scale components rather than to the full field. Examples of feature-based methods include the location-error matching approach of Ebert and McBride (2000) and the morphometric decomposition into scale, location, rotation angle, and intensity differences presented in Micheas et al. (2007).

Field Significance

It has long been of interest in the geophysical sciences to ask whether the differences in two spatial fields (or a collection of such fields) are significantly different. These two spatial fields may correspond to predictions or observations. For example, is the average maximum temperature on a grid over North America for the decade 2001–2010 significantly different from the corresponding average for the decade 1971–1980? One could consider simple pointwise two-sample t tests for the hull hypothesis of mean differences equal to zero at each grid cell. Then a Bonferroni correction of the level of significance, obtained by dividing the desired level by the number of grid cells, could be applied to deal with the multiple testing. However, such a correction leads to an overall test with very low power. Alternatively, one could look at a map of corresponding p-values and qualitatively try to identify regions in which a significant difference is present, which can be effective but lacks rigor.

However, there is not only dependence in time that must be accounted for in any test that considers a sequence of fields (e.g., the effective degrees of freedom would likely be less than the number of time replicates in the presence of positive temporal dependence), but one must also account for the spatial dependence between nearby tests when doing multiple t tests. Historical approaches have attempted to deal with these issues through effective-degrees-of-freedom modifications and Monte Carlo testing (see, for example, Livezey and Chen, 1983; Stanford and Ziemke, 1994; Von Storch and Zwiers, 2002). More recently, expanding on the famous *false discovery rate* (FDR) multiplicity mitigation approach of Benjamini and Hochberg (1995), Shen et al. (2002) developed the so-called *enhanced FDR* (EFDR) approach for spatial field comparison that uses the FDR methodology on a wavelet-based scale decomposition of the spatial fields (which deals with the spatial dependence by carrying out the testing on the decorrelated wavelet coefficients).

As an illustration, consider the difference between the mean SST anomalies in the 1970s and in the 1990s for an area of the Pacific Ocean, as shown in the left panel of Figure 6.9. Visually, it seems clear that the mean SST anomaly in the 1990s was higher than that of the 1970s. However, to check which areas are significantly different, we can run the EFDR procedure on this field of differences and then plot the field corresponding to the wavelets whose coefficients are deemed to be significantly different from zero (at the 5% level). The resulting "field significance" map, shown in the right panel of Figure 6.9, highlights the regions that were significantly warmer or cooler in the 1990s. This procedure was implemented using the **EFDR** R package.

6.4   Model Selection

It is often the case that diagnostic analysis of a model suggests that we consider an alternative model, or that we should use fewer covariates in our regression model. This section is concerned with the question of how to decide which model out of a group of models, say

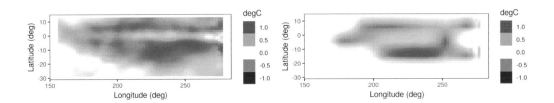

Figure 6.9: Left: Difference between the average SST anomalies in the 1990s and the average SST anomalies in the 1970s. Right: The field significance map of SST anomaly differences that were found to be significantly different from zero at the 5% level. The plot is based on the EFDR procedure and was obtained using the package **EFDR**.

$\{\mathcal{M}_1, \dots, \mathcal{M}_L\}$, is in some sense the "best." We shall assume that all of the models under consideration are reasonable from a scientific perspective, and so the choice is not obvious. First, we note that any of the summaries or score functions discussed above could be used to compare models, for example, using the skill score (6.22). In this section, we focus on more traditional statistical-model-selection approaches, although our presentation is brief. Interested readers can find more details in the excellent overviews of model comparison presented in Gelman et al. (2014), Hooten and Hobbs (2015), and the references therein.

6.4.1 Model Averaging

From a predictive perspective, it may be the case that one obtains better predictions by averaging over several models, rather than focusing on a single model. The formal methodology for doing this is through *Bayesian model averaging*, which provides a probabilistically consistent mechanism for combining posterior distributions (see Hoeting et al., 1999, for an extensive overview). Our presentation follows the concise summary in Hooten and Hobbs (2015).

Suppose we are interested in some vector quantity, $\mathbf{g}$, which can be parameters or predictions of the process or the data, and suppose we have observations, $\mathbf{Z}$, that were used to train the model. Then, for $\ell \in \{1, \dots, L\}$, we can write

$$[\mathbf{g}|\mathbf{Z}] = \sum_{\ell=1}^{L} [\mathbf{g}|\mathbf{Z}, \mathcal{M}_\ell] P(\mathcal{M}_\ell|\mathbf{Z}),$$

where $[\mathbf{g}|\mathbf{Z}, \mathcal{M}_\ell]$ is the posterior distribution of $\mathbf{g}$ given the data and the model $\mathcal{M}_\ell$; and $P(\mathcal{M}_\ell|\mathbf{Z})$ is the posterior probability of the model $\mathcal{M}_\ell$ which, given the data, gives the importance of model $\mathcal{M}_\ell$ among the collection of models. We can obtain the latter distribution from

$$P(\mathcal{M}_\ell|\mathbf{Z}) = \frac{[\mathbf{Z}|\mathcal{M}_\ell] P(\mathcal{M}_\ell)}{\sum_{j=1}^{L} [\mathbf{Z}|\mathcal{M}_j] P(\mathcal{M}_j)}, \tag{6.27}$$

where the prior probabilities for the models, $\{P(\mathcal{M}_j) : j = 1, \ldots, L\}$, have been provided. Often, all the models are assumed equally likely with *a priori* probability $1/L$, but this need not be the case. In (6.27), we also require the marginal data distribution for each model (often called the *integrated likelihood*), $[\mathbf{Z}|\mathcal{M}_\ell]$, which is simply the factor in the denominator in Bayes' rule when one is obtaining the posterior distribution under model $\mathcal{M}_\ell$. That is,

$$[\mathbf{Z}|\mathcal{M}_\ell] = \iint [\mathbf{Z}|\mathbf{Y}, \boldsymbol{\theta}, \mathcal{M}_\ell][\mathbf{Y}|\boldsymbol{\theta}, \mathcal{M}_\ell][\boldsymbol{\theta}|\mathcal{M}_\ell]\mathrm{d}\mathbf{Y}\mathrm{d}\boldsymbol{\theta}, \qquad (6.28)$$

where $[\mathbf{Z}|\mathbf{Y}, \boldsymbol{\theta}, \mathcal{M}_\ell]$ is the data model (likelihood) under model $\mathcal{M}_\ell$; and $[\mathbf{Y}|\boldsymbol{\theta}, \mathcal{M}_\ell]$ and $[\boldsymbol{\theta}|\mathcal{M}_\ell]$ are the process and prior distributions, respectively, under model $\mathcal{M}_\ell$. Unfortunately, (6.28) is typically intractable in BHM settings and cannot be calculated directly. This makes Bayesian model averaging difficult to implement for complex models, although there are various computational approaches used to obtain integrated likelihoods in this setting and in the context of Bayes factors described in Section 6.4.2 (see, for example, Congdon, 2006).

6.4.2   Model Comparison via Bayes Factors

The posterior probability for a given model expressed in (6.27) suggests a way to compare models. In particular, we note that the ratio of two such posteriors (the *posterior odds*) can be written as

$$\frac{p(\mathcal{M}_\ell|\mathbf{Z})}{p(\mathcal{M}_k|\mathbf{Z})} = \frac{[\mathbf{Z}|\mathcal{M}_\ell]P(\mathcal{M}_\ell)}{[\mathbf{Z}|\mathcal{M}_k]P(\mathcal{M}_k)} \equiv B_{\ell,k}(\mathbf{Z})\frac{P(\mathcal{M}_\ell)}{P(\mathcal{M}_k)},$$

where the ratio of the integrated likelihoods, $B_{\ell,k}(\mathbf{Z})$, is known as the *Bayes factor*. It is a constant multiplier (that depends on the data) applied to the prior odds of model $\mathcal{M}_\ell$ relative to model $\mathcal{M}_k$. So, the larger $B_{\ell,k}(\mathbf{Z})$ is, the more support there is for model $\mathcal{M}_\ell$ relative to model $\mathcal{M}_k$. Note that if we take the negative log of the Bayes factor, we obtain the difference of two logarithmic scores (recall (6.20)); using obvious notation,

$$-\log B_{\ell,k} = LS(F_\ell; \mathbf{Z}) - LS(F_k; \mathbf{Z}).$$

6.4.3   Model Comparison via Validation

We can always compare models based on the validation measure that we think is most appropriate for our problem. In this sense, any of the validation measures discussed above might be considered. In spatio-temporal statistics, we most often use a measure of predictive accuracy and typically use an out-of-sample validation or, at least, some type of cross-validation (e.g., using the MSPE or a proper scoring rule as a way to compare models). The logarithmic scoring rule (6.20) is often used in this context. Note that the log predictive density is given by $\log[\mathbf{Z}_p|\mathbf{Z}]$, where $\mathbf{Z}_p$ corresponds to spatio-temporal data that we would like to predict with our model, given data $\mathbf{Z}$ that were used to train the model.

In the context of model selection, we should explicitly denote the model under which this predictive distribution was obtained, namely, $\log[\mathbf{Z}_p|\mathbf{Z}, \mathcal{M}_\ell]$.

As stated previously, when the predictive distribution is Gaussian (which is often assumed in S-T kriging models), it is described by the predictive means, variances, and covariances. Then the negative log predictive density is the LS_{mv} score, which is just the DSS_{mv} score as we defined it in (6.24). More generally, in a BHM context, we can obtain the logarithmic score by averaging over $j = 1, \ldots, N$ MCMC samples from the predictive distribution. That is, up to Monte Carlo error, the log score based on the predictive distribution $[\mathbf{Z}_p|\mathbf{Z}]$ can be obtained as follows:

$$LS_{p,\ell} = -\log\left(\frac{1}{N}\sum_{j=1}^{N}[\mathbf{Z}_p|\mathbf{Z}, \mathbf{Y}^{(j)}, \boldsymbol{\theta}^{(j)}, \mathcal{M}_\ell]\right), \quad \ell = 1, \ldots, L, \qquad (6.29)$$

where $\mathbf{Y}^{(j)}$ and $\boldsymbol{\theta}^{(j)}$ correspond to the jth MCMC sample of the process and parameter components in the ℓth model. Thus, we can compute (6.29) for multiple models, $\ell = 1, \ldots, L$, and use this to select the "best" model(s); with our definition of LS, we prefer models with smaller values of $LS_{p,\ell}$.

As discussed above in Section 6.3.3, we often do not have a hold-out sample to use for validation, so we turn to cross-validation. For example, the K-fold cross-validation estimate of the LS based on the predictive distribution $[\mathbf{Z}_k|\mathbf{Z}^{(-k)}]$ is (up to Monte Carlo error)

$$LS_{cv,\ell} = -\frac{1}{K}\sum_{k=1}^{K}\log\left(\frac{1}{N}\sum_{j=1}^{N}[\mathbf{Z}_k|\mathbf{Z}^{(-k)}, \mathbf{Y}^{(j)}, \boldsymbol{\theta}^{(j)}, \mathcal{M}_\ell]\right), \quad \ell = 1, \ldots, L,$$

where $\mathbf{Z}_k$ corresponds to the components of $\mathbf{Z}$ in the kth hold-out sample. The challenge for many spatio-temporal BHMs is that it can be expensive to perform K-fold cross-validation in the Bayesian setting, since the model has to be fitted K times. As an alternative, we can evaluate the log predictive distribution using data from our training sample and then attempt to correct for the bias associated with using the training sample for both model-parameter estimation and prediction evaluation. The common bias correction methods are often labeled *information criteria* and are discussed briefly in the next subsection.

6.4.4   Information Criteria

Information criteria work in much the same spirit as regularization approaches; that is, they represent a trade-off between bias and variance in the sense that they penalize the bias due to overfitting that can occur when models are evaluated on the same data that were used to train them. This penalty controls for model complexity and favors models that are more parsimonious (see, for example, the discussion in Hooten and Hobbs, 2015).

Perhaps the most famous of the information criteria is the *Akaike information criterion* (AIC). In this case, the parameters, $\boldsymbol{\theta}$, are assumed to be estimated using ML estimation, and the AIC can be defined as

$$AIC(\mathcal{M}_\ell) \equiv -2\log[\mathbf{Z}|\widehat{\boldsymbol{\theta}}, \mathcal{M}_\ell] + 2p_\ell, \qquad (6.30)$$

where notice that $-\log[\mathbf{Z}|\widehat{\boldsymbol{\theta}}, \mathcal{M}_\ell]$ is the LS for model $\mathcal{M}_\ell$, and parameter estimates $\widehat{\boldsymbol{\theta}}$ are ML estimates under model $\mathcal{M}_\ell$ (having integrated out the hidden process $\mathbf{Y}$ to yield $[\mathbf{Z}|\boldsymbol{\theta}, \mathcal{M}_\ell]$). In (6.30), p_ℓ is the number of parameters estimated in model $\mathcal{M}_\ell$ (after integrating out $\mathbf{Y}$). Thus, the LS is penalized by the number of parameters in the model. When comparing two models, the model with the lower AIC is better, which, all other things being equal, favors more parsimonious models. Despite integrating out the process $\mathbf{Y}$, the AIC breaks down when one has random effects and dependence in the model $\mathcal{M}_\ell$, because the number of *effective parameters* is not equal to p_ℓ. Although there are corrections to the AIC that attempt to deal with some of these issues, one must be careful using them in these settings (see, for example, the discussion in Hodges and Sargent, 2001; Overholser and Xu, 2014). In addition, the AIC is not an appropriate criterion for model selection between different BHMs because it depends on ML estimates of parameters, and these parameters have a prior distribution on them. There is no mechanism that we know of to account for general prior distributions when using the AIC.

Another information criterion in common use is the *Bayesian information criterion* (BIC). The BIC is given by

$$BIC(\mathcal{M}_\ell) = -2\log[\mathbf{Z}|\widehat{\boldsymbol{\theta}}, \mathcal{M}_\ell] + \log(m^*)p_\ell, \qquad (6.31)$$

where m^* is the sample size (i.e., the number of spatio-temporal observations) and, as with the AIC, $\widehat{\boldsymbol{\theta}}$ is the ML estimate under $\mathcal{M}_\ell$ and p_ℓ is the number of parameters in the model (with the same caveats as in the AIC case). As with the AIC, we prefer models with smaller BIC values. Note that the BIC formula (6.31) gives larger penalties than the AIC (when $m^* > 7$) and so favors more parsimonious models than AIC. While it is referred to as a "Bayesian" information criterion, it is likewise not appropriate for model selection between different BHMs. Again the BIC relies on ML estimates of parameters and provides no way to adjust the penalty term to account for the effective number of parameters in models with random effects and dependence.

To account for the effective number of parameters in a BHM, Spiegelhalter et al. (2002) proposed the *deviance information criterion* (DIC), given by

$$DIC(\mathcal{M}_\ell) = -2\log[\mathbf{Z}|E(\boldsymbol{\theta}|\mathbf{Z}), \mathcal{M}_\ell] + 2p_\ell^D, \qquad (6.32)$$

where $E(\boldsymbol{\theta}|\mathbf{Z})$ is the posterior expectation of $\boldsymbol{\theta}$ under model $\mathcal{M}_\ell$, and p_ℓ^D is the effective number of parameters, given by

$$p_\ell^D \equiv \overline{D}_\ell - \widehat{D}_\ell. \qquad (6.33)$$

In (6.33), the *estimated model deviance* is $\widehat{D}_\ell = -2\log[\mathbf{Z}|E(\boldsymbol{\theta}|\mathbf{Z}), \mathcal{M}_\ell]$ as in (6.32), and $\overline{D}_\ell$ is the *posterior mean deviance*, which is given by

$$\overline{D}_\ell = \int -2\left(\log[\mathbf{Z}|\boldsymbol{\theta}, \mathcal{M}_\ell]\right)[\boldsymbol{\theta}|\mathbf{Z}, \mathcal{M}_\ell]d\boldsymbol{\theta}.$$

The DIC is fairly simple to calculate in MCMC implementations of BHMs, but it has several well-known limitations, primarily related to the estimate of the effective number of parameters (6.33) and the fact that it is not appropriate for mixture models (see the summary in Hooten and Hobbs, 2015). There are several alternative specifications in the literature that attempt to overcome these limitations.

The *Watanabe–Akaike information criterion (WAIC)* attempts to address some of the limitations of the DIC, and an elementwise (rather than multivariate) form can be written as

$$WAIC(\mathcal{M}_\ell) = -2\sum_{i=1}^{m^*} \log\left(\int [Z_i|\boldsymbol{\theta}, \mathcal{M}_\ell][\boldsymbol{\theta}|\mathbf{Z}, \mathcal{M}_\ell]d\boldsymbol{\theta}\right) + 2p_\ell^w, \qquad (6.34)$$

where the effective number of parameters in (6.34) is given by

$$p_\ell^w = \sum_{i=1}^{m^*} \mathrm{var}_{\theta|Z}(\log[Z_i|\boldsymbol{\theta}, \mathcal{M}_\ell]). \qquad (6.35)$$

There are other ways to define the effective number of parameters in this setting, but Gelman et al. (2014) favor (6.35) because it gives results more similar to LOOCV. Both components of the WAIC can be easily evaluated using the samples from MCMC implementations of BHMs (see, for example, Gelman et al., 2014; Hooten and Hobbs, 2015). The WAIC has several advantages over the DIC for BHM selection (it averages using the posterior predictive distribution of $\boldsymbol{\theta}$ directly, rather than conditioning on a point estimate of the parameters; it has a more realistic effective-number-of-parameters penalty; and it is appropriate both for BHMs and Bayesian mixture models). However, we sound a warning note again in that the elementwise implementation of the WAIC may not be appropriate for dependent processes such as encountered in spatio-temporal modeling (see, for example, Gelman et al., 2014; Hooten and Hobbs, 2015, for further discussion).

Hooten and Hobbs (2015) make the point that there is a similar model-selection approach that may be more appropriate for BHMs with dependent processes. In particular, consider a special case of the so-called *posterior predictive loss (PPL)* approach described by Laud and Ibrahim (1995) and Gelfand and Ghosh (1998). Define

$$PPL(\mathcal{M}_\ell) = \sum_{i=1}^{m^*}(Z_i - E(Z_i|\mathbf{Z}, \mathcal{M}_\ell))^2 + \sum_{i=1}^{m^*}\mathrm{var}(Z_i|\mathbf{Z}, \mathcal{M}_\ell), \qquad (6.36)$$

where $E(Z_i|\mathbf{Z}, \mathcal{M}_\ell)$ and $\mathrm{var}(Z_i|\mathbf{Z}, \mathcal{M}_\ell)$ are the predictive mean and predictive variance, respectively, for the observation Z_i. The PPL given by (6.36) shares with the usual information criteria a first term corresponding to the quality of prediction and a second term penalizing models that are more complex.

> **R tip:** Several R packages used in this book contain functions that help compute or return information criteria from the fitted model. The functions `AIC` and `BIC` can be used to extract the Akaike and Bayesian information criteria, respectively, from the models discussed in Chapter 3 (linear models, generalized linear models, generalized additive models), and the function `inla` in the package **INLA** may be instructed to compute the deviance and Watanabe–Akaike information criteria. Other packages such as **SpatioTemporal**, **FRK**, and **IDE** contain functions to compute the log-likelihood from the fitted model, and then some of the information criteria above could be computed; see Lab 6.1.

6.5   Chapter 6 Wrap-Up

The evaluation of a model through model checking, validation, and selection is a very important step in the model-building process. That said, it is worth making the point here that in spatio-temporal modeling we often have a strong scientific motivation to consider a specific model (e.g., a particular survey design or a particular physical or biological process model). Cressie and Wikle (2011, Chapter 1) and Ver Hoef and Boveng (2015) make the case that in these situations one should focus on building the best single model that is possible rather than carrying out model selection from several models or implementing multi-model inference. Indeed, as we have mentioned several times in this book, with observational data we never select the "true" model, but we can certainly build models that allow us to learn about or predict the spatio-temporal process. This notion of "iterating on a single model" (Ver Hoef and Boveng, 2015) may actually improve our ability to describe the real-world processes of interest, as it allows us to focus more on model checking (diagnostics) and model validation, which may suggest new features of the data about which we were unaware.

This chapter focused on model checking (Section 6.2), model validation (Section 6.3), and model selection (Section 6.4). We discussed how it is difficult to evaluate what we usually care about, the latent process, because we only have noisy and usually incompletely sampled versions of it. Although an OSSE can be used in some cases to evaluate the model with respect to the (simulated) true process of interest, we most often compare predictions obtained from our predictive distribution to various validation data. We typically favor validation data sets that are not used to train the model, and we can mimic such data through cross-validation. We mentioned how there is often a challenge in matching the validation sample with the prediction from our model, in terms of data support, although this was not a topic we covered in detail. We gave some possible spatio-temporal extensions of regression diagnostics and diagnostic plots that could be used for model checking, but we note that this is quite an under-developed area of spatio-temporal statistics.

Validation is the area of model evaluation that has seen the most activity in the spatio-temporal literature, although most of these methods were not developed explicitly for spatio-temporal processes. We are in favor of using proper scoring rules as validation summaries, particularly those that account for the uncertainty included in the predictive distribution. Model selection is a vast topic, and we just touched on some of the basic approaches there. It is worth pointing out again that many of these methods are often not appropriate in fully Bayesian contexts, or when one has dependent random effects. In that sense, there is still a lot of work to be done in developing model-selection approaches for complex spatio-temporal models.

Finally, as we have noted, spatio-temporal statistical models have primarily been used for the purpose of prediction. Disciplines such as meteorology, which have had to develop, improve, and justify predictive (forecast) models publicly on a daily basis for decades, have developed a broader terminology to consider the efficacy of predictive models. In particular, the late Alan Murphy was a pioneer in the formal study of predictive-model performance. In a classic paper, Murphy (1993) gave a list of nine "attributes" to consider when trying to describe the quality of a forecast: *bias, association, accuracy, skill, reliability, resolution, sharpness, discrimination*, and *uncertainty*. In general, his attributes describe three primary aspects of a good prediction: *consistency, quality*, and *value*. Consistency refers to how closely the prediction corresponds to the modeler's prior beliefs or judgement, given his/her understanding of the process and the data; quality corresponds to how well the prediction agrees with observations; and value simply considers if the prediction actually contributes to beneficial decision-making.[1] In statistics, we should consider these issues too, but our subject has primarily focused on bias and accuracy. These other issues are important, and this area offers a wonderful opportunity for researchers to build up this under-developed area in spatio-temporal statistics.

After going through the following Lab, you are invited to go on to the epilogical chapter for some closing remarks about spatio-temporal statistics.

---

[1]See the overview at `http://www.cawcr.gov.au/projects/verification/`

Lab 6.1: Spatio-Temporal Model Validation

In this Lab we consider the validation of two spatio-temporal models that are fitted to the same data set. To show the importance of modeling dynamics, we shall consider the Sydney radar data set and compare predictions obtained using the IDE model to those obtained using the FRK model (which does not incorporate dynamics). We shall carry out validation on data that are held out. The hold-out data set will comprise (i) a block of data spanning two time points, and (ii) a 10% random sample of the data at the other time points. We expect the IDE model to perform particularly well when validating the block of data spanning two points, where information on the dynamics is pivotal for "filling in" the temporal gaps.

For this Lab we use the **IDE** and **FRK** packages for modeling,

```
library("FRK")
library("IDE")
```

the **scoringRules** and **verification** packages for probabilistic validation,

```
library("scoringRules")
library("verification")
```

and the usual packages for handling and plotting spatio-temporal data,

```
library("dplyr")
library("ggplot2")
library("sp")
library("spacetime")
library("STRbook")
library("tidyr")
```

Step 1: Training and Validation Data

First, we load the Sydney radar data set and create a new field `timeHM` that contains the time in an hours:minutes format.

```
data("radar_STIDF", package = "STRbook")
mtot <- length(radar_STIDF)
radar_STIDF$timeHM <- format(time(radar_STIDF), "%H:%M")
```

The initial stage of model verification is to hold out data prior to fitting the model, so that these data can be compared to the predictions once the model is fitted on the retained data. As explained above, we first leave out data at two time points, namely 09:35 and 09:45, by finding the indices of the observations that were made at these times, and then removing them from the complete set of observation indices.

```
valblock_idx <- which(radar_STIDF$timeHM %in% c("09:35",
                                                "09:45"))
obs_idx <- setdiff(1:mtot, valblock_idx)
```

We next leave out 10% of the data at the other time points by randomly sampling 10% of the elements from the remaining observation indices.

```
set.seed(1)
valrandom_idx <- sample(obs_idx,
                        0.1 * length(obs_idx),
                        replace = FALSE) %>% sort()
obs_idx <- setdiff(obs_idx, valrandom_idx)
```

We can now use the indices we have generated above to construct our training data set, a validation data set for the missing time points, and a validation data set corresponding to the data missing at random from the other time points.

```
radar_obs <- radar_STIDF[obs_idx, ]
radar_valblock <- radar_STIDF[valblock_idx, ]
radar_valrandom <- radar_STIDF[valrandom_idx, ]
```

Step 2: Fitting the IDE Model

In Lab 5.2 we fitted the IDE model to the entire data set. Here, instead, we fit the IDE model to the training data set created above. As before, since this computation takes a long time, we can load the results directly from cache.

```
IDEmodel <- IDE(f = z ~ 1,
                data = radar_obs,
                dt = as.difftime(10, units = "mins"),
                grid_size = 41)

fit_results_radar2 <- fit.IDE(IDEmodel,
                              parallelType = 1)


data("IDE_Radar_results2", package = "STRbook")
```

It is instructive to compare the estimated parameters from the full data set in Lab 5.2 to the estimated parameters from the training data set in this Lab. Reassuringly, we see that the intercept, the kernel parameters (which govern the system dynamics), as well as the variance parameters, have similar estimates.

```
## load results with full data set
data("IDE_Radar_results", package = "STRbook")
with(fit_results_radar$IDEmodel, c(get("betahat")[1,1],
                                   unlist(get("k")),
                                   get("sigma2_eps"),
                                   get("sigma2_eta")))
```

```
##          par1   par2   par3   par4   par5   par6
##   0.582  0.135  2.497 -5.488 -1.861 28.384  7.271
```

```
with(fit_results_radar2$IDEmodel, c(get("betahat")[1,1],
                                    unlist(get("k")),
                                    get("sigma2_eps"),
                                    get("sigma2_eta")))
```

```
##            par1    par2    par3    par4    par5     par6
##   0.5735  0.0909  3.6784 -5.2067 -1.8174 28.8660 10.1376
```

Prediction proceeds with the function **predict**. Since we wish to predict at specific locations we now use the argument newdata to indicate where and when the predictions need to be made. In this case we supply newdata with the STIDF objects we constructed above.

```
pred_IDE_block <- predict(fit_results_radar2$IDEmodel,
                          newdata = radar_valblock)
pred_IDE_random <- predict(fit_results_radar2$IDEmodel,
                           newdata = radar_valrandom)
```

Step 3: Fitting the FRK Model

For FRK we need to specify the spatial basis functions and temporal basis functions in order to construct the spatio-temporal basis functions. For the spatial basis functions we specify two resolutions of bisquare functions regularly distributed inside the domain.

```
G_spatial <- auto_basis(manifold = plane(),    # fns on plane
                        data = radar_obs,       # project
                        nres = 2,               # 2 res.
                        type = "bisquare",      # bisquare.
                        regular = 1)            # irregular
```

Type **show_basis**(G_spatial) to visualize the locations and apertures of these basis functions. For the temporal basis functions we regularly place five bisquare functions between 0 and 12 with an aperture of 3.5.

```
t_grid <- matrix(seq(0, 12, length = 5))
G_temporal <- local_basis(manifold = real_line(), # fns on R1
                          type = "bisquare",       # bisquare
                          loc = t_grid,            # centroids
                          scale = rep(3.5, 5))     # aperture par.
```

Type `show_basis`(G_temporal) to visualize these basis functions. Finally, we construct the spatio-temporal basis functions by taking their tensor product.

```
G <- TensorP(G_spatial, G_temporal)  # take the tensor product
```

Next we construct the BAUs. These are regularly placed space-time cubes covering our spatio-temporal domain. The `cellsize` we choose below is one that is similar to that which the **IDE** function constructed when specifying `grid_size` = 41 above. We impose a convex hull as a boundary that is tight around the data points, and not extended.

```
BAUs <- auto_BAUs(manifold = STplane(),   # ST field on plane
                  type = "grid",          # gridded (not "hex")
                  data = radar_obs,       # data
                  cellsize = c(1.65, 2.38, 10), # BAU cell size
                  nonconvex_hull = FALSE, # convex boundary
                  convex = 0,             # no hull extension
                  tunit = "mins")         # time unit is "mins"
BAUs$fs = 1        # fs variation prop. to 1
```

As we did in Lab 4.2, we can take the measurement error to be that estimated elsewhere, in this case by the IDE model. Any remaining residual variation is then attributed to fine-scale variation that is modeled as white noise. Attribution of variation is less critical when validating against observational data, since the total variance is used when constructing prediction intervals.

```
sigma2_eps <- fit_results_radar2$IDEmodel$get("sigma2_eps")
radar_obs$std <- sqrt(sigma2_eps)
```

The function **FRK** is now called to fit the random-effects model using the chosen basis functions and BAUs.

```
S <- FRK(f = z ~ 1,
         BAUs = BAUs,
         data = list(radar_obs), # (list of) data
         basis = G,              # basis functions
         n_EM = 2,               # max. no. of EM iterations
         tol = 0.01)             # tol. on log-likelihood
```

Prediction proceeds using the `predict` function.

```
FRK_pred <- predict(S)
```

Since `predict` predicts over the BAUs, we need to associate each observation in our validation `STIDF`s to a BAU cell. This can be done simply using the function `over`. In the code below, the data frames `df_block_over` and `df_random_over` are data frames containing the predictions and prediction standard errors at the validation locations.

```
df_block_over <- over(radar_valblock, FRK_pred)
df_random_over <- over(radar_valrandom, FRK_pred)
```

Step 4: Organizing Predictions for Further Analysis

Having obtained our predictions and prediction standard errors from the two models, the next step is to combine them into one data frame. We take the hold-out `STIDF` from the two time points, convert it to a data frame, and then put in the FRK and IDE predictions, prediction standard errors on the process, and the prediction standard errors in observation space. We distinguish between the latter two by using the labels `predse` and `predZse`, respectively.

```
radar_valblock_df <- radar_valblock %>%
            data.frame() %>%
            mutate(FRK_pred = df_block_over$mu,
                FRK_predse = df_block_over$sd,
                FRK_predZse = sqrt(FRK_predse^2 +
                            sigma2_eps),
                IDE_pred = pred_IDE_block$Ypred,
                IDE_predse = pred_IDE_block$Ypredse,
                IDE_predZse = sqrt(IDE_predse^2 +
                            sigma2_eps))
```

For plotting purposes, it is also convenient to construct a data frame in long format, where all the predictions are put into the same column, and a second column identifies to which model the prediction corresponds.

```
radar_valblock_df_long <- radar_valblock_df %>%
                    dplyr::select(s1, s2, timeHM, z,
                        FRK_pred, IDE_pred) %>%
                    gather(type, num, FRK_pred, IDE_pred)
```

Construction of `radar_valrandom_df` and `radar_valrandom_df_long` proceeds in identical fashion to the code given above (with `block` replaced with `random`) and is thus omitted.

Step 5: Scoring

Now we have everything in place to start analyzing the prediction errors. We start by simply plotting histograms of the prediction errors to get an initial feel of the distributions of these errors from the two models. As before, we only show the code for the left-out data in `radar_valblock_df_long`.

```
ggplot(radar_valblock_df_long) +
    geom_histogram(aes(z - num, fill = type),
                   binwidth = 2, position = "identity",
                   alpha = 0.4, colour = 'black') + theme_bw()
```

Figure 6.2 shows the resulting distributions. They are relatively similar; however, a close look reveals that the errors from the FRK model have a slightly larger spread, especially for the data at the missing time points. This is a first indication that FRK, and the lack of consideration of dynamics, will be at a disadvantage when predicting the process across time points for which we have no data.

We next look at the correlation between the predictions and the observations, plotted below and shown in Figure 6.10. Again, there does not seem to be much of a difference in the distribution of the errors between the two models when the data are missing at random, but there is a noticeable difference when the data are missing for entire time points. In fact, the correlation between the predictions and observations for the FRK model is, in this case, 0.708, while that for the IDE model is 0.848.

```
ggplot(radar_valblock_df) + geom_point(aes(z, FRK_pred))
ggplot(radar_valblock_df) + geom_point(aes(z, IDE_pred))
```

It is interesting to see the effect of the absence of time points on the quality of the predictions. To this end, we can create a new data frame, which combines the validation data and the predictions, and compute the mean-squared prediction error (MSPE) for each time point.

```
MSPE_time <- rbind(radar_valrandom_df_long,
                   radar_valblock_df_long) %>%
        group_by(timeHM, type) %>%
        dplyr::summarise(MSPE = mean((z - num)^2))
```

The following code plots the evolution of the MSPE as a function of time.

```
ggplot(MSPE_time) +
  geom_line(aes(timeHM, MSPE, colour = type, group = type))
```

The evolution of the MSPE is shown in Figure 6.11, together with vertical dashed lines indicating the time points that were left out when fitting and predicting. It is remarkable to

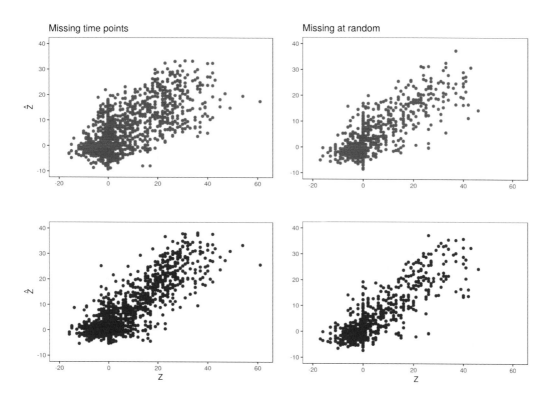

Figure 6.10: Scatter plots of the observations and predictions for the FRK model (red) and the IDE model (blue), when the data are missing for entire time points (left) and at random (right).

note how spatio-temporal FRK, due to its simple descriptive nature, suffers considerably, with an MSPE that is nearly twice that of the IDE model. Note that predictions close to this gap are also severely compromised. The IDE model is virtually unaffected by the missing data, as the trained dynamics are sufficiently informative to describe the evolution of the process at unobserved time points. At time points away from this gap, the MSPEs of the FRK and IDE models are comparable.

The importance of dynamics can be further highlighted by mapping the prediction standard errors at each time point. The plot in Figure 6.12, for which the commands are given below, reveals vastly constrasting spatial structures between the FRK prediction standard errors and the IDE prediction standard errors. Note that at other time points (we only show six adjoining time points) the prediction standard errors given by the two models are comparable.

```
ggplot(rbind(radar_valrandom_df_long, radar_valblock_df_long)) +
  geom_tile(aes(s1, s2, fill= z - num)) +
```

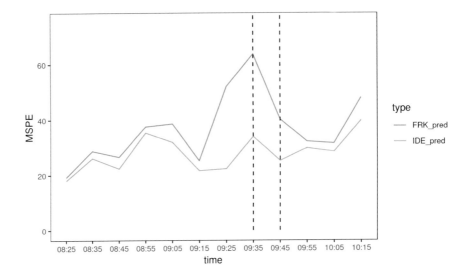

Figure 6.11: MSPE of the FRK predictions (red) and the IDE predictions (blue) as a function of time. The dotted black lines mark the times where no data were available for fitting or predicting.

```
facet_grid(type ~ timeHM) + coord_fixed() +
fill_scale(name = "dBZ") +
theme_bw()
```

Next, we compute some of the cross-validation diagnostics. We consider the bias, the predictive cross-validation (PCV) and the standardized cross-validation (SCV) measures, and the continuous ranked probability score (CRPS). Functions for the first three are simple enough to code from scratch.

```
Bias <- function(x,y) mean(x - y)          # x: val. obs.
PCV <- function(x,y) mean((x - y)^2)        # y: predictions
SCV <- function(x,y,v) mean((x - y)^2 / v)  # v: pred. variances
```

The last one, CRPS, is a bit more tedious to implement, but it is available through the **crps** function of the **verification** package. The function **crps** returns, among other things, a field CRPS containing the average CRPS across all validation observations.

```
## Compute CRPS. s is the pred. standard error
CRPS <- function(x, y, s) verification::crps(x, cbind(y, s))$CRPS
```

Finally, we compute the diagnostics for each of the FRK and IDE models. In the code below, we show how to obtain them for the validation data at the missing time points; those

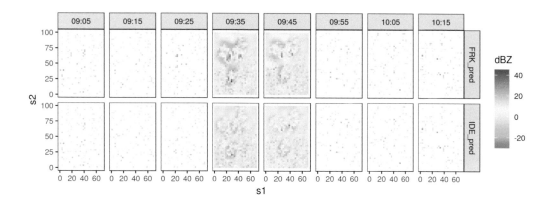

Figure 6.12: Spatial maps of prediction standard errors at the validation-data locations for the two missing time points and the adjoining six time points, based on the FRK model (top row) and the IDE model (bottom row).

for the validation data that are missing at random are obtained in a similar fashion. The diagnostics are summarized in Table 6.2, where it is clear that the IDE model outperforms the FRK model on most of the diagnostics considered here (note, in particular, the PCV for data associated with missing time points). For both models, we note that the SCV and CRPS need to be treated with care in a spatial or spatio-temporal setting, since the errors do exhibit some correlation, which is not taken into account when computing these measures.

```
Diagblock <- radar_valblock_df %>% summarise(
    Bias_FRK = Bias(FRK_pred, z),
    Bias_IDE = Bias(IDE_pred, z),
    PCV_FRK  = PCV(FRK_pred, z),
    PCV_IDE  = PCV(IDE_pred, z),
    SCV_FRK  = SCV(FRK_pred, z, FRK_predZse^2),
    SCV_IDE  = SCV(IDE_pred, z, IDE_predZse^2),
    CRPS_FRK = CRPS(z, FRK_pred, FRK_predZse),
    CRPS_IDE = CRPS(z, IDE_pred, IDE_predZse)
)
```

The multivariate energy score (ES) and variogram score of order p (VS_p) are available in R in the **scoringRules** package. The two functions we shall be using are `es_sample` and `vs_sample`. However, to compute these scores, we first need to simulate forecasts

Table 6.2: Cross-validation diagnostics for the FRK and IDE models fitted to the Sydney radar data set on data that are left out for two entire time intervals (top row) and at random (bottom row). The IDE model fares better for most diagnostics considered here, namely the bias (closer to zero is better), the predictive cross-validation measure (PCV, lower is better), the standardized cross-validation measure (SCV, closer to 1 is better), and the continuous ranked probability score (CRPS, lower is better)

	Bias		PCV		SCV		CRPS	
	FRK	IDE	FRK	IDE	FRK	IDE	FRK	IDE
Missing time points	-0.36	0.61	51.98	29.54	1.37	0.56	3.78	3.02
Missing at random	-0.16	0.02	34.48	27.94	0.79	0.91	3.13	2.76

from the predictive distribution. To do this, we not only need the marginal prediction variances, but also all the prediction covariances. Due to the size of the prediction covariance matrices, multivariate scoring can only be done on at most a few thousand predictions at a time.

For this part of the Lab, we consider the validation data at 09:35 from the Sydney radar data set.

```
radar_val0935 <- subset(radar_valblock,
                        radar_valblock$timeHM == "09:35")
n_0935 <- length(radar_val0935)  # number of validation data
```

To predict with the IDE model and store the covariances, we simply set the argument covariances to TRUE.

```
pred_IDE_block <- predict(fit_results_radar2$IDEmodel,
                          newdata = radar_val0935,
                          covariances = TRUE)
```

To predict with the FRK model and store the covariances, we also set the argument covariances to TRUE.

```
FRK_pred_block <- predict(S,
                          newdata = radar_val0935,
                          covariances = TRUE)
```

The returned objects are lists that contain the predictions in the item newdata and the covariances in an item Cov. Now, both es_sample and vs_sample are designed to compare a *sample* of forecasts to data, and therefore we need to simulate some realizations from the predictive distribution before calling these functions.

Recalling Lab 5.1, one of the easiest ways to simulate from a Gaussian random vector $\mathbf{x}$ with mean $\boldsymbol{\mu}$ and covariance matrix $\boldsymbol{\Sigma}$ is to compute the lower Cholesky factor of $\boldsymbol{\Sigma}$, call this $\mathbf{L}$, and then to compute

$$\mathbf{Z}_{\text{sim}} = \boldsymbol{\mu} + \mathbf{Le},$$

where $\mathbf{e} \sim iid\ Gau(\mathbf{0}, \mathbf{I})$. In our case, $\boldsymbol{\mu}$ contains the estimated intercept plus the predictions, while $\mathbf{L}$ is the lower Cholesky factor of whatever covariance matrix was returned in `Cov` with the measurement-error variance, σ_ϵ^2, added onto the diagonal (since we are validating against observations, and not process values). Recall that we have set σ_ϵ^2 to be the same for the FRK and the IDE models.

```
Veps <- diag(rep(sigma2_eps, n_0935))
```

Now the Cholesky factors of the predictive covariance matrices for the IDE and FRK models are given by the following commands.

```
L_IDE <- t(chol(pred_IDE_block$Cov + Veps))
L_FRK <- t(chol(FRK_pred_block$Cov + Veps))
```

The intercepts estimated by both models are given by the following commands.

```
IntIDE <- coef(fit_results_radar2$IDEmodel)
IntFRK <- coef(S)
```

We can generate 100 simulations at once by adding on the mean component (intercept plus prediction) to 100 realizations simulated using the Cholesky factor as follows.

```
nsim <- 100
E <- matrix(rnorm(n_0935*nsim), n_0935, nsim)
Sims_IDE <- IntIDE + pred_IDE_block$newdata$Ypred + L_IDE %*% E
Sims_FRK <- IntFRK + FRK_pred_block$newdata$mu + L_FRK %*% E
```

In Figure 6.13 we show one of the simulations for both the FRK and the IDE model, together with the validation data, at time point 09:35. Note how the IDE model is able to capture more structure in the predictions than the FRK model.

```
## Put into long format
radar_val0935_long <- cbind(data.frame(radar_val0935),
                    IDE = Sims_IDE[,1],
                    FRK = Sims_FRK[,1]) %>%
                gather(type, val, z, FRK, IDE)

## Plot
gsims <- ggplot(radar_val0935_long) +
    geom_tile(aes(s1, s2, fill = val)) +
  facet_grid(~ type) + theme_bw() + coord_fixed() +
  fill_scale(name = "dBZ")
```

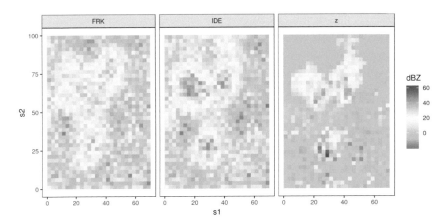

Figure 6.13: One of the 100 simulations from the predictive distribution of the FRK model (left) and the IDE model (center), and the data (not used to train the model, right) at 09:35.

We now compute the ES for both models by supplying the data and the simulations in matrix form to `es_sample`.

```
es_sample(radar_val0935$z, dat = as.matrix(Sims_IDE))
```

```
## [1] 145
```

```
es_sample(radar_val0935$z, dat = as.matrix(Sims_FRK))
```

```
## [1] 208
```

As with all proper scoring rules, lower is better, and we clearly see in this case that the IDE model has a lower ES than that for the FRK model for these validation data. For VS_p, we also need to specify weights. Here we follow the example given in the help file of **vs_sample** and set $w_{ij} = 0.5^{d_{ij}}$, where d_{ij} is the distance between the ith and jth prediction locations.

```
distances <- radar_val0935 %>%
            coordinates() %>%
            dist() %>%
            as.matrix()
weights <- 0.5^distances
```

The function **vs_sample** is then called in a similar way to **es_sample**, but this time specifying the weights and the order (we chose $p = 1$).

```
vs_sample(radar_val0935$z, dat = as.matrix(Sims_IDE),
          w = weights, p = 1)
```

```
## [1] 67423
```

```
vs_sample(radar_val0935$z, dat = as.matrix(Sims_FRK),
          w = weights, p = 1)
```

```
## [1] 79028
```

As expected, we find that the IDE model has a lower VS_1 than the FRK model. Thus, the IDE model in this case has provided better probabilistic predictions than the FRK model, both marginally and jointly.

Step 6: Model Comparison

We conclude this Lab by evaluating the Akaike information criterion (AIC) and Bayesian information criterion (BIC) for the two models. Recall that the AIC and BIC of a model $\mathcal{M}_\ell$ with p_l parameters estimated with m^* data points are given by

$$AIC(\mathcal{M}_\ell) = -2\log p(\mathbf{Z}|\widehat{\boldsymbol{\theta}}, \mathcal{M}_\ell) + 2p_l,$$
$$BIC(\mathcal{M}_\ell) = -2\log p(\mathbf{Z}|\widehat{\boldsymbol{\theta}}, \mathcal{M}_\ell) + \log(m^*)p_l.$$

For both AIC and BIC, we need the log-likelihood of the model at the estimated parameters. For the models we consider, these can be extracted using the function `loglik` in **FRK** and the negative of the function `negloglik` supplied with the `IDE` object.

```
loglikFRK <- FRK:::loglik(S)
loglikIDE <- -fit_results_radar2$IDEmodel$negloglik()
```

Before we can compute the AIC and BIC for our models, we first need to find out how many parameters were estimated. For the IDE model, the intercept, two variance parameters (one for measurement error and one for the temporal invariant disturbance term) and four kernel parameters were estimated, for a total of seven parameters. For the FRK model, the intercept, four variance parameters (one for measurement error, one for fine-scale variation, and one for each resolution of the basis functions) and four length-scale parameters (one spatial and one temporal length-scale for each resolution) were estimated, for a total of nine parameters.

```
pIDE <- 7
pFRK <- 9
```

The total number of data points used to fit the two models is

```
m <- length(radar_obs)
```

We now find the AIC and BIC for both models.

```
## Initialize table
Criteria <- data.frame(AIC = c(0, 0), BIC = c(0, 0),
                           row.names = c("FRK", "IDE"))

## Compute criteria
Criteria["FRK", "AIC"] <- -2*loglikFRK + 2*pFRK
Criteria["IDE", "AIC"] <- -2*loglikIDE + 2*pIDE
Criteria["FRK", "BIC"] <- -2*loglikFRK + pFRK*log(m)
Criteria["IDE", "BIC"] <- -2*loglikIDE + pIDE*log(m)
Criteria

##        AIC    BIC
## FRK 65980  66045
## IDE 45701  45751
```

Both the AIC and BIC are much smaller for the IDE model than for the FRK model. When the difference in the criteria is so large (in this case around 10,000), it safe to conclude that one model is a much better representation of the data. Combined with the other visualizations and diagnostics, we can conclude that the IDE model is preferrable to the FRK model for modeling and predicting with the Sydney radar data set.

As a final note, as discussed in Section 6.4.4, the AIC and BIC are not really appropriate for model selection in the presence of dependent random effects as the effective number of parameters in such settings is more than the number of parameters describing the fixed effects and covariance functions, and less than this number plus the number of basis-function coefficients (due to dependence; e.g., Hodges and Sargent, 2001; Overholser and Xu, 2014). Excluding the number of basis functions (i.e., the number of random effects) when computing the AIC and BIC clearly results in optimistic criteria; other measures such as the conditional AIC (e.g., Overholser and Xu, 2014) or the DIC, WAIC, and PPL are more suitable for such problems (see Section 6.4.4).

Pergimus (Epilogue)

These are our last words in the book, but that doesn't mean this is the end. This epilogical chapter's title is meant to convey an open-endedness to our project. Our Latin sources tell us that *pergimus* means "let's go forward" or "let's continue to progress," derived from the verb *pergere*. Here's an opportunity for you, the reader, to move beyond the previous six chapters and develop your own statistical approaches to the analysis of spatio-temporal data. You now have a sense for the motivations, main concepts, and practicalities behind spatio-temporal statistics, and the R Labs have given you an important "hands-on" perspective.

We hope you've seen enough to want more than what is in our book. A stepping-off point for more theory and methods might be in the pages of Chapters 6–9 of Cressie and Wikle (2011); and you can find more and more applications in the literature, most recently where the spatio-temporal models fitted are non-Gaussian, nonlinear, and multivariate. We expect that by the time our book comes out, new applications and software for spatio-temporal statistics will have appeared, and we hope you'll be motivated yourself to contribute.

We've tried to emphasize that spatio-temporal data are ubiquitous in the real, complex, messy world, and making sense of them depends on accounting for spatio-temporal dependencies. In the past, it's been difficult to handle the complexity of such data, the hidden processes behind them, and the sheer size of many of the data sets. Yet the principles of good statistical practice still apply – they're just a bit more involved! We should still explore our data through visualization and quantitative summaries; we should still try to build parsimonious models; we should add complexity to our models only when necessary; and we still need to evaluate our inferences through simulation and (cross-)validation. Then, after making all necessary modifications to the model, we go through the modeling–evaluation cycle again!

There are several challenges that are particular to spatio-temporal statistics. The obvious one is how to accommodate the complex dependencies that are typically present in spatio-temporal data. This is often exacerbated by the curse of dimensionality – that is, we may have a lot of data and/or are interested in predicting at a lot of locations in space and time. It's often worse when we're data-rich in one of the dimensions (e.g., space) but data-poor in the other (e.g., time, or vice versa). These challenges can be met by focusing on parsimonious parameterizations, for example when parameterizing spatio-temporal covariance functions in the descriptive approach or propagator matrices in the dynamic approach. In the latter case, using mechanistic processes to motivate parsimonious dynamic models has proven very useful.

In both cases, a very effective strategy is to treat scientifically interpretable parameters as random processes (e.g., spatial stochastic processes) at a lower level in a hierarchical statistical model. We've also seen that if we're not careful about how our models are parameterized, we can run into serious computational roadblocks. One of the most helpful solutions comes through basis-function expansions, where the modeling effort is typically redirected towards specifying multivariate-time-series models for the random coefficients of the basis functions.

Finally, we've presented some approaches to model evaluation (checking, validation, and selection) for models fitted to spatio-temporal data. However, this is very much an open area of research, and there's no "one way" to go about it. Nor should there be: using the analogy of a medical professional trying to evaluate a patient's health status, such evaluation comes from running a battery of diagnostics.

We've entered an interesting time where statistical applications are increasingly using machine-learning methods to answer all sorts of questions. All the rage at the time of writing are "deep learning" methods based on deep models, which are quite complicated but, as noted earlier, essentially hierarchical. Statistical and machine-learning versions of these models share many things in common, such as requiring a lot of training data and prior information, substantial regularization (smoothing), and high-performance computing. The biggest difference to date is that machine-learning methods don't always provide estimates of uncertainty or account for uncertainties in inputs and outputs. In the near future, we expect there will be substantially more cross-fertilization between these two paradigms, leading to new avenues of research and development in spatio-temporal modeling. This is an interesting and exciting place to be, at the intersection of statistics and the data-oriented disciplines in science, technology, engineering, and mathematics (STEM) that loosely define "data science."

We believe that the statistical methods presented in this book provide a good practical foundation for much of spatio-temporal statistics, although there are many things that we didn't cover – not because they are less important, but mainly because of space and time limitations (pun intended!). For example, some of the topics on our "should've but didn't" list are:

- spatio-temporal point processes
- spatio-temporal random sets
- continuous-time spatio-temporal processes
- spatio-temporal extremes
- multivariate spatio-temporal processes
- spatio-temporal design of sampling networks
- spatio-temporal change-of-support (resolution, alignment, scale)

- small-area panel data from surveys
- more details on estimation, computation, and implementation (especially in "big data" situations)
- more R examples of Bayesian spatio-temporal statistical analyses.

It's time to take a break, but let's continue to progress … and we invite you to share your progress and check up on ours through the book's website: `https://spacetimewithr.org`.

Appendices

A   Some Useful Matrix-Algebra Definitions and Properties

For the sake of completeness, we provide some definitions and properties of vectors and matrices that are needed to understand many of the formulas and equations in this book. Readers who are already familiar with matrix algebra can skip this section. Readers who would like more detail than the bare minimum presented here can find them in books on matrix algebra or multivariate statistics (e.g., Johnson and Wichern, 1992; Schott, 2017).

Vectors and matrices. In this book we denote a *vector* (a column of numbers) by a bold letter (Latin or Greek); for example,

$$\mathbf{a} = \begin{bmatrix} a_1 \\ a_2 \\ \vdots \\ a_p \end{bmatrix}$$

represents a p-dimensional vector, and $\mathbf{a}' = [a_1, a_2, \ldots, a_p]$ or $(a_1, a_2, \ldots, a_p)$ is its p-dimensional transpose.

We also denote a *matrix* (an array of numbers) by bold upper-case letters (Latin or Greek); for example,

$$\mathbf{A} = \begin{bmatrix} a_{11} & a_{12} & \cdots & a_{1n} \\ a_{21} & a_{22} & \cdots & a_{2n} \\ \vdots & \vdots & & \vdots \\ a_{p1} & a_{p2} & \cdots & a_{pn} \end{bmatrix}$$

is a $p \times n$ matrix, and $a_{k\ell}$ corresponds to the element in the kth row and ℓth column; sometimes it is also written as $\{a_{k\ell}\}$. The matrix transpose, $\mathbf{A}'$, is then an $n \times p$ matrix given by

$$\mathbf{A}' = \begin{bmatrix} a_{11} & a_{21} & \cdots & a_{p1} \\ a_{12} & a_{22} & \cdots & a_{p2} \\ \vdots & \vdots & & \vdots \\ a_{1n} & a_{2n} & \cdots & a_{pn} \end{bmatrix}.$$

We often consider a special matrix known as the *identity matrix*, denoted $\mathbf{I}_n$, which is an $n \times n$ diagonal matrix with ones along the main diagonal (i.e., $a_{ii} = 1$ for $i = 1, \dots, n$) and zeros for all of the off-diagonal elements (i.e., $a_{ij} = 0$, for $i \neq j$). It is sometimes the case that the dimensional subscript (in this case, n) is left off if the context is clear.

Finally, note that a vector can be thought of as a special case of a $p \times n$ matrix, where either $p = 1$ or $n = 1$.

Matrix addition. Matrix addition is defined for two matrices that have the same dimension. Then, given $p \times n$ matrices $\mathbf{A}$ and $\mathbf{B}$, with elements $\{a_{k\ell}\}$ and $\{b_{k\ell}\}$ for $k = 1, \dots, p$ and $\ell = 1, \dots, n$, respectively, the elements of the matrix sum, $\mathbf{C} = \{c_{k\ell}\} = \mathbf{A} + \mathbf{B}$, are given by

$$c_{k\ell} = a_{k\ell} + b_{k\ell}, \quad k = 1, \dots, p; \; \ell = 1, \dots, n.$$

Scalar multiplication. Consider an arbitrary scalar, c, and the $p \times n$ matrix $\mathbf{A}$. Scalar multiplication by a matrix then gives a new matrix in which each element of the matrix $\mathbf{A}$ is multiplied individually by the scalar c. Specifically, $c\mathbf{A} = \mathbf{A}c = \mathbf{G}$, where each element of $\mathbf{G} = \{g_{k\ell}\}$ is given by $g_{k\ell} = ca_{k\ell}$, for $k = 1, \dots, p$ and $\ell = 1, \dots, n$.

Matrix subtraction. As with matrix addition, matrix subtraction is defined for two matrices that have the same dimension. Consider the two $p \times n$ matrices $\mathbf{A}$ and $\mathbf{B}$, with elements $\{a_{k\ell}\}$ and $\{b_{k\ell}\}$, for $k = 1, \dots, p$ and $\ell = 1, \dots, n$, respectively. The matrix difference between $\mathbf{A}$ and $\mathbf{B}$ is then given by

$$\mathbf{C} = \{c_{k\ell}\} = \mathbf{A} - \mathbf{B} = \mathbf{A} + (-1)\mathbf{B},$$

where it can be seen that the elements of $\mathbf{C}$ are given by $c_{k\ell} = a_{k\ell} - b_{k\ell}$, for $k = 1, \dots, p$ and $\ell = 1, \dots, n$. Thus, matrix subtraction is just a combination of matrix addition and scalar multiplication (by -1).

Matrix multiplication. The product of the $p \times n$ matrix $\mathbf{A}$ and $n \times m$ matrix $\mathbf{B}$ is given by the $p \times m$ matrix $\mathbf{C}$, where $\mathbf{C} = \{c_{kj}\} = \mathbf{AB}$, with

$$c_{kj} = \sum_{\ell=1}^{n} a_{k\ell} b_{\ell j}, \quad k = 1, \dots, p; \; j = 1, \dots, m.$$

Thus, for the matrix product $\mathbf{AB}$ to exist, the number of columns in $\mathbf{A}$ must equal the number of rows in $\mathbf{B}$; so $\mathbf{C}$ always has the number of rows that are in $\mathbf{A}$ and the number of columns that are in $\mathbf{B}$.

Orthogonal matrix. A square $p \times p$ matrix $\mathbf{A}$ is said to be *orthogonal* if $\mathbf{AA}' = \mathbf{A}'\mathbf{A} = \mathbf{I}_p$.

Vector inner product. As a special case of matrix multiplication, consider two vectors, $\mathbf{a}$ and $\mathbf{b}$, both of length p. The *inner product* of $\mathbf{a}$ and $\mathbf{b}$ is given by the scalar $\mathbf{a}'\mathbf{b} = \mathbf{b}'\mathbf{a} \equiv \sum_{k=1}^{p} a_k b_k$.

Vector outer product. For another special case of matrix multiplication, consider a p-dimensional vector $\mathbf{a}$ and a q-dimensional vector $\mathbf{b}$. The *outer product* $\mathbf{ab}'$ is given by the $p \times q$ matrix

$$\mathbf{ab}' \equiv \begin{bmatrix} a_1 b_1 & a_1 b_2 & \cdots & a_1 b_q \\ a_2 b_1 & a_2 b_2 & \cdots & a_2 b_q \\ \vdots & \vdots & & \vdots \\ a_p b_1 & a_p b_2 & \cdots & a_p b_q \end{bmatrix}.$$

Note that (in general) $\mathbf{ab}' \neq \mathbf{b}'\mathbf{a}$.

Kronecker product. Consider two matrices, an $n_a \times m_a$ matrix, $\mathbf{A}$, and an $n_b \times m_b$ matrix, $\mathbf{B}$. The Kronecker product of $\mathbf{A}$ and $\mathbf{B}$ is given by the $n_a n_b \times m_a m_b$ matrix $\mathbf{A} \otimes \mathbf{B}$ defined as

$$\mathbf{A} \otimes \mathbf{B} \equiv \begin{bmatrix} a_{11}\mathbf{B} & \cdots & a_{1m_a}\mathbf{B} \\ \vdots & \vdots & \vdots \\ a_{n_a 1}\mathbf{B} & \cdots & a_{n_a m_a}\mathbf{B} \end{bmatrix}.$$

If $\mathbf{A}$ is $n_a \times n_a$ and $\mathbf{B}$ is $n_b \times n_b$, the inverse and determinant of the Kronecker product can be expressed in terms of the Kronecker product of the inverses and determinants of the individual matrices, respectively:

$$(\mathbf{A} \otimes \mathbf{B})^{-1} = \mathbf{A}^{-1} \otimes \mathbf{B}^{-1},$$

$$|\mathbf{A} \otimes \mathbf{B}| = |\mathbf{A}|^{n_b} |\mathbf{B}|^{n_a}.$$

Euclidean norm. Consider the p-dimensional real-valued vector $\mathbf{a} = [a_1, a_2, \ldots, a_p]'$. The Euclidean norm is simply the Euclidean distance in p-dimensional space, given by

$$\|\mathbf{a}\| \equiv \sqrt{\mathbf{a}'\mathbf{a}} \equiv \sqrt{\sum_{k=1}^{p} a_k^2}.$$

Symmetric matrix. A matrix $\mathbf{A}$ is said to be *symmetric* if $\mathbf{A}' = \mathbf{A}$.

Diagonal matrix. Consider the $p \times p$ matrix $\mathbf{A}$. The (main) diagonal elements of this matrix are given by the vector $[a_{11}, a_{22}, \ldots, a_{pp}]'$. Sometimes it is helpful to use a shorthand notation to construct a matrix with specific elements of a vector on the main diagonal and zeros for all other elements. For example,

$$\text{diag}(b_1, b_2, \ldots, b_q) \equiv \begin{bmatrix} b_1 & 0 & 0 & \cdots & 0 \\ 0 & b_2 & 0 & \cdots & 0 \\ \vdots & \vdots & \ddots & \vdots & \vdots \\ 0 & 0 & 0 & \cdots & b_q \end{bmatrix}.$$

Trace of a matrix. Let $\mathbf{A}$ be a $p \times p$ square matrix. We then define the *trace* of this matrix, denoted $\mathrm{trace}(\mathbf{A})$ (or $\mathrm{tr}(\mathbf{A})$) as the sum of the diagonal elements of $\mathbf{A}$; that is,

$$\mathrm{trace}(\mathbf{A}) = \sum_{k=1}^{p} a_{kk}.$$

Non-negative-definite and positive-definite matrices. Consider a $p \times p$ symmetric and real-valued matrix, $\mathbf{A}$. If, for any non-zero real-valued vector $\mathbf{x}$, the scalar given by the *quadratic form* $\mathbf{x}'\mathbf{A}\mathbf{x}$ is non-negative, we say $\mathbf{A}$ is a *non-negative-definite* matrix. Similarly, if $\mathbf{x}'\mathbf{A}\mathbf{x}$ is strictly positive for any $\mathbf{x} \neq \mathbf{0}$, we say that $\mathbf{A}$ is a *positive-definite* matrix.

Matrix inverse. Consider the $p \times p$ square matrix, $\mathbf{A}$. If it exists, the matrix $\mathbf{B}$ such that $\mathbf{A}\mathbf{B} = \mathbf{B}\mathbf{A} = \mathbf{I}_p$ is known as the *inverse matrix* of $\mathbf{A}$, and it is denoted by $\mathbf{A}^{-1}$. Thus, $\mathbf{A}^{-1}\mathbf{A} = \mathbf{A}\mathbf{A}^{-1} = \mathbf{I}_p$. If the inverse exists, we say that the matrix is *invertible*. Not every square matrix has an inverse, but every positive-definite matrix is invertible (and, the inverse matrix is also positive-definite).

Matrix square root. Let $\mathbf{A}$ be a $p \times p$ positive-definite matrix. Then there exists a matrix $\mathbf{B}$ such that $\mathbf{A} = \mathbf{B}\mathbf{B} \equiv \mathbf{B}^2$ and we say that $\mathbf{B}$ is the *matrix square root* of $\mathbf{A}$ and denote it by $\mathbf{A}^{1/2}$. The matrix square root of a positive-definite matrix is also positive-definite and we can write the inverse matrix as $\mathbf{A}^{-1} = \mathbf{A}^{-1/2}\mathbf{A}^{-1/2}$, where $\mathbf{A}^{-1/2}$ is the inverse of $\mathbf{A}^{1/2}$.

Spectral decomposition. Let $\mathbf{A}$ be a $p \times p$ symmetric matrix of real values. This matrix can be decomposed as

$$\mathbf{A} = \sum_{k=1}^{p} \lambda_k \boldsymbol{\phi}_k \boldsymbol{\phi}_k' = \boldsymbol{\Phi}\boldsymbol{\Lambda}\boldsymbol{\Phi}',$$

where $\boldsymbol{\Lambda} = \mathrm{diag}(\lambda_1, \ldots, \lambda_p)$, $\boldsymbol{\Phi} = [\boldsymbol{\phi}_1, \ldots, \boldsymbol{\phi}_p]$, and $\{\lambda_k\}$ are called the *eigenvalues* that are associated with the *eigenvectors*, $\{\boldsymbol{\phi}_k\}$, $k = 1, \ldots, p$, which are orthogonal (i.e., $\boldsymbol{\Phi}\boldsymbol{\Phi}' = \boldsymbol{\Phi}'\boldsymbol{\Phi} = \mathbf{I}_p$). Note that for a symmetric non-negative-definite matrix $\mathbf{A}$, $\lambda_k \geq 0$, and for a symmetric positive-definite matrix $\mathbf{A}$, $\lambda_k > 0$, for all $k = 1, \ldots, p$. The matrix square root and its inverse can be written as $\mathbf{A}^{1/2} = \boldsymbol{\Phi}\,\mathrm{diag}(\lambda_1^{1/2}, \ldots, \lambda_p^{1/2})\boldsymbol{\Phi}'$ and $\mathbf{A}^{-1/2} = \boldsymbol{\Phi}\,\mathrm{diag}(\lambda_1^{-1/2}, \ldots, \lambda_p^{-1/2})\boldsymbol{\Phi}'$, respectively.

Singular value decomposition (SVD). Let $\mathbf{A}$ be a $p \times n$ matrix of real values. Then the matrix $\mathbf{A}$ can be decomposed as $\mathbf{A} = \mathbf{U}\mathbf{D}\mathbf{V}'$, where $\mathbf{U}$ and $\mathbf{V}$ are $p \times p$ and $n \times n$ orthogonal matrices, respectively. In addition, the $p \times n$ matrix $\mathbf{D}$ contains all zeros except for the (k, k)th non-negative elements, $\{d_k : k = 1, 2, \ldots, \min(p, n)\}$, which are known as *singular values*.

B General Smoothing Kernels

Consider data $\{Z_i : i = 1, \ldots, m\}$, which we can write as a vector, $\mathbf{Z} = (Z_1, \ldots, Z_m)'$. Now, a homogeneously linear (smoothing) predictor for $\mathbf{Z}$ can always be written as $\widehat{\mathbf{Z}} = \mathbf{HZ}$, where the ith row of the $m \times m$ matrix $\mathbf{H}$, sometimes referred to as the influence matrix, corresponds to smoothing weights for the prediction, $\widehat{Z}_i$; that is,

$$\widehat{Z}_i = \sum_{j=1}^{m} h_{ij} Z_j,$$

where h_{ij} corresponds to the (i, j)th element of $\mathbf{H}$ and, by definition, the elements of $\mathbf{H}$ do *not* depend on $\mathbf{Z}$. Note that both the kernel and regression predictors given in Section 3.1 and Section 3.2, respectively, are linear predictors of this form. In the case of the kernel predictors, h_{ij} corresponds to the kernel evaluated at location i and j. For the regression case, $\mathbf{H} = \mathbf{X}(\mathbf{X}'\mathbf{X})^{-1}\mathbf{X}'$ (sometimes called the "hat" matrix in books on regression). The difference is that, in general, under a kernel model, $\mathbf{H}$ gives more weight to locations that are near to each other, whereas standard regression matrices do not necessarily do so, although so-called local linear regression approaches do (see, for example, James et al., 2013). There are several useful properties of the general linear smoothing matrix, $\mathbf{H}$, used in the linear predictor. First, as we have noted, if one has m observations but they are statistically dependent, then there are effectively fewer than m degrees of freedom (e.g., some of the information is redundant due to the dependence). Specifically, the *effective degrees of freedom* in the sample of m observations are given by the trace of the matrix $\mathbf{H}$,

$$df_{\text{eff}} = \text{tr}(\mathbf{H}) = \sum_{i=1}^{m} h_{ii}.$$

Another important property of linear predictors of this form is that we can obtain the LOOCV estimate (see Technical Note 3.1) without actually having to refit the model. That is, in the case of evaluating the MSPE, the LOOCV statistic is given by

$$CV_{(m)} = \frac{1}{m} \sum_{i=1}^{m} (Z_i - \widehat{Z}_i^{(-i)})^2 = \frac{1}{m} \sum_{i=1}^{m} \left(\frac{Z_i - \widehat{Z}_i}{1 - h_{ii}} \right)^2, \tag{B.1}$$

and the so-called *generalized cross-validiation* statistic is given by replacing the denominator in the right-hand side of (B.1) by $(1 - \text{tr}(\mathbf{H})/m)$.

In cases where *regularization* is considered in the context of the linear predictor (e.g., when we wish to shrink the parameters toward zero by using, for example, a *ridge regression* (L_2-norm) penalty; see Technical Note 3.4), we can write $\mathbf{H} = \mathbf{X}(\mathbf{X}'\mathbf{X} + \mathbf{R})^{-1}\mathbf{X}'$ (with $\mathbf{R} = \lambda\mathbf{I}$ in the ridge-regression case), and the effective degrees of freedom and LOOCV properties are still valid (see James et al., 2013). As discussed in Technical Note 3.4, a lasso (L_1-norm) penalty can also be used for regularization, but the smoothing kernel has no closed form in this case.

C   Estimation and Prediction for Dynamic Spatio-Temporal Models

Estimation and prediction for linear dynamic spatio-temporal models (DSTMs) with Gaussian errors can sometimes be done using methods developed for state-space models (when there are many more temporal observations than spatial locations). In particular, after conditioning on parameter estimates, the hidden (state) process can be predicted using a Kalman filter or smoother, and the parameters might be estimated using an expectation-maximization (EM) algorithm or a Markov chain Monte Carlo (MCMC) algorithm. This appendix illustrates, first, a method-of-moments estimation approach that is common in vector autoregession modeling in time series, and second, a detailed description of parameter estimation and prediction of the process in linear DSTMs with Gaussian errors using the Kalman filter, Kalman smoother, and the EM algorithm.

C.1   Estimation in Vector Autoregressive Spatio-Temporal Models via the Method of Moments

In traditional vector autoregressive (VAR) time-series applications, the autoregressive process is assumed to correspond directly to the data-generating process (i.e., there is no separate data model and process model). In the spatio-temporal context this implies a model such as

$$\mathbf{Z}_t = \mathbf{M}\mathbf{Z}_{t-1} + \boldsymbol{\eta}_t, \quad \boldsymbol{\eta}_t \sim Gau(\mathbf{0}, \mathbf{C}_\eta), \tag{C.1}$$

for $t = 1, \ldots, T$, where we assume that $\mathbf{Z}_0$ is known and recall that $\mathbf{Z}_t = (Z_t(\mathbf{s}_1), \ldots, Z_t(\mathbf{s}_m))'$. Estimation of the matrices $\mathbf{M}$ and $\mathbf{C}_\eta$ can be obtained via maximum likelihood, least squares, or the method of moments (see Lütkepohl, 2005, Chapter 3). We illustrate the latter here.

For simplicity, we assume $\{\mathbf{Z}_t\}$ has mean zero and is second-order stationary in time. If we post-multiply both sides of (C.1) by $\mathbf{Z}'_{t-1}$ and take the expectation, we get,

$$E(\mathbf{Z}_t \mathbf{Z}'_{t-1}) = \mathbf{M}E(\mathbf{Z}_{t-1}\mathbf{Z}'_{t-1}),$$

which we write as

$$\mathbf{C}_z^{(1)} = \mathbf{M}\mathbf{C}_z^{(0)}. \tag{C.2}$$

Recall from Chapter 2 that $\mathbf{C}_z^{(\tau)}$ is the lag-τ spatial covariance matrix for $\{\mathbf{Z}_t\}$. Now, (C.2) implies that

$$\mathbf{M} = \mathbf{C}_z^{(1)}(\mathbf{C}_z^{(0)})^{-1}. \tag{C.3}$$

Similarly, if we post-multiply (C.1) by $\mathbf{Z}'_t$ and take expectations, we can show that

$$\mathbf{C}_\eta = \mathbf{C}_z^{(0)} - \mathbf{M}\mathbf{C}_z^{(1)'} = \mathbf{C}_z^{(0)} - \mathbf{C}_z^{(1)}(\mathbf{C}_z^{(0)})^{-1}\mathbf{C}_z^{(1)'}. \tag{C.4}$$

It follows that the method-of-moments estimators (where empirical moments are equated with theoretical moments) of (C.3) and (C.4) are given by

$$\widehat{\mathbf{M}} = \widehat{\mathbf{C}}_z^{(1)}(\widehat{\mathbf{C}}_z^{(0)})^{-1}, \tag{C.5}$$

$$\widehat{\mathbf{C}}_\eta = \widehat{\mathbf{C}}_z^{(0)} - \widehat{\mathbf{C}}_z^{(1)}(\widehat{\mathbf{C}}_z^{(0)})^{-1}\widehat{\mathbf{C}}_z^{(1)\prime}. \tag{C.6}$$

In (C.6), the empirical lag-τ covariance matrices, $\widehat{\mathbf{C}}_z^{(\tau)}$, are calculated as shown in (2.4). Note that T needs to be larger than the dimension of $\mathbf{Z}_t$ to ensure that $\widehat{\mathbf{C}}_z^{(0)}$ is invertible.

As we have said throughout this book, we prefer to consider DSTMs that have a separate data and process model. Estimation for these models is described below in Appendix C.2. So, what is the benefit of the method-of-moments approach in the context of DSTMs? In cases where the signal-to-noise ratio is high, the estimates given by (C.5) and (C.6) can provide reasonable estimates for exploratory data analysis. We illustrate an example using method-of-moments estimation in Lab 5.3. Specifically, assume that we project the spatial-mean-centered data onto orthogonal basis functions, $\mathbf{\Phi}$: $\boldsymbol{\alpha}_t = \mathbf{\Phi}'(\mathbf{Z}_t - \widehat{\boldsymbol{\mu}})$. We then assume that the projected data come from the model, $\boldsymbol{\alpha}_t = \mathbf{M}\boldsymbol{\alpha}_{t-\tau} + \boldsymbol{\eta}_t$, and we obtain estimates $\widehat{\mathbf{M}}$ and $\widehat{\mathbf{C}}_\eta$ based on the projected data. One can then produce forecasts such as $\widehat{\boldsymbol{\alpha}}_{T+\tau} = \widehat{\mathbf{M}}\widehat{\boldsymbol{\alpha}}_T$, with estimated forecast covariance matrix, $\widehat{\mathbf{C}}_\alpha = \widehat{\mathbf{M}}\widehat{\mathbf{C}}_\alpha^{(0)}\widehat{\mathbf{M}}' + \widehat{\mathbf{C}}_\eta$, where $\widehat{\mathbf{C}}_\alpha^{(0)}$ is the empirical estimate of $E(\boldsymbol{\alpha}_t\boldsymbol{\alpha}_t')$. To obtain a forecast for $\widehat{\mathbf{Z}}_{T+\tau}$, one would have to multiply the forecast $\widehat{\boldsymbol{\alpha}}_{T+\tau}$ by the basis-function matrix and add back the spatial mean: $\widehat{\mathbf{Z}}_{T+\tau} = \widehat{\boldsymbol{\mu}} + \mathbf{\Phi}\widehat{\boldsymbol{\alpha}}_{T+\tau}$. The forecast covariance matrix is then approximated by $\widehat{\mathbf{C}}_Z = \mathbf{\Phi}\widehat{\mathbf{C}}_\alpha\mathbf{\Phi}'$, where we have ignored the truncation and measurement error when projecting onto the basis functions. Although this procedure is somewhat *ad hoc*, it is simple and can give a quick forecast. More importantly, the parameter estimates in this procedure would be used as starting values in the state-space EM algorithm described in Appendix C.2. This is demonstrated in the second portion of Lab 5.3.

For completeness, note that when one makes the assumption that the initial spatial data vector $\mathbf{Z}_0$ is known, it can be shown that, conditional on $\mathbf{Z}_0$, maximum likelihood, least squares, and method-of-moments estimation all give equivalent estimates, $\widehat{\mathbf{M}}$ and $\widehat{\mathbf{C}}_\eta$ (see, for example, Harvey, 1993, Section 7.4).

C.2 Prediction and Estimation in Fully Parameterized Linear DSTMs

Traditionally, from the data model,

$$\mathbf{Z}_t = \mathbf{H}_t\mathbf{Y}_t + \boldsymbol{\varepsilon}_t, \quad \boldsymbol{\varepsilon}_t \sim Gau(\mathbf{0}, \mathbf{C}_{\epsilon,t}), \tag{C.7}$$

and from the process model,

$$\mathbf{Y}_t = \mathbf{M}\mathbf{Y}_{t-1} + \boldsymbol{\eta}_t, \quad \boldsymbol{\eta}_t \sim Gau(\mathbf{0}, \mathbf{C}_\eta), \tag{C.8}$$

we obtain a hierarchical model (HM). Note that we have assumed here that there is no additive offset in the data model and that the process has mean zero to simplify the exposition.

Next we can perform prediction on the hidden process via the Kalman filter and Kalman smoother if the parameter matrices are all known. In practice these are not known, and estimates are sometimes used in their place, which is an empirical hierarchical model (EHM) approach. Note that although in general $\mathbf{M}$ could depend on time (and hence would be written as $\mathbf{M}_t$), we consider the simpler time-invariant case here.

Sequential Prediction of the Process via Kalman Filtering and Smoothing

In Chapter 1 we discussed the notions of smoothing, filtering, and forecasting. Before we show the filtering and smoothing distributions and algorithms, we need to define some notation and terms. In particular, let $\mathbf{w}_{c:d} \equiv \{\mathbf{w}_c, \ldots, \mathbf{w}_d\}$, for the generic vector $\mathbf{w}_t$ at times $t \in \{c, c+1, \ldots, d-1, d\}$. Then we define the *forecasting distribution* to be the distribution of $\mathbf{Y}_t$ given all of the observations that occur before time t, namely, $[\mathbf{Y}_t|\mathbf{Z}_{1:t-1}]$. We also define the *filtering distribution* to be the distribution of $\mathbf{Y}_t$ given all of the observations up to and including time t, namely, $[\mathbf{Y}_t|\mathbf{Z}_{1:t}]$. Finally, we define the *smoothing distribution* to be the distribution of $\mathbf{Y}_t$ given all the observations before, including, and after time t, namely, $[\mathbf{Y}_t|\mathbf{Z}_{1:T}]$, for $1 \leq t \leq T$.

The forecasting distribution is of most interest when one would like to predict the process one time step into the future; the filtering distribution is typically most useful when one seeks to "filter out" observation error from the true process as data come along sequentially (e.g., in real time); and the smoothing distribution is most useful when one retrospectively wants to smooth out the observation errors for any time in the entire observation period. Now, consider the following notation for the conditional expectations of the forecast and filtering distributions, respectively: $\mathbf{Y}_{t|t-1} \equiv E[\mathbf{Y}_t|\mathbf{Z}_{1:t-1}]$ and $\mathbf{Y}_{t:t} \equiv E[\mathbf{Y}_t|\mathbf{Z}_{1:t}]$. Similarly, define the conditional covariance matrices for the forecast error and filtering error distributions, respectively, as: $\mathbf{P}_{t|t-1} \equiv E[(\mathbf{Y}_t - \mathbf{Y}_{t|t-1})(\mathbf{Y}_t - \mathbf{Y}_{t|t-1})'|\mathbf{Z}_{1:t-1}]$ and $\mathbf{P}_{t|t} \equiv E[(\mathbf{Y}_t - \mathbf{Y}_{t|t})(\mathbf{Y}_t - \mathbf{Y}_{t|t})'|\mathbf{Z}_{1:t}]$.

In the case of linear Gaussian data models and process models given by (C.7) and (C.8), the forecast and filtering distributions can be found analytically by using standard conditional expectation/variance relationships and Bayes' Rule, respectively. In particular, the forecast and filtering distributions are denoted, respectively, by

$$\mathbf{Y}_t|\mathbf{Z}_{1:t-1} \sim Gau(\mathbf{Y}_{t|t-1}, \mathbf{P}_{t|t-1}),$$

and

$$\mathbf{Y}_t|\mathbf{Z}_{1:t} \sim Gau(\mathbf{Y}_{t|t}, \mathbf{P}_{t|t}),$$

and they can be found through the famous *Kalman filter algorithm* given in Algorithm C.1. Thus, given the initial conditions $\mathbf{Y}_{0|0} \equiv \boldsymbol{\mu}_0$ and $\mathbf{P}_{0|0} \equiv \mathbf{C}_0$ and the parameter matrices, $\{\mathbf{H}_t\}_{t=1}^T$, $\{\mathbf{C}_{\epsilon,t}\}_{t=1}^T$, $\mathbf{M}$, and $\mathbf{C}_\eta$, one can iterate sequentially between the forecast and filtering steps to obtain these distributions for all times $t = 1, \ldots, T$.

Algorithm C.1: Kalman Filter

Set initial conditions: $\mathbf{Y}_{0|0} = \boldsymbol{\mu}_0$ and $\mathbf{P}_{0|0} = \mathbf{C}_0$
for $t = 1$ to T **do**

   1. Forecast distribution step:

      (a) Obtain $\mathbf{Y}_{t|t-1} = \mathbf{M}\mathbf{Y}_{t-1|t-1}$
      (b) Obtain $\mathbf{P}_{t|t-1} = \mathbf{C}_\eta + \mathbf{M}\mathbf{P}_{t-1|t-1}\mathbf{M}'$

   2. Filtering distribution step:

      (a) Obtain the Kalman gain, $\mathbf{K}_t \equiv \mathbf{P}_{t|t-1}\mathbf{H}_t'(\mathbf{H}_t\mathbf{P}_{t|t-1}\mathbf{H}_t' + \mathbf{C}_{\epsilon,t})^{-1}$
      (b) Obtain $\mathbf{Y}_{t|t} = \mathbf{Y}_{t|t-1} + \mathbf{K}_t(\mathbf{Z}_t - \mathbf{H}_t\mathbf{Y}_{t|t-1})$
      (c) Obtain $\mathbf{P}_{t|t} = (\mathbf{I} - \mathbf{K}_t\mathbf{H}_t)\mathbf{P}_{t|t-1}$

end for

Recall that the smoothing distribution considers the distribution of the process at time t given *all* of the observations regardless of whether they come before, during, or after time t. This smoothing distribution is denoted by

$$\mathbf{Y}_t|\mathbf{Z}_{1:T} \sim \ Gau(\mathbf{Y}_{t|T}, \mathbf{P}_{t|T})$$

and, if one saves the results from the Kalman filter, this can be obtained for all t by the *Kalman smoother algorithm* (also known as the Rauch–Tung–Striebel smoother) given in Algorithm C.2.

Algorithm C.2: Kalman Smoother

Obtain $\{\mathbf{Y}_{t|t-1}, \mathbf{P}_{t|t-1}\}_{t=1}^{T}$ and $\{\mathbf{Y}_{t|t}, \mathbf{P}_{t|t}\}_{t=0}^{T}$ from the Kalman filter algorithm (Algorithm C.1)
for $t = T - 1$ to 0 **do**

   1. Obtain $\mathbf{J}_t \equiv \mathbf{P}_{t|t}\,\mathbf{M}'\,\mathbf{P}_{t+1|t}^{-1}$
   2. Obtain $\mathbf{Y}_{t|T} = \mathbf{Y}_{t|t} + \mathbf{J}_t(\mathbf{Y}_{t+1|T} - \mathbf{Y}_{t+1|t})$
   3. Obtain $\mathbf{P}_{t|T} = \mathbf{P}_{t|t} + \mathbf{J}_t(\mathbf{P}_{t+1|T} - \mathbf{P}_{t+1|t})\mathbf{J}_t'$

end for

Parameter Estimation via the EM Algorithm

The state-space approach discussed above in terms of the Kalman filter and smoother makes the assumption that the parameter matrices in the data and process models are known. This is unrealistic in most cases, and one must use the data to estimate these parameters; that is, the HM being used is an EHM. One of the most popular (and effective) ways to do this in the state-space time-series case is through the EM algorithm (recall the general EM algorithm presented in Algorithm 4.1).

The state-space version of the EM algorithm, originally developed by Shumway and Stoffer (1982), denotes by $\mathbf{Z}_{1:T}$ the observations and the unobservable latent process, and by $\mathbf{Y}_{0:T}$ the "missing data." Denote the parameters by $\mathbf{\Theta} \equiv \{\boldsymbol{\mu}_0, \mathbf{C}_0, \mathbf{C}_\eta, \mathbf{C}_\epsilon, \mathbf{M}\}$, where we assume typically that the observation matrices, $\{\mathbf{H}_t\}$, are all known. We assume that the initial distribution is given by $\mathbf{Y}_{0|0} \sim Gau(\boldsymbol{\mu}_0, \mathbf{C}_0)$, and we further assume here (for simplicity) that $\mathbf{C}_\epsilon$ corresponds to the $m \times m$ measurement-error covariance matrix for all possible observation locations (thus, $\mathbf{C}_{\epsilon,t} = \mathbf{C}_\epsilon$, for all t, so $m_t = m$ and we assume no missing observations at each time point). The EM algorithm is then based on the complete-data likelihood given by $[\mathbf{Z}_{1:T}, \mathbf{Y}_{0:T}|\mathbf{\Theta}] = \left(\prod_{t=1}^{T}[\mathbf{Z}_t|\mathbf{Y}_t]\right)\left(\prod_{t=1}^{T}[\mathbf{Y}_t|\mathbf{Y}_{t-1}]\right)[\mathbf{Y}_0]$, which again makes use of the conditional independencies in the data model and the Markov property of the process model. The EM algorithm for a linear DSTM, presented in Algorithm C.3, makes use of the Kalman smoother algorithm to evaluate both the E-step and the M-step. Note that, in addition to running the Kalman smoother at each iteration of the algorithm, we also have to obtain the so-called "lagged-one smoother" variance–covariance matrix, $\mathbf{P}_{t,t-1|T} \equiv E((\mathbf{Y}_t - \mathbf{Y}_{t|T})(\mathbf{Y}_{t-1} - \mathbf{Y}_{t-1|T})'|\mathbf{Z}_{1:T})$, for $t = T, T-1, \ldots$. This is accomplished by the so-called *lag-one covariance smoother*, which is part of Algorithm C.3. Convergence can be assessed by considering parameter changes and/or changes to the log complete-data likelihood (i.e., see (C.9) in Algorithm C.3). Typically, in the linear DSTM case, one considers the latter because there are a large number of parameters. An example of using the EM algorithm for a linear DSTM is given in Lab 5.3.

Algorithm C.3: Linear DSTM E-M Algorithm

Choose initial condition covariance matrix, $\mathbf{C}_0$
Choose starting values: $\widehat{\mathbf{\Theta}}^{(0)} = \{\widehat{\boldsymbol{\mu}}_0^{(0)}, \widehat{\mathbf{C}}_\eta^{(0)}, \widehat{\mathbf{C}}_\epsilon^{(0)}, \widehat{\mathbf{M}}^{(0)}\}$
repeat $i = 1, 2, \ldots$

    1. E-step:

- Use $\widehat{\mathbf{\Theta}}^{(i-1)}$ in the Kalman smoother (Algorithm C.2) to obtain $\{\mathbf{Y}_{t|T}^{(i-1)}, \mathbf{P}_{t|T}^{(i-1)}\}$
- Use Kalman smoother output to obtain the lag-one covariance smoother estimates

- Calculate $\mathbf{P}_{T,T-1|T}^{(i-1)} = (\mathbf{I} - \mathbf{K}_T^{(i-1)}\mathbf{H}_T)\mathbf{M}^{(i-1)}\mathbf{P}_{T-1|T-1}^{(i-1)}$
- **for** $t = T, T-1, \ldots, 2$ **do**

$$
\begin{aligned}
\mathbf{P}_{t-1,t-2|T}^{(i-1)} &= \mathbf{P}_{t-1|t-1}^{(i-1)}\mathbf{J}_{t-2}^{(i-1)\prime} + \mathbf{J}_{t-1}^{(i-1)\prime}(\mathbf{P}_{t,t-1|T}^{(i-1)} \\
&\quad - \mathbf{M}^{(i-1)}\mathbf{P}_{t-1|t-1}^{(i-1)})\mathbf{J}_{t-2}^{(i-1)\prime}
\end{aligned}
$$

- **end for**
- Calculate $\mathbf{S}_{00} \equiv \sum_{t=1}^{T}(\mathbf{P}_{t-1|T}^{(i-1)} + \mathbf{Y}_{t-1|T}^{(i-1)}\mathbf{Y}_{t-1|T}^{(i-1)\prime})$
- Calculate $\mathbf{S}_{11} \equiv \sum_{t=1}^{T}(\mathbf{P}_{t|T}^{(i-1)} + \mathbf{Y}_{t|T}^{(i-1)}\mathbf{Y}_{t|T}^{(i-1)\prime})$
- Calculate $\mathbf{S}_{10} \equiv \sum_{t=1}^{T}(\mathbf{P}_{t,t-1|T}^{(i-1)} + \mathbf{Y}_{t|T}^{(i-1)}\mathbf{Y}_{t-1|T}^{(i-1)\prime})$

2. M-step:

- Update: $\widehat{\boldsymbol{\mu}}_0^{(i)} = \mathbf{Y}_{0|T}^{(i-1)}$
- Update: $\widehat{\mathbf{M}}^{(i)} = \mathbf{S}_{10}\mathbf{S}_{00}^{-1}$
- Update: $\widehat{\mathbf{C}}_\eta^{(i)} = (1/T)(\mathbf{S}_{11} - \mathbf{S}_{10}\mathbf{S}_{00}^{-1}\mathbf{S}_{10}')$
- Update:

$$
\begin{aligned}
\widehat{\mathbf{C}}_\epsilon^{(i)} &= \frac{1}{T}\sum_{t=1}^{T}((\mathbf{Z}_t - \mathbf{H}_t\mathbf{Y}_{t|T}^{(i-1)})(\mathbf{Z}_t - \mathbf{H}_t\mathbf{Y}_{t|T}^{(i-1)})' \\
&\quad + \mathbf{H}_t\mathbf{P}_{t|T}^{(i-1)}\mathbf{H}_t')
\end{aligned}
$$

until convergence (typically, based on differences in $-2\ln(L(\boldsymbol{\Theta}|\mathbf{Z}_{1:T}, \mathbf{Y}_{0:T}))$ as calculated in (C.9):

$$
\begin{aligned}
-2\ln(L(\boldsymbol{\Theta}^{(i)}|\mathbf{Z}_{1:T}, \mathbf{Y}_{0:T}^{(i)})) &= \ln(|\widehat{\mathbf{C}}_0^{(i)}|) + (\mathbf{Y}_{0|T}^{(i)} - \widehat{\boldsymbol{\mu}}_0^{(i)})'\widehat{\mathbf{C}}_0^{-1(i)}(\mathbf{Y}_{0|T}^{(i)} - \widehat{\boldsymbol{\mu}}_0^{(i)}) \\
&\quad + T\ln(|\widehat{\mathbf{C}}_\eta^{(i)}|) + \sum_{t=1}^{T}(\mathbf{Y}_{t|T}^{(i)} - \widehat{\mathbf{M}}^{(i)}\mathbf{Y}_{t-1|T}^{(i)})'\widehat{\mathbf{C}}_\eta^{-1(i)}(\mathbf{Y}_{t|T}^{(i)} - \widehat{\mathbf{M}}^{(i)}\mathbf{Y}_{t-1|T}^{(i)}) \\
&\quad + T\ln(|\widehat{\mathbf{C}}_\epsilon^{(i)}|) + \sum_{t=1}^{T}(\mathbf{Z}_t - \mathbf{H}_t\mathbf{Y}_{t|T}^{(i)})'\widehat{\mathbf{C}}_\epsilon^{-1(i)}(\mathbf{Z}_t - \mathbf{H}_t\mathbf{Y}_{t|T}^{(i)}). \quad \text{(C.9)}
\end{aligned}
$$

Uncertainty estimates are less easily obtained for the parameter estimates than they are for the state-process estimates, but they can be obtained through considering the inverse of the associated asymptotic information matrix or by parametric bootstrap methods. Unfortunately, obtaining uncertainty estimates even for the state-process estimates is not often done in practice and, as discussed in the comments motivating DSTMs in Section 5.2.3, it can be problematic because of the potential for explosive behavior by some of the transition matrices whose parameters are within the joint confidence region.

More flexible inference for DSTMs can be accomplished by the fully hierarchical Bayesian hierarchical model (BHM); see Section 4.5.2 as well as Cressie and Wikle (2011, Chapter 8). These BHM implementations are often problem-specific, and they are often best implemented directly in R or in a so-called probabilistic programming language (e.g., Stan, WinBugs, JAGS). For an example, see the Gibbs sampler MCMC algorithm (corresponding to a BHM) to predict Mediterranean surface winds implemented in Appendix E.

C.3   Estimation for Non-Gaussian and Nonlinear DSTMs

In principle, the filtering and smoothing methods presented in Appendix C.2 can be generalized to the setting of non-Gaussian and nonlinear DSTMs (e.g., particle filters and smoothers, ensemble Kalman filters; see Chapter 8 of Cressie and Wikle, 2011). However, in the high-dimensional settings with deep BHMs with complicated parameter-dependence structures, one typically has to consider fully Bayesian implementations. As mentioned above, these implementations are often programmed "from scratch" rather than from particular R packages. As an example, see the BHM based on a linear DSTM with Gaussian error given in Appendix E.

D   Mechanistically Motivated Dynamic Spatio-Temporal Models

As discussed in Section 5.3, it can be quite useful to parameterize DSTMs by considering transition matrices that are motivated by a mechanistic model. Here, we show the details of how one can do this with a partial differential equation (PDE) and an integro-difference equation (IDE).

D.1   Example of a Process Model Motivated by a PDE: Finite Differences

Consider the case where the parameters a, b, u, and v in (5.14) vary with space and denote the two-dimensional spatial location by the vector $\mathbf{s} = (x, y)'$. Then the PDE is

$$\frac{\partial Y}{\partial t} = \frac{\partial}{\partial x}\left(a(x, y)\frac{\partial Y}{\partial x}\right) + \frac{\partial}{\partial y}\left(b(x, y)\frac{\partial Y}{\partial y}\right) + u(x, y)\frac{\partial Y}{\partial x} + v(x, y)\frac{\partial Y}{\partial y}. \quad \text{(D.1)}$$

If we consider this process on a regular two-dimensional grid and employ a standard centered finite difference in space and a forward difference in time for (D.1), we obtain a lagged nearest-neighbor relationship given by

$$
\begin{aligned}
Y_t(x,y) \;=\; & \theta_{p,1}(x,y)Y_{t-\Delta_t}(x,y) + \theta_{p,2}(x,y)Y_{t-\Delta_t}(x+\Delta_x,y) \\
& + \theta_{p,3}(x,y)Y_{t-\Delta_t}(x-\Delta_x,y) + \theta_{p,4}(x,y)Y_{t-\Delta_t}(x,y+\Delta_y) \\
& + \theta_{p,5}(x,y)Y_{t-\Delta_t}(x,y-\Delta_y),
\end{aligned}
\tag{D.2}
$$

where Δ_t is a time-discretization constant, Δ_x and Δ_y are spatial-discretization constants, and the θs are defined as

$$
\theta_{p,1}(x,y) = \left[\frac{-2a(x,y)\Delta_t}{\Delta_x^2} + \frac{-2b(x,y)\Delta_t}{\Delta_y^2}\right] + 1,
$$

$$
\theta_{p,2}(x,y) = \frac{a(x+\Delta_x,y)\Delta_t}{4\Delta_x^2} - \frac{a(x-\Delta_x,y)\Delta_t}{4\Delta_x^2} + \frac{a(x,y)\Delta_t}{\Delta_x^2} + \frac{u(x,y)\Delta_t}{2\Delta_x},
$$

$$
\theta_{p,3}(x,y) = \frac{-a(x+\Delta_x,y)\Delta_t}{4\Delta_x^2} + \frac{a(x-\Delta_x,y)\Delta_t}{4\Delta_x^2} + \frac{a(x,y)\Delta_t}{\Delta_x^2} - \frac{u(x,y)\Delta_t}{2\Delta_x},
$$

$$
\theta_{p,4}(x,y) = \frac{b(x,y+\Delta_y)\Delta_t}{4\Delta_y^2} - \frac{b(x,y-\Delta_y)\Delta_t}{4\Delta_y^2} + \frac{b(x,y)\Delta_t}{\Delta_y^2} + \frac{v(x,y)\Delta_t}{2\Delta_y},
$$

$$
\theta_{p,5}(x,y) = \frac{-b(x,y+\Delta_y)\Delta_t}{4\Delta_y^2} + \frac{b(x,y-\Delta_y)\Delta_t}{4\Delta_y^2} + \frac{b(x,y)\Delta_t}{\Delta_y^2} - \frac{v(x,y)\Delta_t}{2\Delta_y}.
$$

Thus, we see that the finite differences suggest that the neighbors of location (x,y) at the previous time (i.e., locations $(x-\Delta_x,y)$, $(x+\Delta_x,y)$, $(x,y-\Delta_y)$, and $(x,y+\Delta_y)$), as well as the location (x,y) itself, play a role in the transition from one time to the next. Note the role of the spatially varying parameters. Let $\mathbf{Y}_t$ be the process evaluated at all interior grid points at time t, and assume the process is defined to be 0 on the boundary (for ease of presentation). Then one can write (D.2) as $\mathbf{Y}_t = \mathbf{M}\mathbf{Y}_{t-\Delta_t}$, where $\mathbf{M}$ is parameterized with the elements of $\{\theta_{p,i}(x,y),\ i=1,\ldots,5\}$. Assume first for simplicity that there is no advection (i.e., $u(x,y)=0$, $v(x,y)=0$ for all locations) and the diffusion coefficients in the x- and y-directions are equal (i.e., $a(x,y)=b(x,y)$). Then, it can be shown that the transition operator $\mathbf{M}$ is still asymmetric if the diffusion coefficients vary with space. If the diffusion coefficients are constant in space and equal (i.e., $a=b$), then transition-operator asymmetry is only due to the advection component.

The type of diffusion represented in (D.1) is typically called "Fickian" diffusion. Similar finite-difference discretizations of other diffusion representations (e.g., so-called "ecological diffusion," $\nabla^2(a(x,y)Y)$) lead to different formulations of the parameters $\boldsymbol{\theta}_p$ (in terms of the coefficients $a(x,y)$), but they still correspond to a five-diagonal sparse transition operator $\mathbf{M}$. Thus, in the context of a linear DSTM process model, we typically allow the parameters $\boldsymbol{\theta}_p$ to be spatially explicit random processes with the possible addition of covariates, rather than model the specific diffusion equation coefficients (e.g., $a(x,y)$, $b(x,y)$, $u(x,y)$, and $v(x,y)$ in (D.1)) directly. (Although, one can certainly do this, and the different diffusions correspond to different scientific interpretations.) A last point to make is that different types of finite-difference discretizations lead to different parameterizations (e.g., a higher-order spatial discretization leads to larger neighborhoods, and higher-order temporal differences lead to higher-order Markov models.)

D.2   Example of a Process Model Motivated by a PDE: Spectral

This section considers a natural basis-function (i.e., spectral) approach to motivating a DSTM from a mechanistic model.

Consider a simple one-dimensional spatial version of the advection–diffusion PDE in (5.14), and denote the spatial index by x,

$$\frac{\partial Y(x,t)}{\partial t} = a\frac{\partial^2 Y(x,t)}{\partial x^2} - b\frac{\partial Y(x,t)}{\partial x}, \tag{D.3}$$

where in this example the advection and diffusion coefficients (b and a, respectively) are assumed to be constant. Now consider the solution as a superposition of Fourier functions (i.e., sines and cosines),

$$Y_t(x) = \sum_{j=1}^{J}[\alpha_{j,t}(1)\cos(\omega_j x) + \alpha_{j,t}(2)\sin(\omega_j x)], \tag{D.4}$$

where $\omega_j = 2\pi j/|D_x|$ is the spatial frequency of a sinusoid with spatial wave number $j = 1,\ldots,J$ in the spatial domain D_x (and $|D_x|$ corresponds to the length of the spatial domain). For simplicity of exposition, we do not include a constant term in the expansion (D.4). For the n spatial locations of interest and n_α Fourier basis functions, we let $\mathbf{Y}_t = \mathbf{\Phi}\boldsymbol{\alpha}_t$, where $\mathbf{Y}_t$ is an n_α-dimensional vector, $\mathbf{\Phi}$ is an $n_\alpha \times n_\alpha$ matrix consisting of Fourier basis functions, and $\boldsymbol{\alpha}_t$ contains the associated n expansion coefficients, where $n_\alpha = 2J$.

The deterministic solution of (D.3) gives formulas for $\alpha_{j,t}(1), \alpha_{j,t}(2)$, which are exponentially decaying sinusoids in time:

$$\begin{aligned}
\alpha_{j,t}(1) &= \exp(-a\omega_j^2 t)\sin(b\omega_j t), \\
\alpha_{j,t}(2) &= \exp(-a\omega_j^2 t)\cos(b\omega_j t), \quad j = 1,\ldots,J.
\end{aligned}$$

In this case, the time evolution is given by,

$$\boldsymbol{\alpha}_{j,t+\Delta_t} = \mathbf{M}_j\boldsymbol{\alpha}_{j,t}, \quad j = 1,\ldots,J,$$

where $\boldsymbol{\alpha}_{j,t} \equiv (\alpha_{j,t}(1)\ \alpha_{j,t}(2))'$ and

$$\mathbf{M}_j = \begin{bmatrix} e^{-a\omega_j^2\Delta_t}\cos\{\omega_j\Delta_t\} & e^{-a\omega_j^2\Delta_t}\sin\{\omega_j\Delta_t\} \\ -e^{-a\omega_j^2\Delta_t}\sin\{\omega_j\Delta_t\} & e^{-a\omega_j^2\Delta_t}\cos\{\omega_j\Delta_t\} \end{bmatrix}.$$

This motivates the $2J$-dimensional linear DSTM process model,

$$\boldsymbol{\alpha}_t = \mathbf{M}_\alpha \boldsymbol{\alpha}_{t-1} + \boldsymbol{\eta}_t \,,$$

where $\boldsymbol{\alpha}_t \equiv (\boldsymbol{\alpha}'_{1,t} \ \cdots \ \boldsymbol{\alpha}'_{J,t})'$ for $\boldsymbol{\alpha}_{j,t} = (\alpha_{j,t}(1), a_{j,t}(2))'$, $\boldsymbol{\eta}_t = (\boldsymbol{\eta}'_{1,t}, \ldots, \boldsymbol{\eta}'_{J,t})'$ for $\boldsymbol{\eta}_{j,t} = (\eta_{j,t}(1), \eta_{j,t}(2))'$, $\mathbf{M}_\alpha$ is a $2J \times 2J$ block diagonal matrix with blocks $\{\mathbf{M}_j, j = 1, \ldots, J\}$, and we have assumed that $\Delta_t = 1$. This then suggests block-diagonal parameterizations where the 2×2 coefficients associated with each set of Fourier functions are unknown and must be estimated (e.g., via a Bayesian hierarchical model). The result is a very sparse representation for the transition matrix, $\mathbf{M}_\alpha$, when J is large.

D.3 Example of a Process Model Motivated by an IDE

The stochastic IDE framework discussed in Chapter 5 naturally motivates a DSTM process model. Consider a decomposition similar to (5.24), where we let the process $\{\widetilde{Y}_t(\mathbf{s}) : \mathbf{s} \in D_s\}$ be decomposed as

$$\widetilde{Y}_t(\mathbf{s}) = \mathbf{x}_t(\mathbf{s})'\boldsymbol{\beta} + Y_t(\mathbf{s}) + \nu_t(\mathbf{s}) \,,$$

where $\{Y_t(\mathbf{s}) : \mathbf{s} \in D_s\}$ is assumed to be a dynamical process, and $\nu_t(\mathbf{s})$ is a non-dynamical process in the sense that it does not exhibit Markovian temporal dependence. Now, we assume that $\{Y_t(\mathbf{s})\}$ follows a stochastic IDE model as in (5.9). That is, for $\mathbf{s} \in D_s$,

$$Y_t(\mathbf{s}) \;=\; \int_{D_s} m(\mathbf{s}, \mathbf{x}; \boldsymbol{\theta}_p)\, Y_{t-1}(\mathbf{x})\mathrm{d}\mathbf{x} + \eta_t(\mathbf{s}) \,, \tag{D.5}$$

where $m(\mathbf{s}, \mathbf{x}; \boldsymbol{\theta}_p)$ is the transition kernel over the domain D_s, and $\boldsymbol{\theta}_p$ are kernel parameters.

As in (5.15), we assume that the dynamical process can be expanded in terms of n_α basis functions, $\{\phi_i(\mathbf{s}) : i = 1, \ldots, n_\alpha\}$. That is,

$$Y_t(\mathbf{s}) \;=\; \sum_{i=1}^{n_\alpha} \phi_i(\mathbf{s})\alpha_{i,t} \,. \tag{D.6}$$

Now, we can also expand the transition kernel in terms of these basis functions (although we could use different basis functions in general; see, for example Cressie and Wikle, 2011, Chapter 7):

$$m(\mathbf{s}, \mathbf{x}; \boldsymbol{\theta}_p) \;=\; \sum_{j=1}^{n_\alpha} \phi_j(\mathbf{x})b_j(\mathbf{s}; \boldsymbol{\theta}_p). \tag{D.7}$$

Substituting (D.6) and (D.7) into (D.5) and, for the sake of simplicity, adding the assumption that the basis functions are orthonormal,

$$\int_{D_s} \phi_i(\mathbf{x})\phi_j(\mathbf{x})\mathrm{d}\mathbf{x} \;=\; \begin{cases} 1, & i = j, \\ 0, & i \neq j, \end{cases}$$

we can show that

$$
\begin{aligned}
Y_t(\mathbf{s}) &= \sum_{i=1}^{n_\alpha} b_i(\mathbf{s}; \boldsymbol{\theta}_p) \alpha_{i,t-1} + \eta_t(\mathbf{s}) \\
&= \mathbf{b}'(\mathbf{s}; \boldsymbol{\theta}_p) \boldsymbol{\alpha}_{t-1} + \eta_t(\mathbf{s}),
\end{aligned}
$$

where $\mathbf{b}(\mathbf{s}; \boldsymbol{\theta}_p) \equiv (b_1(\mathbf{s}; \boldsymbol{\theta}_p), \ldots, b_{n_\alpha}(\mathbf{s}; \boldsymbol{\theta}_p))'$ and $\boldsymbol{\alpha}_t \equiv (\alpha_{1,t}, \ldots, \alpha_{n_\alpha,t})'$. Note also that $\mathbf{Y}_t = \boldsymbol{\Phi} \boldsymbol{\alpha}_t$, where $\boldsymbol{\Phi}$ is an $n \times n_\alpha$ basis-function matrix,

$$
\boldsymbol{\Phi} \equiv \begin{pmatrix} \boldsymbol{\phi}(\mathbf{s}_1)' \\ \vdots \\ \boldsymbol{\phi}(\mathbf{s}_n)' \end{pmatrix},
$$

and $\boldsymbol{\phi}(\mathbf{s}_i) \equiv (\phi_1(\mathbf{s}_i), \ldots, \phi_{n_\alpha}(\mathbf{s}_i))'$, for $i = 1, \ldots, n$. Now, define the $n \times n_\alpha$ matrix

$$
\mathbf{B} \equiv \begin{pmatrix} \mathbf{b}(\mathbf{s}_1; \boldsymbol{\theta}_p)' \\ \vdots \\ \mathbf{b}(\mathbf{s}_n; \boldsymbol{\theta}_p)' \end{pmatrix}.
$$

Then, for all n process locations, we can write

$$
\begin{aligned}
\boldsymbol{\alpha}_t &= (\boldsymbol{\Phi}'\boldsymbol{\Phi})^{-1}\boldsymbol{\Phi}'\mathbf{B}\boldsymbol{\alpha}_{t-1} + (\boldsymbol{\Phi}'\boldsymbol{\Phi})^{-1}\boldsymbol{\Phi}'\boldsymbol{\eta}_t \\
&= \boldsymbol{\Phi}'\mathbf{B}\boldsymbol{\alpha}_{t-1} + \boldsymbol{\Phi}'\boldsymbol{\eta}_t \\
&\equiv \mathbf{M}_\alpha \boldsymbol{\alpha}_{t-1} + \tilde{\boldsymbol{\eta}}_t,
\end{aligned}
$$

where the second equality is due to orthonormality (i.e., $\boldsymbol{\Phi}'\boldsymbol{\Phi} = \mathbf{I}$), $\tilde{\boldsymbol{\eta}}_t \equiv \boldsymbol{\Phi}'\boldsymbol{\eta}_t$ is the n_α-dimensional noise process, and the $n_\alpha \times n_\alpha$ propagator matrix is given by $\mathbf{M}_\alpha \equiv \boldsymbol{\Phi}'\mathbf{B}$.

The truncated expansion (D.6) leads to a lower-dimensional dynamical process (since $n_\alpha \ll n$). In principle, we still have to estimate the $n \times n_\alpha$ matrix $\mathbf{B}$ and the covariance matrix associated with $\tilde{\boldsymbol{\eta}}_t$. However, the IDE formulation allows the kernel $m(\mathbf{s}, \mathbf{x}; \boldsymbol{\theta}_p)$ to be parameterized parsimoniously. In some cases, one can select $\{\phi_j(\mathbf{s})\}$ to ensure that the expansion coefficients for the kernel can be specified analytically in terms of its parameters. For example, letting $\{\phi_j(\cdot)\}$ in (D.7) be Fourier basis functions allows one to parameterize the kernel in terms of its characteristic function. This can facilitate a BHM parameterization that allows kernel asymmetry and scale parameters to vary in space.

Lab 5.1 gives an introduction to the implementation of the IDE in one-dimensional space. Lab 5.2 then gives an implementation of a DSTM motivated by the stochastic IDE model to generate nowcasts for weather radar images.

E Case Study: Physical-Statistical Bayesian Hierarchical Model for Predicting Mediterranean Surface Winds

In this section we present a specific and detailed example of how to develop a physically motivated bivariate spatio-temporal model for the purpose of predicting near-surface wind fields in the Mediterranean Sea. The implementation of this model in R is given below.

Consider a simple analytical model for the surface wind known as the *Rayleigh friction equations* (e.g., Stevens et al., 2002):

$$\frac{\partial u}{\partial t} = fv - \frac{1}{\rho_0}\frac{\partial P}{\partial x} - \gamma u,$$

$$\frac{\partial v}{\partial t} = -fu - \frac{1}{\rho_0}\frac{\partial P}{\partial y} - \gamma v,$$

where u and v are the east–west and north–south components of the wind, respectively (recall that winds are vectors with a magnitude and direction that can be decomposed into x (east–west) and y (north–south) coordinates); f is the Coriolis parameter; ρ_0 is a reference atmospheric density; P is the sea-level pressure; and γ is the Rayleigh friction parameter. Note that u, v, and P are functions of time and space. As in Section D.1, simple forward differencing in time with $\Delta_t = 1$, and centered differencing in space, give the analogous discretized form of these equations:

$$u_{t+1}(i,j) = u_t(i,j) + \Delta_t\left\{fv_t(i,j) - \frac{1}{\rho_0}\left(\frac{P_t(i+1,j) - P_t(i-1,j)}{2\Delta_x}\right) - \gamma u_t(i,j)\right\},$$
$$\tag{E.1}$$

$$v_{t+1}(i,j) = v_t(i,j) + \Delta_t\left\{-fu_t(i,j) - \frac{1}{\rho_0}\left(\frac{P_t(i,j+1) - P_t(i,j-1)}{2\Delta_y}\right) - \gamma v_t(i,j)\right\},$$
$$\tag{E.2}$$

where $u_t(i,j)$, $v_t(i,j)$, and $P_t(i,j)$ are discretized wind components and pressure, respectively, at grid location (i,j) and time t, and Δ_t, Δ_x, and Δ_y are the time-, x-, and y-discretization constants, respectively. Note that this is a multivariate linear system, with each component of the wind conditioned on the past values of that component, the other component, and the difference (gradient) in pressure.

Now a simple *statistical* process model based on these equations can be written in vector form as:

$$\mathbf{u}_{t+1} = \theta_{uu}\mathbf{u}_t + \theta_{uv}\mathbf{v}_t + \theta_{up}\mathbf{D}_x\mathbf{P}_t + \boldsymbol{\eta}_{u,t}, \tag{E.3}$$

$$\mathbf{v}_{t+1} = \theta_{vv}\mathbf{v}_t + \theta_{vu}\mathbf{u}_t + \theta_{vp}\mathbf{D}_y\mathbf{P}_t + \boldsymbol{\eta}_{v,t}, \tag{E.4}$$

for $t = 1, \ldots, T - 1$, where $\mathbf{v}_t$ and $\mathbf{u}_t$ are n_g-dimensional ($n_g = n_x \times n_y$) vectors of the discretized u and v components, n_x and n_y being the number of grid locations in the x- and y-directions on the prediction grid; $\mathbf{P}_t$ is an $n_e = (n_x + 2) \times (n_y + 2)$-dimensional vector of surface pressure values on an expanded grid (which in our example will come from data); $\mathbf{D}_x$ and $\mathbf{D}_y$ are $n_g \times n_e$ matrix operators that give the centered difference in the x and y directions, respectively; and $\boldsymbol{\eta}_{u,t} \sim iid\ Gau(0, \sigma_u^2 \mathbf{I})$ and $\boldsymbol{\eta}_{v,t} \sim iid\ Gau(0, \sigma_v^2 \mathbf{I})$ are residual error processes. Although we could specify the θ-values in (E.3) and (E.4) according to the discretization constants and f and γ in (E.1) and (E.2), we instead allow them to be unknown and random here (see below) and include the additive error terms to adapt to the data and to reflect the fact that the Rayleigh friction equations are a pretty rough approximation for reality. More complicated versions of this model are given in Milliff et al. (2011) and Cressie and Wikle (2011, Chapter 9) to account for spatio-temporal dependent errors as well as a random pressure process.

We have two sources of data on Mediterranean surface winds (as described in Section 2.1): gridded analysis wind and pressure data from the European Center for Medium Range Weather Forecasting (ECMWF) (these observations are complete in space and time); and higher-resolution satellite observations of near-surface winds over the ocean from the polar-orbiting QuikSCAT scatterometer (these observations are irregular in space and time). The BHM is given by a data model, a process model, and a parameter model. For modeling the Mediterranean wind data, these are defined as follows.

Data model. For $t = 1, \ldots, T$, assume

$$
\begin{align}
\mathbf{E}_{u,t}|\mathbf{u}_t, \sigma_e^2 &\sim \quad indep.\ Gau(\mathbf{H}_e \mathbf{u}_t, \sigma_e^2 \mathbf{I}), & \text{(E.5)} \\
\mathbf{E}_{v,t}|\mathbf{v}_t, \sigma_e^2 &\sim \quad indep.\ Gau(\mathbf{H}_e \mathbf{v}_t, \sigma_e^2 \mathbf{I}), & \text{(E.6)} \\
\mathbf{S}_{u,t}|\mathbf{u}_t, \sigma_s^2 &\sim \quad indep.\ Gau(\mathbf{H}_{s,t} \mathbf{u}_t, \sigma_s^2 \mathbf{I}), & \text{(E.7)} \\
\mathbf{S}_{v,t}|\mathbf{v}_t, \sigma_s^2 &\sim \quad indep.\ Gau(\mathbf{H}_{s,t} \mathbf{v}_t, \sigma_s^2 \mathbf{I}), & \text{(E.8)}
\end{align}
$$

where $\mathbf{E}_{u,t}, \mathbf{E}_{v,t}$ are n_e-vectors of ECMWF observations at time t, with associated $n_e \times n_g$ incidence matrix $\mathbf{H}_e$; and $\mathbf{S}_{u,t}, \mathbf{S}_{v,t}$ are $n_{s,t}$-dimensional vectors of QuikSCAT observations at time t, with associated $n_{s,t} \times n_g$ incidence matrices, $\mathbf{H}_{s,t}$. (Note that there are different numbers of QuikSCAT observations at each time, and there can be times for which there are no QuikSCAT observations.)

Process model. For $t = 1, \ldots, T - 1$, assume

$$
\mathbf{u}_{t+1}|\mathbf{u}_t, \mathbf{v}_t, \mathbf{P}_t, \theta_{uu}, \theta_{uv}, \theta_{up}, \sigma_u^2 \sim indep.\ Gau(\theta_{uu}\mathbf{u}_t + \theta_{uv}\mathbf{v}_t + \theta_{up}\mathbf{D}_x\mathbf{P}_t, \sigma_u^2\mathbf{I}), \text{(E.9)}
$$
$$
\mathbf{v}_{t+1}|\mathbf{v}_t, \mathbf{u}_t, \mathbf{P}_t, \theta_{vv}, \theta_{vu}, \theta_{vp}, \sigma_v^2 \sim indep.\ Gau(\theta_{vv}\mathbf{v}_t + \theta_{vu}\mathbf{u}_t + \theta_{vp}\mathbf{D}_y\mathbf{P}_t, \sigma_v^2\mathbf{I}), \text{(E.10)}
$$

where we have pressure observations, $\mathbf{P}_t$, from the ECMWF data within the Mediterranean wind data. We also need to specify the process's initial conditions at time $t = 1$. Assume

$$\mathbf{u}_1 | \boldsymbol{\mu}_{u,1}, \sigma_{u,1}^2 \sim Gau(\boldsymbol{\mu}_{u,1}, \sigma_{u,1}^2 \mathbf{I}), \quad \text{(E.11)}$$

$$\mathbf{v}_1 | \boldsymbol{\mu}_{v,1}, \sigma_{v,1}^2 \sim Gau(\boldsymbol{\mu}_{v,1}, \sigma_{v,1}^2 \mathbf{I}). \quad \text{(E.12)}$$

Parameter model. All of the process-model parameters are assumed to be independent and their distributions are given by

$$\theta_{ab} | \mu_{ab}, \sigma_{ab}^2 \sim Gau(\mu_{ab}, \sigma_{ab}^2), \quad \text{(E.13)}$$

for $ab = \{uu, vv, uv, vu, up, vp\}$. Further,

$$\sigma_a^2 | q_a, r_a \sim IG(q_a, r_a), \quad \text{(E.14)}$$

for $a = \{u, v\}$ (where $IG(q_a, r_a)$ is the *inverse gamma* distribution with shape parameter q_a and rate parameter r_a).

Hyperparameters (fixed and specified). The following hyperparameters are specified based on scientific assumptions or to correspond to "vague" prior distributions: $\sigma_e^2, \sigma_s^2, \{\mu_{ab}, \sigma_{ab}^2 : ab = uu, vv, uv, vu, up, vp\}, \{q_a, r_a : a = u, v\}, \boldsymbol{\mu}_{u,1}, \boldsymbol{\mu}_{v,1}, \sigma_{u,1}^2, \sigma_{v,1}^2$. Specific values are given in the R example that follows.

Gibbs sampler. The BHM presented above is amenable to a Gibbs sampler MCMC implementation because all the full conditional distributions are available in closed form (see Cressie and Wikle, 2011, Chapter 8, for details on how to derive full conditional distributions for spatio-temporal models). Recall that from Algorithm 4.2 that the Gibbs sampler simply cycles through the full conditional distributions, sampling each variable given the most recent samples. The Gibbs sampler for the BHM of the Mediterranean winds data is outlined in Algorithm E.1, where the equation numbers correspond to the full conditional distributions presented in the next section.

Algorithm E.1 Gibbs Sampler for BHM of Mediterranean winds data set

Select hyperparameters: $\sigma_e^2, \sigma_s^2, \{\mu_{ab}, \sigma_{ab}^2 : ab = uu, vv, uv, vu, up, vp\}, \{q_a, r_a : a = u, v\}, \boldsymbol{\mu}_{u,1}, \boldsymbol{\mu}_{v,1}, \sigma_{u,1}^2, \sigma_{v,1}^2$

Select initial values: $\{\mathbf{u}_t^{(0)} : t = 2, \ldots, T\}, \{\mathbf{v}_t^{(0)} : t = 1, \ldots, T\}, \theta_{uu}^{(0)}, \theta_{vv}^{(0)}, \theta_{uv}^{(0)}, \theta_{vu}^{(0)}, \theta_{up}^{(0)}, \theta_{vp}^{(0)}, \sigma_u^{2(0)}, \sigma_v^{2(0)}$

for $i = 1, 2, \ldots, N_{\text{gibbs}}$ **do**

---

1. using (E.15), **sample from**

$$\mathbf{u}_1^{(i)}|\mathbf{v}_1^{(i-1)},\mathbf{u}_2^{(i-1)},\mathbf{v}_2^{(i-1)},\theta_{vu}^{(i-1)},\theta_{uu}^{(i-1)},\theta_{uv}^{(i-1)},\theta_{vv}^{(i-1)},\theta_{vp}^{(i-1)},\theta_{up}^{(i-1)},\sigma_u^{2(i-1)},\sigma_v^{2(i-1)}$$

2. using (E.16) for $t = 2, \ldots, T-1$, **sample from**

$$\mathbf{u}_t^{(i)}|\mathbf{u}_{t-1}^{(i)},\mathbf{u}_{t+1}^{(i-1)},\mathbf{v}_{t-1}^{(i-1)},\mathbf{v}_t^{(i-1)},\mathbf{v}_{t+1}^{(i-1)},\theta_{vu}^{(i-1)},\theta_{uu}^{(i-1)},\theta_{uv}^{(i-1)},\theta_{vv}^{(i-1)},$$
$$\theta_{vp}^{(i-1)},\theta_{up}^{(i-1)},\sigma_u^{2(i-1)},\sigma_v^{2(i-1)}$$

3. using (E.17), **sample from**

$$\mathbf{u}_T^{(i)}|\mathbf{u}_{T-1}^{(i)},\mathbf{v}_{T-1}^{(i-1)},\theta_{uu}^{(i-1)},\theta_{uv}^{(i-1)},\theta_{up}^{(i-1)},\sigma_u^{2(i-1)}$$

4. using (E.18), **sample from**

$$\mathbf{v}_1^{(i)}|\mathbf{u}_1^{(i)},\mathbf{u}_2^{(i)},\mathbf{v}_2^{(i-1)},\theta_{uv}^{(i-1)},\theta_{uu}^{(i-1)},\theta_{vu}^{(i-1)},\theta_{vv}^{(i-1)},\theta_{up}^{(i-1)},\theta_{vp}^{(i-1)},\sigma_u^{2(i-1)},\sigma_v^{2(i-1)}$$

5. using (E.19) for $t = 2, \ldots, T-1$, **sample from**

$$\mathbf{v}_t^{(i)}|\mathbf{u}_t^{(i)},\mathbf{u}_{t-1}^{(i)},\mathbf{u}_{t+1}^{(i)},\mathbf{v}_{t-1}^{(i)},\mathbf{v}_{t+1}^{(i-1)},\theta_{vu}^{(i-1)},\theta_{uu}^{(i-1)},\theta_{uv}^{(i-1)},\theta_{vv}^{(i-1)},\theta_{vp}^{(i-1)},\theta_{up}^{(i-1)},\sigma_u^{2(i-1)},\sigma_v^{2(i-1)}$$

6. using (E.20), **sample from**

$$\mathbf{v}_T^{(i)}|\mathbf{u}_{T-1}^{(i)},\mathbf{v}_{T-1}^{(i)},\theta_{vv}^{(i-1)},\theta_{vu}^{(i-1)},\theta_{vp}^{(i-1)},\sigma_v^{2(i-1)}$$

7. using (E.21), **sample from**

$$\theta_{uu}^{(i)}|\{\mathbf{u}_t^{(i)} : t = 1, \ldots, T\},\{\mathbf{v}_t^{(i)} : t = 1, \ldots, T\},\theta_{uv}^{(i-1)},\theta_{up}^{(i-1)},\sigma_u^{2(i-1)}$$

8. using (E.22), **sample from**

$$\theta_{vv}^{(i)}|\{\mathbf{u}_t^{(i)} : t = 1, \ldots, T\},\{\mathbf{v}_t^{(i)} : t = 1, \ldots, T\},\theta_{vu}^{(i-1)},\theta_{vp}^{(i-1)},\sigma_v^{2(i-1)}$$

9. using (E.23), **sample from**

$$\theta_{uv}^{(i)}|\{\mathbf{u}_t^{(i)} : t = 1, \ldots, T\},\{\mathbf{v}_t^{(i)} : t = 1, \ldots, T\},\theta_{uu}^{(i)},\theta_{up}^{(i-1)},\sigma_u^{2(i-1)}$$

10. using (E.24), **sample from**

$$\theta_{vu}^{(i)}|\{\mathbf{u}_t^{(i)} : t = 1, \ldots, T\},\{\mathbf{v}_t^{(i)} : t = 1, \ldots, T\},\theta_{vv}^{(i)},\theta_{vp}^{(i-1)},\sigma_v^{2(i-1)}$$

11. using (E.25), **sample from**

$$\theta_{up}^{(i)}|\{\mathbf{u}_t^{(i)} : t = 1, \ldots, T\},\{\mathbf{v}_t^{(i)} : t = 1, \ldots, T\},\theta_{uu}^{(i)},\theta_{uv}^{(i)},\sigma_u^{2(i-1)}$$

12. using (E.26), **sample from**

$$\theta_{vp}^{(i)}|\{\mathbf{u}_t^{(i)} : t = 1, \ldots, T\},\{\mathbf{v}_t^{(i)} : t = 1, \ldots, T\},\theta_{vv}^{(i)},\theta_{vu}^{(i)},\sigma_v^{2(i-1)}$$

13. using (E.27), **sample from**

$$\sigma_u^{2(i)}|\{\mathbf{u}_t^{(i)} : t = 1, \ldots, T\},\{\mathbf{v}_t^{(i)} : t = 1, \ldots, T\},\theta_{uu}^{(i)},\theta_{uv}^{(i)},\theta_{up}^{(i)}$$

14. using (E.28), **sample from**

$$\sigma_v^{2(i)}|\{\mathbf{u}_t^{(i)} : t = 1, \ldots, T\},\{\mathbf{v}_t^{(i)} : t = 1, \ldots, T\},\theta_{vv}^{(i)},\theta_{vu}^{(i)},\theta_{vp}^{(i)}$$

end for

Full conditional distributions. Readers be warned that this material is very technical! The full conditional distributions for the Gibbs sampler presented in Algorithm E.1 are included here for advanced readers. For more examples in the spatio-temporal context, see Cressie and Wikle (2011, Chapter 8), and for other examples see Gelman et al. (2014). In the representation to follow, $[equation\ number]_t$ corresponds to the distribution associated with the equation number above, where the variable on the left side of the conditioning symbol is given at time t. When referring to the parameter model, (E.13) and (E.14), the notation $[equation\ number]_{ab}$ and $[equation\ number]_a$ correspond to the specific parameter distribution given by $ab = \{uu, vv, uv, vu, up, vp\}$ or $a = \{u, v\}$, respectively.

- Full conditional distribution for $\mathbf{u}_1$:

$$
\begin{aligned}
[\mathbf{u}_1|\cdot] &\propto [E.5]_1 \times [E.7]_1 \times [E.9]_2 \times [E.10]_2 \times [E.11] \\
\mathbf{u}_1|\cdot &\sim Gau(\mathbf{A}_{u,1}\mathbf{b}_{u,1}, \mathbf{A}_{u,1}),
\end{aligned} \tag{E.15}
$$

where

$$
\begin{aligned}
\mathbf{A}_{u,1} &\equiv \left(\mathbf{H}'_e\mathbf{H}_e/\sigma_e^2 + \mathbf{H}'_{s,1}\mathbf{H}_{s,1}/\sigma_s^2 + \theta_{vu}^2\mathbf{I}/\sigma_v^2 + \theta_{uu}^2\mathbf{I}/\sigma_u^2 + \mathbf{I}/\sigma_{u,1}^2\right)^{-1}, \\
\mathbf{b}_{u,1} &\equiv \left(\mathbf{E}'_{u,1}\mathbf{H}_e/\sigma_e^2 + \mathbf{S}'_{u,1}\mathbf{H}_{s,1}/\sigma_s^2 + (\mathbf{v}_2 - \mathbf{c}_{v,1})'\theta_{vu}/\sigma_v^2 \right. \\
&\quad \left. + (\mathbf{u}_2 - \mathbf{c}_{u,1})'\theta_{uu}/\sigma_u^2 + \boldsymbol{\mu}'_{u,1}/\sigma_{u,1}^2\right)',
\end{aligned}
$$

with

$$
\begin{aligned}
\mathbf{c}_{u,1} &\equiv \theta_{uv}\mathbf{v}_1 + \theta_{up}\mathbf{D}_x\mathbf{P}_1, \\
\mathbf{c}_{v,1} &\equiv \theta_{vv}\mathbf{v}_1 + \theta_{vp}\mathbf{D}_y\mathbf{P}_1.
\end{aligned}
$$

- Full conditional distribution for $\mathbf{u}_t, t = 2, \ldots, T-1$:

$$
\begin{aligned}
[\mathbf{u}_t|\cdot] &\propto [E.5]_t \times [E.7]_t \times [E.10]_{t+1} \times [E.9]_{t+1} \times [E.9]_t \\
\mathbf{u}_t|\cdot &\sim Gau(\mathbf{A}_{u,t}\mathbf{b}_{u,t}, \mathbf{A}_{u,t}),
\end{aligned} \tag{E.16}
$$

where

$$
\begin{aligned}
\mathbf{A}_{u,t} &\equiv \left(\mathbf{H}'_e\mathbf{H}_e/\sigma_e^2 + \mathbf{H}'_{s,t}\mathbf{H}_{s,t}/\sigma_s^2 + \theta_{vu}^2\mathbf{I}/\sigma_v^2 + \theta_{uu}^2\mathbf{I}/\sigma_u^2 + \mathbf{I}/\sigma_u^2\right)^{-1}, \\
\mathbf{b}_{u,t} &\equiv \left(\mathbf{E}'_{u,t}\mathbf{H}_e/\sigma_e^2 + \mathbf{S}'_{u,t}\mathbf{H}_{s,t}/\sigma_s^2 + (\mathbf{v}_{t+1} - \mathbf{c}_{v,t})'\theta_{vu}/\sigma_v^2 + (\mathbf{u}_{t+1} - \mathbf{c}_{u,t})'\theta_{uu}/\sigma_u^2 \right. \\
&\quad \left. + (\mathbf{c}_{u,t-1} + \theta_{uu}\mathbf{u}_{t-1})'/\sigma_u^2\right)',
\end{aligned}
$$

with

$$
\begin{aligned}
\mathbf{c}_{v,t} &\equiv \theta_{vv}\mathbf{v}_t + \theta_{vp}\mathbf{D}_y\mathbf{P}_t, \\
\mathbf{c}_{u,t} &\equiv \theta_{uv}\mathbf{v}_t + \theta_{up}\mathbf{D}_x\mathbf{P}_t, \\
\mathbf{c}_{u,t-1} &\equiv \theta_{uv}\mathbf{v}_{t-1} + \theta_{up}\mathbf{D}_x\mathbf{P}_{t-1}.
\end{aligned}
$$

- Full conditional distribution for $\mathbf{u}_T$:

$$
\begin{aligned}
[\mathbf{u}_T|\cdot] &\propto [E.5]_T \times [E.7]_T \times [E.9]_T \\
\mathbf{u}_T|\cdot &\sim Gau(\mathbf{A}_{u,T}\mathbf{b}_{u,T}, \mathbf{A}_{u,T}),
\end{aligned}
\tag{E.17}
$$

where

$$
\begin{aligned}
\mathbf{A}_{u,T} &\equiv \left(\mathbf{H}_e'\mathbf{H}_e/\sigma_e^2 + \mathbf{H}_{s,T}'\mathbf{H}_{s,T}/\sigma_s^2 + \mathbf{I}/\sigma_u^2\right)^{-1}, \\
\mathbf{b}_{u,T} &\equiv (\mathbf{E}_{u,T}'\mathbf{H}_e/\sigma_e^2 + \mathbf{S}_{u,T}'\mathbf{H}_{s,T}/\sigma_s^2 + (\mathbf{c}_{u,T-1} + \theta_{uu}\mathbf{u}_{T-1})'/\sigma_u^2)',
\end{aligned}
$$

with

$$
\mathbf{c}_{u,T-1} \equiv \theta_{uv}\mathbf{v}_{T-1} + \theta_{up}\mathbf{D}_x\mathbf{P}_{T-1}.
$$

- Full conditional distribution for $\mathbf{v}_1$:

$$
\begin{aligned}
[\mathbf{v}_1|\cdot] &\propto [E.6]_1 \times [E.8]_1 \times [E.10]_2 \times [E.9]_2 \times [E.12] \\
\mathbf{v}_1|\cdot &\sim Gau(\mathbf{A}_{v,1}\mathbf{b}_{v,1}, \mathbf{A}_{v,1}),
\end{aligned}
\tag{E.18}
$$

where

$$
\begin{aligned}
\mathbf{A}_{v,1} &\equiv \left(\mathbf{H}_e'\mathbf{H}_e/\sigma_e^2 + \mathbf{H}_{s,1}'\mathbf{H}_{s,1}/\sigma_s^2 + \theta_{uv}^2\,\mathbf{I}/\sigma_u^2 + \theta_{vv}^2\,\mathbf{I}/\sigma_v^2 + \mathbf{I}/\sigma_{v,1}^2\right)^{-1}, \\
\mathbf{b}_{v,1} &\equiv \big(\mathbf{E}_{v,1}'\mathbf{H}_e/\sigma_e^2 + \mathbf{S}_{v,1}'\mathbf{H}_{s,1}/\sigma_s^2 + (\mathbf{u}_2 - \mathbf{c}_{u,1})'\theta_{uv}/\sigma_u^2 \\
&\quad + (\mathbf{v}_2 - \mathbf{c}_{v,1})'\theta_{vv}/\sigma_v^2 + \boldsymbol{\mu}_{v,1}'/\sigma_{v,1}^2\big)',
\end{aligned}
$$

with

$$
\begin{aligned}
\mathbf{c}_{v,1} &\equiv \theta_{vu}\mathbf{u}_1 + \theta_{vp}\mathbf{D}_y\mathbf{P}_1, \\
\mathbf{c}_{u,1} &\equiv \theta_{uu}\mathbf{u}_1 + \theta_{up}\mathbf{D}_x\mathbf{P}_1.
\end{aligned}
$$

- Full conditional distribution for $\mathbf{v}_t$, $t = 2, \ldots, T-1$:

$$
\begin{aligned}
[\mathbf{v}_t|\cdot] &\propto [E.6]_t \times [E.8]_t \times [E.9]_{t+1} \times [E.10]_{t+1} \times [E.10]_t \\
\mathbf{v}_t|\cdot &\sim Gau(\mathbf{A}_{v,t}\mathbf{b}_{v,t}, \mathbf{A}_{v,t}),
\end{aligned}
\tag{E.19}
$$

where

$$
\begin{aligned}
\mathbf{A}_{v,t} &\equiv \left(\mathbf{H}_e'\mathbf{H}_e/\sigma_e^2 + \mathbf{H}_{s,t}'\mathbf{H}_{s,t}/\sigma_s^2 + \theta_{uv}^2\,\mathbf{I}/\sigma_u^2 + \theta_{vv}^2\,\mathbf{I}/\sigma_v^2 + \mathbf{I}/\sigma_v^2\right)^{-1}, \\
\mathbf{b}_{v,t} &\equiv (\mathbf{E}_{v,t}'\mathbf{H}_e/\sigma_e^2 + \mathbf{S}_{v,t}'\mathbf{H}_{s,t}/\sigma_s^2 + (\mathbf{u}_{t+1} - \mathbf{c}_{u,t})'\theta_{uv}/\sigma_u^2 + (\mathbf{v}_{t+1} - \mathbf{c}_{v,t})'\theta_{vv}/\sigma_v^2 \\
&\quad + (\mathbf{c}_{v,t-1} + \theta_{vv}\mathbf{v}_{t-1})'/\sigma_v^2)',
\end{aligned}
$$

with

$$
\begin{aligned}
\mathbf{c}_{u,t} &\equiv \theta_{uu}\mathbf{u}_t + \theta_{up}\mathbf{D}_x\mathbf{P}_t, \\
\mathbf{c}_{v,t} &\equiv \theta_{vu}\mathbf{u}_t + \theta_{vp}\mathbf{D}_y\mathbf{P}_t, \\
\mathbf{c}_{v,t-1} &\equiv \theta_{vu}\mathbf{u}_{t-1} + \theta_{vp}\mathbf{D}_y\mathbf{P}_{t-1}.
\end{aligned}
$$

- Full conditional distribution for $\mathbf{v}_T$:

$$[\mathbf{v}_T|\cdot] \quad \propto \quad [E.6]_T \times [E.8]_T \times [E.10]_T$$
$$\mathbf{v}_T|\cdot \quad \sim \quad Gau(\mathbf{A}_{v,T}\mathbf{b}_{v,T}, \mathbf{A}_{v,T}), \qquad (E.20)$$

where

$$\mathbf{A}_{v,T} \equiv \left(\mathbf{H}'_e\mathbf{H}_e/\sigma_e^2 + \mathbf{H}'_{s,T}\mathbf{H}_{s,T}/\sigma_s^2 + \mathbf{I}/\sigma_v^2\right)^{-1},$$
$$\mathbf{b}_{v,T} \equiv (\mathbf{E}'_{v,T}\mathbf{H}_e/\sigma_e^2 + \mathbf{S}'_{v,T}\mathbf{H}_{s,T}/\sigma_s^2 + (\mathbf{c}_{v,T-1} + \theta_{vv}\mathbf{v}_{T-1})'/\sigma_v^2)',$$

with

$$\mathbf{c}_{v,T-1} \equiv \theta_{vu}\mathbf{u}_{T-1} + \theta_{vp}\mathbf{D}_y\mathbf{P}_{T-1}.$$

- Full conditional distribution for θ_{uu}:

$$[\theta_{uu}|\cdot] \quad \propto \quad \prod_{t=1}^{T-1}[E.9]_{t+1} \times [E.13]_{(uu)}$$
$$\theta_{uu}|\cdot \quad \sim \quad Gau(A_{uu}b_{uu}, A_{uu}), \qquad (E.21)$$

where

$$A_{uu} \equiv \left(\sum_{t=1}^{T-1}\mathbf{u}'_t\mathbf{u}_t/\sigma_u^2 + 1/\sigma_{uu}^2\right)^{-1},$$
$$b_{uu} \equiv \sum_{t=1}^{T-1}(\mathbf{u}_{t+1} - \mathbf{k}_{v,t})'\mathbf{u}_t/\sigma_u^2 + \mu_{uu}/\sigma_{uu}^2,$$

with

$$\mathbf{k}_{v,t} \equiv \theta_{uv}\mathbf{v}_t + \theta_{up}\mathbf{D}_x\mathbf{P}_t.$$

- Full conditional distribution for θ_{vv}:

$$[\theta_{vv}|\cdot] \quad \propto \quad \prod_{t=1}^{T-1}[E.10]_{t+1} \times [E.13]_{(vv)}$$
$$\theta_{vv}|\cdot \quad \sim \quad Gau(A_{vv}b_{vv}, A_{vv}), \qquad (E.22)$$

where

$$A_{vv} \equiv \left(\sum_{t=1}^{T-1}\mathbf{v}'_t\mathbf{v}_t/\sigma_v^2 + 1/\sigma_{vv}^2\right)^{-1},$$
$$b_{vv} \equiv \sum_{t=1}^{T-1}(\mathbf{v}_{t+1} - \mathbf{k}_{u,t})'\mathbf{v}_t/\sigma_v^2 + \mu_{vv}/\sigma_{vv}^2,$$

with

$$\mathbf{k}_{u,t} \equiv \theta_{vu}\mathbf{u}_t + \theta_{vp}\mathbf{D}_y\mathbf{P}_t.$$

- Full conditional distribution for θ_{uv}:

$$
\begin{aligned}
[\theta_{uv}|\cdot] &\propto \prod_{t=1}^{T-1} [E.9]_{t+1} \times [E.13]_{(uv)} \\
\theta_{uv}|\cdot &\sim Gau(A_{uv}b_{uv}, A_{uv}),
\end{aligned}
\tag{E.23}
$$

where

$$
A_{uv} \equiv \left(\sum_{t=1}^{T-1} \mathbf{v}_t' \mathbf{v}_t / \sigma_u^2 + 1/\sigma_{uv}^2 \right)^{-1},
$$

$$
b_{uv} \equiv \sum_{t=1}^{T-1} (\mathbf{u}_{t+1} - \mathbf{k}_{u,t})' \mathbf{v}_t / \sigma_u^2 + \mu_{uv}/\sigma_{uv}^2,
$$

with

$$
\mathbf{k}_{u,t} \equiv \theta_{uu}\mathbf{u}_t + \theta_{up}\mathbf{D}_x\mathbf{P}_t.
$$

- Full conditional distribution for θ_{vu}:

$$
\begin{aligned}
[\theta_{vu}|\cdot] &\propto \prod_{t=1}^{T-1} [E.10]_{t+1} \times [E.13]_{(vu)} \\
\theta_{vu}|\cdot &\sim Gau(A_{vu}b_{vu}, A_{vu}),
\end{aligned}
\tag{E.24}
$$

where

$$
A_{vu} \equiv \left(\sum_{t=1}^{T-1} \mathbf{u}_t' \mathbf{u}_t / \sigma_v^2 + 1/\sigma_{vu}^2 \right)^{-1},
$$

$$
b_{vu} \equiv \sum_{t=1}^{T-1} (\mathbf{v}_{t+1} - \mathbf{k}_{v,t})' \mathbf{u}_t / \sigma_v^2 + \mu_{vu}/\sigma_{vu}^2,
$$

with

$$
\mathbf{k}_{v,t} \equiv \theta_{vv}\mathbf{v}_t + \theta_{vp}\mathbf{D}_y\mathbf{P}_t.
$$

- Full conditional distribution for θ_{up}:

$$
\begin{aligned}
[\theta_{up}|\cdot] &\propto \prod_{t=1}^{T-1} [E.9]_{t+1} \times [E.13]_{(up)} \\
\theta_{up}|\cdot &\sim Gau(A_{up}b_{up}, A_{up}),
\end{aligned}
\tag{E.25}
$$

where

$$A_{up} \equiv \left(\sum_{t=1}^{T-1} (\mathbf{D}_x \mathbf{P}_t)'(\mathbf{D}_x \mathbf{P}_t)/\sigma_u^2 + 1/\sigma_{up}^2 \right)^{-1},$$

$$b_{up} \equiv \sum_{t=1}^{T-1} (\mathbf{u}_{t+1} - \mathbf{k}_{u,t})' \mathbf{D}_x \mathbf{P}_t/\sigma_u^2 + \mu_{up}/\sigma_{up}^2,$$

with

$$\mathbf{k}_{u,t} \equiv \theta_{uu} \mathbf{u}_t + \theta_{uv} \mathbf{v}_t.$$

- Full conditional distribution for θ_{vp}:

$$[\theta_{vp}|\cdot] \quad \propto \quad \prod_{t=1}^{T-1} [E.10]_{t+1} \times [E.13]_{(vp)}$$

$$\theta_{vp}|\cdot \quad \sim \quad Gau(A_{vp} b_{vp}, A_{vp}), \tag{E.26}$$

where

$$A_{vp} \equiv \left(\sum_{t=1}^{T-1} (\mathbf{D}_y \mathbf{P}_t)'(\mathbf{D}_y \mathbf{P}_t)/\sigma_v^2 + 1/\sigma_{vp}^2 \right)^{-1},$$

$$b_{vp} \equiv \sum_{t=1}^{T-1} (\mathbf{v}_{t+1} - \mathbf{k}_{v,t})' \mathbf{D}_y \mathbf{P}_t/\sigma_v^2 + \mu_{vp}/\sigma_{vp}^2,$$

with

$$\mathbf{k}_{v,t} \equiv \theta_{vv} \mathbf{u}_t + \theta_{vu} \mathbf{u}_t.$$

- Full conditional distribution for σ_u^2:

$$[\sigma_u^2|\cdot] \quad \propto \quad \prod_{t=1}^{T-1} [E.9]_{t+1} \times [E.14]_{(u)}$$

$$\sigma_u^2|\cdot \quad \sim \quad IG(q_{\text{new},u}, r_{\text{new},u}), \tag{E.27}$$

where

$$q_{\text{new},u} = q_u + (T-1)n_g/2,$$

$$r_{\text{new},u} = \left(\frac{1}{r_u} + \frac{1}{2} \sum_{t=1}^{T-1} (\mathbf{u}_{t+1} - \mathbf{k}_{u,t})'(\mathbf{u}_{t+1} - \mathbf{k}_{u,t}) \right)^{-1},$$

with

$$\mathbf{k}_{u,t} \equiv \theta_{uu} \mathbf{u}_t + \theta_{uv} \mathbf{v}_t + \theta_{up} \mathbf{D}_x \mathbf{P}_t.$$

- Full conditional distribution for σ_v^2:

$$[\sigma_v^2|\cdot] \quad \propto \quad \prod_{t=1}^{T-1} [E.10]_{t+1} \times [E.14]_{(v)}$$

$$\sigma_v^2|\cdot \quad \sim \quad IG(q_{\text{new},v}, r_{\text{new},v}), \tag{E.28}$$

where

$$q_{\text{new},v} = q_v + (T-1)n_g/2,$$

$$r_{\text{new},v} = \left(\frac{1}{r_v} + \frac{1}{2} \sum_{t=1}^{T-1} (\mathbf{v}_{t+1} - \mathbf{k}_{v,t})'(\mathbf{v}_{t+1} - \mathbf{k}_{v,t}) \right)^{-1},$$

with

$$\mathbf{k}_{v,t} \equiv \theta_{vv}\mathbf{v}_t + \theta_{vu}\mathbf{u}_t + \theta_{vp}\mathbf{D}_y\mathbf{P}_t.$$

Implementation in `R`

`R` Preliminaries

We will need the **Matrix** package because the BHM Gibbs sampler uses sparse matrices, and **ggquiver** and **ggmap** to make "quiver" plots of the wind vectors on a map of the Mediterranean region.

```
library("Matrix")
library("ggmap")
library("ggquiver")
library("STRbook")
```

The functions needed for this case study are provided with **STRbook**. Two functions designed to work for this specific application are `Medwind_BHM_preproc` and `Medwind_BHM`, which we describe in more detail below. Their purpose is to show that this realistic, complex, science-motivated spatio-temporal BHM can be analyzed in `R` using the dynamical approach described in Chapter 5. It is worth browsing through the code of these functions to see how it is implemented (visit `https://github.com/andrewzm/STRbook`).

Preprocessing the Data and Model Setup

The function `Medwind_BHM_preproc` is a preprocessor function that takes the following as arguments:

- `Edat`: A list of four items

- `ECMWFxylocs`: Data frame containing the (x, y) coordinates on which the wind vectors and pressures are defined

- `EUdat`: Data frame containing the east–west (u) component of the ECMWF wind vector (in units of m/s) in time-wide format

- `EVdat`: Data frame containing the north–south (v) component of the ECMWF wind vector (in units of m/s) in time-wide format

- `EPdat`: Data frame containing the ECMWF atmospheric pressure (in pascals (Pa)) in time-wide format.

- `Sdat`: A list of three items

  - `Sxylocs`: A list of items (one per time point) containing the spatial locations (for each time point) of the scatterometer data

  - `SUdat`: The east–west (u) component of the QuikSCAT wind vector (in units of m/s)

  - `SVdat`: The north–south (v) component of the QuikSCAT wind vector (in units of m/s)

- `Predlocs`: Data frame containing the (x, y) coordinates of the spatial prediction grid.

- `Inparm`: Other parameters for the Gibbs sampler, discussed further below.

The data objects required for this application, `Edat`, `Sdat`, and `Predlocs`, can be loaded from **STRbook** as follows:

```
data("Medwind_data")
```

The data here correspond to 28 time periods from 00:00 UTC on 29 January 2005 to 18:00 UTC on 04 February 2005 (every 6 hours). The ECMWF analysis winds and pressure are on a $0.5° \times 0.5°$ grid and the QuikSCAT scatterometer winds are polar-orbiting satellite observations at a finer resolution (see the description in Milliff et al., 2011). There are typically no QuikSCAT observations in the prediction domain considered here at 00:00 UTC and 12:00 UTC. The prediction grid consists of 1035 ($n_y = 23$, $n_x = 45$) grid locations with a $0.5°$ spacing; this prediction grid coincides with the interior grid points of the ECMWF domain; see Section 2.1 for more details.

The preprocessor argument `Inparm` is a list of several parameters. These parameters are associated with the prediction grid, the distance to search for data near prediction-grid locations, and hyperparameters for the parameter model (prior) distributions. In particular, the item `gspdeg` is the prediction-grid spacing in degrees, and `srad` and `erad` correspond to the distance (in degrees) to search for ECMWF and QuikSCAT data locations, respectively, centered on a prediction-grid location. So `srad = 0.5` would mean that we would identify all QuikSCAT observations within 0.25 degrees of a prediction-grid location. The values of hx and hy correspond to the average longitudinal and latitudinal spacing (in meters), respectively, between the prediction-grid locations. With regard to the fixed hyperparameters in Inparm, the variables s2e and s2s are the measurement-error variances for the ECMWF and QuikSCAT wind data (σ_e^2 and σ_s^2), respectively, as given in the data-model equations (E.5)–(E.8). These are assumed to be known (Milliff et al., 2011). The variables mu_pri and s2_pri corresponding to the normal-distribution-prior mean (μ_{ab}) and variance (σ_{ab}^2) for the θ parameters, given in model (E.13), are also fixed (at 0 and 10^6, respectively, corresponding to a vague prior for θ_{ab}). IGshape and IGrate are the prior shape (q_a) and rate (r_a) parameters, respectively, for an inverse gamma prior on the process-model error variances given in (E.14); these are fixed at 1 and 1, respectively, corresponding to a vague prior for σ_a^2.

```
##      parameters to control the grid, data search, and priors
##
  preprocInput = list(
    gspdeg = .5,
    srad = .5,
    erad = .5,
    hx = 2*19.42865*1000,
    hy = 2*27.75*1000,
    s2s = 1,
    s2e = 10,
    mu_pri = 0,
    s2_pri = 10^6,
    IGshape = 1,
    IGrate = 1
  )
```

The function **Medwind_BHM_preproc** takes the data and other parameters and then builds the data-model incidence matrices ($\mathbf{H}_e$, $\mathbf{H}_{s,t}$) given in equations (E.5)–(E.8), and the difference operator matrices ($\mathbf{D}_x$, $\mathbf{D}_y$) given in equations (E.3) and (E.4), respectively. The returned list (denoted below as Mpre) contains the lists Mdata (data), Mgrid (grid), Mpriors (prior hyperparameters), and Mstrt (MCMC starting values). This list is used in the Gibbs sampler given in the next section. Recall that the Gibbs sampler is an MCMC algorithm that produces samples from the posterior distribution of all the "unknowns" given the data.

```
Mpre <- Medwind_BHM_preproc(Edat = Edat,
                            Sdat = Sdat,
                            Predlocs = Predlocs,
                            Inparm = preprocInput)
```

Running the Gibbs Sampler

This section provides the commands necessary to implement the Gibbs sampler presented in Algorithm E.1. We specify the parameters that control the number of Gibbs sampler iterations (`ngibbs`), the number of burn-in samples (`nburn`), and the number of iterations to save in memory (`nreal`).

```
GibbsInput = list(ngibbs = 10000,
                  nburn = 1000,
                  nreal = 10)
```

The Gibbs sampler for this problem takes the arguments `GibbsInput`, defined as above, and the output of the pre-processor, `Mpre`.

```
set.seed(1)  # ensure reproducibility
Mout <- Medwind_BHM(GibbsInput, Mpre)
```

The algorithm can take quite a long time to run in order to obtain a reasonable number of iterations. For this Lab, the output can be loaded directly from **STRbook**, if desired, as follows.

```
data("Medwind_Gibbs_output")
```

The Gibbs sampler function **Medwind_BHM** returns a list that includes posterior means, posterior standard deviations, and `nreal` realizations for the u and v wind components, as well as all of the iterations for the θ parameters and the process-model variances. Specifically, in this case, the list contains the following items:

- `uS`: posterior mean of the u components (1035 locations × 28 time points)

- `vS`: posterior mean of the v components (1035 locations × 28 time points)

- `uSTD`: posterior standard deviation of the u components (1035 locations × 28 time points)

- `vSTD`: posterior standard deviation of the v components (1035 locations × 28 time points)

- `uSreal`: `nreal` realizations of all 1035 × 28 locations/time points for the u components (list)

- `vSreal`: `nreal` realizations of all 1035 × 28 locations/time points for the v components (list)

- `theta_xxS` (xx = `uu,vv,uv,vu,up,vp`): `ngibbs` samples for the θ parameters

- `s2uS, s2vS`: `ngibbs` samples for the variance parameters.

We reiterate that both the pre-processor and the Gibbs sampler functions are specifically designed for this BHM fitted to the Mediterranean winds data. They would need substantial modification for different BHMs fitted to different data sets using different space-time grids. Advanced readers could examine and modify the code contained in the **STRbook** package for their applications.

Examining the Model Output

It is customary to plot the Gibbs sampler output against iteration number of the Gibbs sampler, to provide visual evidence that the samples have reasonably converged. More formal diagnostics for MCMC convergence can be found in the **coda** package. In Figure E.1, we plot the post burn-in samples for θ_{up} and σ_u^2 as a demonstration of the code and the graphics.

```
indx <- (GibbsInput$nburn+1):GibbsInput$ngibbs #plot post "burn-in"

p1 <- ggplot(data.frame(indx = indx,
                        theta_up = Mout$theta_upS[indx]),
             aes(x = indx, y = theta_up)) + geom_line() +
          ylab(expression(theta[up])) + theme_bw()

p2 <- ggplot(data.frame(indx = indx, s2u = Mout$s2uS[indx]),
             aes(x = indx, y = s2u)) + geom_line() +
          ylab(expression(sigma[vu]^2)) + theme_bw()
```

For inference on the model's parameters, we are usually interested in their marginal posterior distributions. In the top-left panel of Figure E.2, we plot the posterior distribution for θ_{up} and give the code below; coding for the other parameters in the process-model equations (E.3) and (E.4) proceeds in a similar fashion.

```
p1 <- ggplot(data.frame(theta_up = Mout$theta_upS[indx]),
             aes(x = theta_up)) + geom_density() +
          geom_vline(aes(xintercept = mean(theta_up)),
          color = "red", linetype = "dashed", size = 1) +
          xlab(expression(theta[up])) + theme_bw()
```

Finally, the main goal in this case study was to fuse the ECMWF and QuikSCAT wind observations to generate a posterior probability distribution on wind speeds. In Figure E.3, we plot the posterior mean and a posterior realization quiver plot for the winds for 06:00 UTC on 01 February 2005, using the following code.

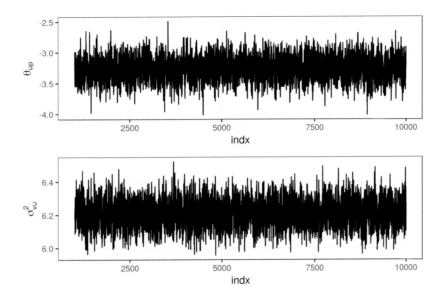

Figure E.1: Post burn-in trace plots for MCMC samples from the posterior distribution, plotted against iteration number `indx`. Top: MCMC samples for θ_{up}. Bottom: MCMC samples for σ_{vu}^2.

```
## Extract mean
u14 <- Mout$uS[, 14]                #time 14 is Feb 1, 2005 06 UTC
v14 <- Mout$vS[, 14]

## Extract realization
u14r <- Mout$uSreal[[5]][, 14]      #consider the 5th realization
v14r <- Mout$vSreal[[5]][, 14]

## Get map using get_map
lat <- c(34, 45)
long <- c(-6, 16)
bbox <- make_bbox(long, lat, f = 0.05)
bb2 <- get_map(bbox, maptype = "watercolor", source = "stamen")

## Create grid on which to plot
xg <- Mpre$Mgrid$Mgridxylocs[, 1]
yg <- Mpre$Mgrid$Mgridxylocs[, 2]
c2 <- expand.grid(x = seq(long[1], long[2], 0.5),
                  y = seq(lat[1], lat[2], 0.5))

## Plot the posterior mean and realization
```

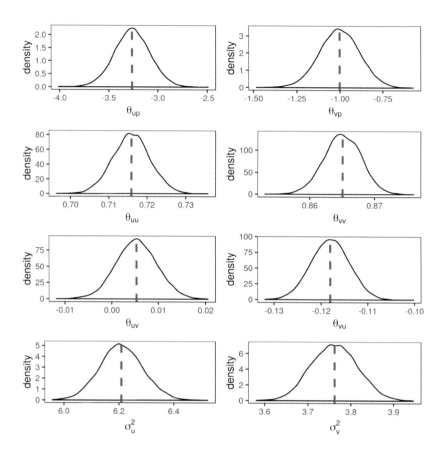

Figure E.2: The marginal posterior distributions (after applying a kernel smoother to the MCMC samples) for the process-model parameters. The red dashed line shows the posterior mean for each distribution.

```
p1 <- ggmap(bb2) + geom_quiver(data = c2,
              aes(x = xg, y = yg, u = u14, v = v14),
              vecsize = 1.5)
p2 <- ggmap(bb2) + geom_quiver(data = c2,
              aes(x = xg, y = yg, u = u14r, v = v14r),
              vecsize = 1.5)
```

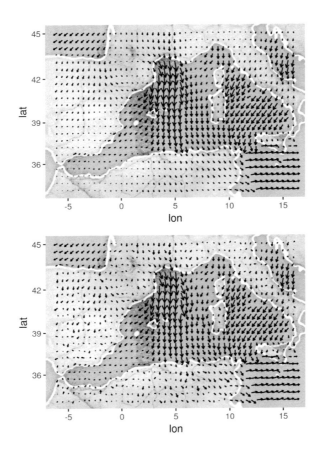

Figure E.3: Quiver plots derived from the posterior distribution of winds for 06:00 UTC on 01 February 2005. Top: Posterior mean. Bottom: A single realization from the posterior distribution.

F   Case Study: Quadratic Echo State Networks for Sea Surface Temperature Long-Lead Prediction

Recall from Section 5.4 that recurrent neural networks (RNNs) were developed in the engineering and machine-learning literature to accommodate time-dependent cycles and sequences as well as the concept of "memory" in a neural network. But, like the statistical GQN (also discussed in Section 5.4), RNNs have an extremely high-dimensional parameter space and can be difficult to fit. In contrast, the echo state network (ESN) is a type of RNN that considers sparsely connected hidden layers that allow for sequential interactions, yet specifies (remarkably) most of the parameters ("weights") to be randomly generated and then fixed, with only the parameters that connect the hidden layer to the response being estimated (see the overview in Jaeger, 2007). We consider a modification of the ESN in this case study. Note that although the models presented here are relatively simple to implement, the notational burden and (especially) the machine-learning jargon can take some getting used to.

A simple representation of an ESN is given by the following hierarchical model for data vector $\mathbf{Z}_t$ (assumed to be m-dimensional here):

$$\mathbf{Z}_t = g_o(\mathbf{V}\mathbf{h}_t), \tag{F.1}$$

$$\mathbf{h}_t = g_h(\mathbf{W}\mathbf{h}_{t-1} + \mathbf{U}\mathbf{x}_t). \tag{F.2}$$

In data model (F.1) and process model (F.2), $\mathbf{h}_t$ is an n-dimensional vector of latent ("hidden") states, $\mathbf{x}_t$ is a p-dimensional input vector, $\mathbf{V}$ is an $m \times n$ output-parameter weight matrix, $\mathbf{W}$ is an $n \times n$ hidden-process-evolution-parameter weight matrix, $\mathbf{U}$ is an $n \times p$ input-parameter weight matrix, and $g_o(\cdot)$ and $g_h(\cdot)$ are so-called "activation functions" (e.g., identity, softmax, hyperbolic tangent). The hidden-state model (F.2) is sometimes called a "reservoir." This reservoir is key to this modeling framework in that the parameter weight matrices in (F.2), $\mathbf{W}$ and $\mathbf{U}$, are sparse (only 1–10% of the parameters are non-zero) with non-zero elements *chosen at random* and fixed (for details, see the example that follows). This means that only the output weights (in $\mathbf{V}$) are estimated, substantially reducing the estimation burden. In most applications, $g_o(\cdot)$ is the identity function, and $\mathbf{V}$ can be estimated with regression methods that include regularization, such as a ridge regression or a lasso penalty (see Section 3.4). These models work surprisingly well for forecasting central tendency and classification, but they are limited for inference and uncertainty quantification. Notice that there are no error terms in this model!

McDermott and Wikle (2017) modified the basic ESN algorithm for use with spatio-temporal data to include quadratic nonlinear outputs, so-called "embedding inputs" (see below), and reservoir parameter uncertainty by considering an ensemble (bootstrap) sample of forecasts. Their quadratic echo state network (QESN), for $t = 1, \ldots, T$, is given by:

response: $\mathbf{Y}_t = \mathbf{V}_1 \mathbf{h}_t + \mathbf{V}_2 \mathbf{h}_t^2 + \boldsymbol{\epsilon}_t, \quad \text{for} \quad \boldsymbol{\epsilon}_t \sim Gau(\mathbf{0}, \sigma_\epsilon^2 \mathbf{I});$ (F.3)

hidden state: $\mathbf{h}_t = g_h \left(\dfrac{\nu}{|\lambda_w|} \mathbf{W} \mathbf{h}_{t-1} + \mathbf{U} \tilde{\mathbf{x}}_t \right);$ (F.4)

parameters: $\mathbf{W} = [w_{i,\ell}]_{i,\ell} : w_{i,\ell} = \gamma_{i,\ell}^w \, Unif(-a_w, a_w) + (1 - \gamma_{i,\ell}^w) \, \delta_0,$ (F.5)

$\mathbf{U} = [u_{i,j}]_{i,j} : u_{i,j} = \gamma_{i,j}^u \, Unif(-a_u, a_u) + (1 - \gamma_{i,j}^u) \, \delta_0,$ (F.6)

$\gamma_{i,\ell}^w \sim Bern(\pi_w),$ (F.7)

$\gamma_{i,j}^u \sim Bern(\pi_u),$ (F.8)

where $\mathbf{Y}_t$ is the n_y-dimensional response vector at time t; $\mathbf{h}_t$ is the n_h-dimensional hidden-state vector;

$$\tilde{\mathbf{x}}_t = [\mathbf{x}_t', \mathbf{x}_{t-\tau*}', \mathbf{x}_{t-2\tau*}', \ldots, \mathbf{x}_{t-m\tau*}']'$$ (F.9)

is the $n_{\tilde{x}} = (m+1)n_x$-dimensional "embedding input" vector, containing lagged values (embeddings) of the inputs $\{\mathbf{x}_t\}$ for time periods $t - \tau*$ through $t - m\tau*$, where the quantity $\tau*$ is the embedding lag (a positive integer, often set equal to the forecast lead time); and $Bern(\cdot)$ denotes the Bernoulli distribution. As in the basic ESN above, $\mathbf{W}$ is the $n \times n$ hidden-process-evolution weight matrix, $\mathbf{U}$ is the $n \times p$ input weight matrix, and $\mathbf{V}_1, \mathbf{V}_2$ are the $n \times n_h$ linear and quadratic output weight matrices, respectively. Furthermore, δ_0 is a Kronecker delta function at zero, λ_w corresponds to the largest eigenvalue of $\mathbf{W}$ (i.e., the "spectral radius" of $\mathbf{W}$), and ν is a spectral-radius control parameter. The "activation function" $g_h(\cdot)$ (a hyperbolic tangent function in our application below) controls the nonlinearity of the hidden-state evolution. The only parameters that are estimated in this model are $\mathbf{V}_1, \mathbf{V}_2$, and σ_ϵ^2 from (F.3), for which we require a ridge-regression penalty parameter, r_v (see Technical Note 3.4). Importantly, note that the matrices $\mathbf{W}$ and $\mathbf{U}$ are simulated from mixture distributions of small values (uniformly sampled in the range $(-a_w, a_w)$ and $(-a_u, a_u)$, respectively) with, respectively, $(1 - \pi_w)$ and $(1 - \pi_u)$ elements set equal to zero on average. After being sampled, these parameters are assumed to be fixed and known. Typically, these weight matrices are very sparse (e.g., of the order of 1–10% non-zeros). The hyperparameters, $\{\nu, n_h, r_v, \pi_w, \pi_u, a_w, a_u\}$, are usually chosen by cross-validation.

As is the case in most traditional ESN applications, the QESN model does not have an explicit mechanism to quantify uncertainty in the process or in the parameters. This is a bit troubling given that the reservoir weight matrices $\mathbf{W}$ and $\mathbf{U}$ are not estimated, but are chosen at random. We would expect that the model is likely to behave differently with a different set of weight matrices. This is especially true when the number of hidden units is fairly small. Although traditional ESN models typically have a very large number of hidden units, which tends to give more stable predictions, it can be desirable to have many different forecasts using a smaller number of hidden units. This provides flexibility in that it prevents overfitting, allows the various forecasts to behave as a "committee of relatively weak learners," and gives a more realistic sense of the prediction uncertainty for out-of-sample forecasts. Thus, we could generate an ensemble or bootstrap sample of forecasts.

As shown in McDermott and Wikle (2017), this ensemble approach can be implemented straightforwardly with the QESN model using Algorithm F.1.

Algorithm F.1: Ensemble QESN Algorithm

Initialize: Select tuning parameters $\{\nu, n_h, r_v, \pi_w, \pi_u, a_w, a_u\}$ (e.g., by cross-validation with a standalone QESN)

for $k = 1$ to K **do**

    1. Simulate $\mathbf{W}^{(k)}$, $\mathbf{U}^{(k)}$ using (F.5) and (F.6) and initialize $\mathbf{h}_1^{(k)}$

    2. Calculate $\{\mathbf{h}_t^{(k)} : t = 1, \ldots, T\}$ using (F.4)

    3. Use ridge regression to estimate $\mathbf{V}_1^{(k)}$, $\mathbf{V}_2^{(k)}$, and σ_ϵ^2

    4. Calculate out-of-sample forecasts $\{\widehat{\mathbf{Y}}_t^{(k)} : t = T + 1, \ldots, T + \tau\}$, where τ is the forecast lead time (requires calculating $\{\widehat{\mathbf{h}}_t^{(k)} : t = T + 1, \ldots, T + \tau\}$ from (F.4))

end for

Use ensemble of forecasts $\{\widehat{\mathbf{Y}}_t^{(k)} : t = T + 1, \ldots, T + \tau; k = 1, \ldots, K\}$ to calculate moments, prediction intervals, etc.

Implementation in R

In what follows, we provide a demonstration of the ensemble QESN model applied to long-lead forecasting of sea-surface temperature using the SST data set.

Ensemble QESN Model Data Preparation

To prepare the data, we need **ggplot2**, **dplyr**, **STRbook**, and **tidyr**.

```
library("ggplot2")
library("dplyr")
library("STRbook")
library("tidyr")
```

The functions needed for this case study are provided with **STRbook**. Our purpose here is to show that this nonlinear DSTM can be implemented in R fairly easily. If readers are interested in adapting these functions to their own applications, it is worth browsing through the functions to see how the code is implemented (visit `https://github.com/andrewzm/STRbook`).

We first load the SST data set. This time we shall use the data up to October 1996 as training data and perform out-of-sample six-month forecasts from April 1997 to July 1999.

```
data("SSTlandmask")
data("SSTlonlat")
data("SSTdata")
delete_rows <- which(SSTlandmask == 1)    # find land values
SSTdataA <- SSTdata[-delete_rows, ]       # remove land values
```

In this application, we shall evaluate the forecast in terms of the time series corresponding to the average of the SST anomalies in the so-called Niño 3.4 region (defined to be the region of the tropical Pacific Ocean contained by 5°S–5°N, 170°W–120°W).

```
## find grid locations corresponding to Nino 3.4 region;
## note, 190 - 240 deg E longitude corresponds
## to 170 - 120 deg W longitude
nino34Index <- which(SSTlonlat[,2] <= 5 & SSTlonlat[, 2] >= -5 &
                     SSTlonlat[,1] >= 190 & SSTlonlat[, 1] <= 240)
```

The object `SSTdataA` is a 2261×399 matrix in time-wide format. In the code below, we save the number of spatial locations in the variable `nspatial`. Of the 399 time points, we only need 322 for training, the number of months between (and including) January 1970 and October 1996. We define a six-month-ahead forecast by specifying `tau = 6`.

```
nspatial <- nrow(SSTdataA)    # number of spat. locations
TrainLen <- 322               # no. of months to Oct 1996
tau <- 6                      # forecast lead time (months)
```

We train the ESN on time series associated with the first ten EOFs extracted from the SST (training) data. The following code follows closely what was done in Labs 2.3 and 5.3.

```
n <- 10                                   # number of EOFs to retain
Z <- t(SSTdataA[, 1:TrainLen])            # data matrix
spat_mean <- apply(SSTdataA, 1, mean)         # spatial mean
Zspat_detrend <- Z - outer(rep(1, TrainLen), # detrend data
                    spat_mean)
Zt <- 1/sqrt(TrainLen - 1)*Zspat_detrend      # normalize
E <- svd(Zt)                  # SVD
PHI <- E$v[, 1:n]             # 10 EOF spatial basis functions
TS <- t(SSTdataA) %*% PHI     # project data onto basis functions
                              # for PC time series
```

Now we need to create the training and validation data sets. Both data sets will need input data and output data. Since we are setting up the ESN for six-months-ahead forecasting, as input we use the PC time series (see Section 2.4.3) lagged by six months with respect to the output. For example, the PC time-series values at January 1970 are inputs ($\mathbf{x}_t$) to forecast the SST (output) in July 1970. For prediction, we consider forecasting at ten three-month intervals starting from October 1996 (we chose three-month intervals to improve the visualization, but one can forecast each month if desired).

```
## training set
xTrainIndex <- 1:(TrainLen - tau) # training period ind. for input
yTrainIndex <- (tau+1):(TrainLen) # shifted period ind. for output
xTrain <- TS[xTrainIndex, ]       # training input time series
yTrain <- TS[yTrainIndex, ]       # training output time series

## test set: forecast every three months
xTestIndex <- seq(TrainLen, by = 3, length.out = 10)
yTestIndex <- xTestIndex+tau       # test output indices
xTest <- TS[xTestIndex,]           # test input data
yTest <- TS[yTestIndex,]           # test output data
testLen <- nrow(xTest)             # number of test cases
```

Ensemble QESN Model Implementation

We first have to make some model choices and set some parameters to run the ensemble QESN model. For model details and terminology, see the description above.

```
quadInd <- TRUE   # include both quadratic and linear output terms
                  # if FALSE, then include only linear terms
ensembleLen <- 500  # number of ensemble members (i.e., QESN runs)
```

The next set of parameters can be trained by cross-validation or out-of-sample validation (see McDermott and Wikle, 2017). For simplicity, we use the values obtained in that paper (which considered a similar long-lead SST forecasting application) here. The model arguments required as input are: wWidth, which corresponds to the parameter a_w that specifies the range of the uniform distribution for the $\mathbf{W}$ weight matrix parameters in (F.5); similarly, uWidth, which corresponds to the parameter a_u that specifies the range of the uniform distribution for the $\mathbf{U}$ matrix in (F.6); piW, which corresponds to π_w in (F.7), the probability of a non-zero $\mathbf{W}$ weight; piU, which corresponds to π_u, the probability of non-zero $\mathbf{U}$ weight parameter in (F.8); curNh, which corresponds to n_h, the number of hidden units; curNu, which corresponds to ν, the spectral radius of the $\mathbf{W}$ matrix; curM, which corresponds to m, the number of lags (embeddings) of input vectors to use; tauEMB, which corresponds to the embedding lag ($\tau*$ in (F.9)); and curRV, which corresponds to r_v, the ridge-regression parameter associated with the estimation of the output matrices, $\mathbf{V}_1$ and $\mathbf{V}_2$ in (F.3).

```
wWidth <- .10          # W-weight matrix, uniform dist "width" param.
uWidth <- .10          # U-weight matrix, uniform dist "width" param.
piW <- .10             # sparseness parameter for W-weight matrix
piU <- .10             # sparseness parameter for U-weight matrix
curNh <- 120           # number of hidden units
curNu <- .35           # scaling parameter for W-weight matrix
curM <- 4              # number of embeddings
tauEMB <- 6            # embedding lag
curRV <- .01           # output ridge regression parameter
```

Now we use the function **createEmbedRNNData** to create a data object containing responses and embedding matrix inputs (see equation (F.9)) for the training and prediction data sets (note that the responses and inputs are scaled by their respective standard deviations, as is common in the ESN literature). The function takes as inputs variables defined above: curM, the number of embedding lags; tauEMB, the embedding lag; tau, the forecast lead time; yTrain, the training output time series; TS, the input time series associated with the projection of the data onto the EOFs; and xTestIndex, which identifies the indices for the input data corresponding to the test periods.

```
## standardize and create embedding matrices
DataObj <- createEmbedRNNData(curM, tauEMB, tau, yTrain, TS,
                              xTestIndex)
```

The returned object, DataObj, is a list containing inputs and training data in the format required to train the ESN. We now need to create a parameter object that contains the parameters to be used in constructing the ESN. The function we use initializes the vectors associated with: the embedding matrix $\tilde{\mathbf{x}}_t$ in (F.9); the hidden state $\mathbf{h}_t$ in (F.4); and the ridge-regression matrix, $r_v\mathbf{I}$ (as defined in Technical Note 3.4).

```
setParObj <- setParsEESN(curRV ,curNh, n, curM, quadInd)
```

We save the forecasts in a three-dimensional array, with the first dimension indexing the ensemble number, the second dimension indexing the forecast time point, and the third dimension indexing the EOF number. We also create a second three-dimensional array with the first two dimensions the same, and the third dimension indexing spatial location.

```
fmatESNFin <- array(NA, c(ensembleLen, testLen, n))
fmatESNFinFull <- array(NA,c(ensembleLen, testLen, nspatial))
```

We are now ready to run the ensemble of QESN models to obtain forecasts. For each ensemble, we run the function **genResR**, which takes arguments defined previously as input: curNh, the number of hidden units; wWidth and uWidth, the uniform distribution sampling range for **W** and **U**, respectively; piW and piU, the probabilities of non-zeros in **W** and **U**, respectively; curNu, the spectral-radius parameter; quadInd, the indicator on whether to include the quadratic output weights or not; DataObj, the embedding input matrices; setParObj, the initializations corresponding to the hidden state vectors and the ridge-regression matrices; and testLen, the number of test cases.

```
for(iEnsem in 1:ensembleLen) {
  ## Run the QESN model for a single ensemble
  QESNOutObj = genResR(nh = curNh,
                       wWidth = wWidth,
                       uWidth = uWidth,
                       piW = piW,
                       piU = piU,
                       nuESN = curNu,
                       quadInd = quadInd,
                       DataObj = DataObj,
                       setParObj = setParObj,
                       testLen = testLen)

  ## save forecasts for the reduced dimension output
  fmatESNFin[iEnsem, , ] <- t(QESNOutObj$unScaledForecasts)

  ## forecasts for the full spatial field
  fmatESNFinFull[iEnsem, , ] <- fmatESNFin[iEnsem, , ] %*% t(PHI)
}
```

Post-Processing the Ensemble QESN Output

In this section we focus on post-processing the ensemble QESN output for the Niño 3.4 region. To assess whether or not we have the correct coverage of the prediction intervals, we consider 95% (pointwise) prediction intervals.

```
alpha <- .05   # alpha-level of 1-alpha pred. intervals (P.I.s)
lwPI <- alpha/2
```

In the following code, we calculate the mean and the lower/upper boundaries of the 95% prediction interval for the Niño 3.4 region (across the whole ensemble of realizations from the predictive distribution).

```
nino34AvgPreds <- nino34LwPI <- nino34UpPI <- rep(NA, testLen)
for(i in 1:testLen){
  nino34AvgPreds[i] <- fmatESNFinFull[,i,nino34Index] %>%
                    mean()
  nino34LwPI[i] <- fmatESNFinFull[, i, nino34Index] %>%
                rowMeans() %>%
```

```
                          quantile(lwPI)
    nino34UpPI[i] <- fmatESNFinFull[,i,nino34Index] %>%
                          rowMeans() %>%
                          quantile(1 - lwPI)
 }
 nino34_results <- data.frame(AvgPres = nino34AvgPreds,
                              LwPI = nino34LwPI,
                              UpPI = nino34UpPI)
```

These predictive-distribution summaries can be compared to the average SST at the prediction month, which we calculate as follows:

```
nino34_results$AvgObs <- SSTdata[nino34Index, yTestIndex] %>%
             colMeans()
```

Finally, we allocate the prediction-month labels to the data frame which, for this example, is achieved as follows.

```
nino34_results$date <- seq(as.Date("1997-04-01"),
                           length.out = 10, by = "3 months")
```

Plotting Results for Forecasts in the Niño 3.4 Region

In this last section, we focus our plots on the results for spatially averaged SST anomalies over the Niño 3.4 region for every third month (to make the plot less cluttered). Although we skip months and present the spatial average for ease of visualization, we note that the full spatial fields could easily be plotted for any of the forecast months, as shown in Lab 5.3. In Figure F.1, we plot the prediction and prediction intervals for the spatial average alongside the spatial average of the observations by month. The following code produces this figure.

```
gresults <- ggplot(nino34_results) +
    geom_line(aes(x = date, y = AvgObs)) +
    geom_ribbon(aes(x = date, ymin = LwPI, ymax = UpPI),
                alpha = 0.1, fill = "black") +
    geom_line(aes(x = date, y = AvgPres), col = "red") +
    ylab(expression(paste("Ni", tilde(n), "o 3.4 Index"))) +
    xlab("Month") + theme_bw()
```

Although pointwise predition intervals are informative, it can also be helpful to plot the trajectories of individual forecasts from the QESN model. This can be done by first assigning a number (say, the first 15) of ensemble trajectories to the data frame and then putting the data frame into long format using `gather`. The following code produces Figure F.2.

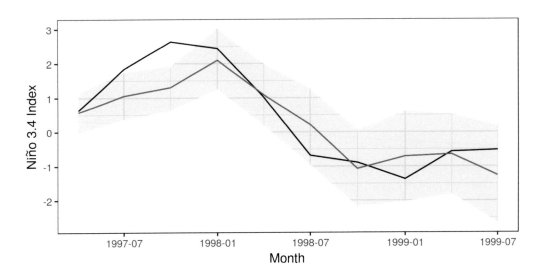

Figure F.1: Out-of-sample six-month forecasts of SST anomalies averaged over the Niño 3.4 region from April 1997 to July 1999 (every three months). The black line shows the truth and the red line shows the average of the ensemble of QESN forecasts. The point-wise 95% prediction intervals from the ensemble of QESN forecasts is shown with light-gray shading.

```r
## Compute the spatial average over Nino3.4 for each ensemble
for(i in 1:15)
  nino34_results[paste0("Sim",i)] <-
       rowMeans(fmatESNFinFull[i,,nino34Index])

## Convert to long data frame
nino34_results_long <- nino34_results %>%
                   dplyr::select(-AvgPres, -LwPI,
                                 -UpPI, -AvgObs) %>%
                   gather(SimNum, SSTindex, -date)

## Plot
gresults2 <- ggplot(nino34_results_long) +
  geom_line(data = nino34_results, aes(x = date, y = AvgObs)) +
  geom_line(aes(x = date, y = SSTindex, group = SimNum,
                linetype = SimNum, colour = SimNum)) +
  ylab(expression(paste("Ni", tilde(n), "o 3.4 Index"))) +
  xlab("Month") + theme_bw() +  theme(legend.position="none")
```

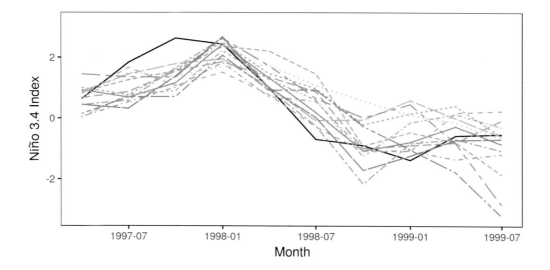

Figure F.2: Out-of-sample six-month forecasts of SST anomalies averaged over the Niño 3.4 region from April 1997 through July 1999 (every three months). The black line shows the truth, and other lines show the first 15 ensemble members from the ensemble of QESN model forecasts.

List of R packages

This book would of course not have been possible without the free availability of R (R Core Team, 2018) and some excellent packages. The book itself was compiled using **knitr** v1.20 (Xie, 2015) while running R v3.4.4. All other R packages that were at some point used throughout the book, together with a brief description of how they were used, are listed below.

Package	Reference	How used
animation v2.5	Xie (2013)	spatio-temporal visualizations
ape v5.1	Paradis et al. (2004)	Moran's I tests in space and space-time using `Moran.I`
broom v0.5.0	Robinson and Hayes (2018)	casting the summary outputs of tests into data frames using `tidy`
CCA v1.2	González and Déjean (2012)	canonical correlation analysis of spatio-temporal data using `cancor`
devtools v1.13.6	Wickham and Chang (2017)	installing **STRbook**
dplyr v0.7.6	Wickham et al. (2017)	data-wrangling spatio-temporal data – in particular filtering, sorting, selecting variables and creating new variables
EFDR v0.1.1	Zammit-Mangion and Huang (2015)	carrying out field significance tests with EFDR
expm v0.999-2	Goulet et al. (2017)	raising matrices to a power using `%^%`
fields v9.6	Nychka et al. (2015)	computing distances using `rdist` and plotting using `image.show`
FRK v0.2.2	Zammit-Mangion (2018a)	modeling spatio-temporal data with spatio-temporal basis functions
ggplot2 v3.0.0	Wickham (2016)	visualizing data by plotting facets of line, contour, or raster plots
ggmap v2.6.1	Kahle and Wickham (2013)	plotting of regional maps

ggquiver v0.1.0	O'Hara-Wild (2017)	visualizing directional data with quivers
gridExtra v2.3	Auguie (2016)	arranging **ggplot2** plots into a grid using `grid.arrange` or `arrangeGrob`
gstat v1.1-6	Gräler et al. (2016)	inverse distance weighting, fitting spatio-temporal semivariograms, and spatio-temporal kriging
IDE v0.2.0	Zammit-Mangion (2018b)	modeling spatio-temporal data with integro-difference-equation models
INLA v18.07.12	Lindgren and Rue (2015)	modeling non-Gaussian spatio-temporal data with a separable model
lattice v0.20-35	Sarkar (2008)	surface plots using `wireframe`
leaps v3.0	Lumley (2017)	stepwise regression using `regsubsets`
lmtest v0.9-36	Zeileis and Hothorn (2002)	Durbin–Watson tests using `dwtest`
maps v3.3.0	Becker and Wilks (2017)	reading in state boundaries for plotting using `map_data`
maptools v0.9-3	Bivand and Lewin-Koh (2017).	reading in data from shapefiles using `readShapePoly`
Matrix v1.2-14	Bates and Maechler (2017)	handling objects of class `Matrix`
mgcv v1.8-23	Wood (2017)	modeling non-Gaussian spatio-temporal data using generalized additive models using `gamm`
nlme v3.1-131	Pinheiro et al. (2017)	fitting linear models with correlated error using `gls`
plyr v1.8.4	Wickham (2011)	filling in missing columns during row binding using `rbind.fill`
purrr v0.2.5	Henry and Wickham (2017)	carrying out tests on groups of data using `map`
RColorBrewer v1.1-2	Neuwirth (2014)	generating a color palette using `brewer.pal`
scoringRules v0.9.5	Jordan et al. (2017a)	multivariate probabilistic validation using `es_sample` and `vs_sample`
sp v1.3-1	Bivand et al. (2013)	creating and handling spatial objects such as `SpatialPoints` or `SpatialPolygons`
spacetime v1.2-2	Pebesma (2012)	creating and handling spatio-temporal objects such as `STIDF` or `STFDF`

SpatialVx v0.6-3	Gilleland (2018)	field-matching methods for assessing predictions
SpatioTemporal v1.1.7	Lindstrom et al. (2013)	modeling spatio-temporal data with temporal basis functions
SpecsVerification v0.5-2	Siegert (2017)	plotting verification rank histograms
stargazer v5.2.2	Hlavac (2015)	generating LaTeX tables from the results of standard tests
STRbook v0.1.0	Zammit-Mangion (2018c)	companion package for this book, containing several data sets and helper functions. Needs to be installed using **devtools** through `install_github`("andrewzm/STRbook")
tidyr v0.8.1	Wickham and Henry (2017)	data-wrangling spatio-temporal data – in particular, going from space-wide or time-wide to long formats, and nesting and unnesting data frames
verification v1.42	NCAR – Research Applications Laboratory (2015)	probabilistic validation using the continuous ranked probability score with the function `crps`
xtable v1.8-2	Dahl (2016)	generating LaTeX tables from data frames

References

Auguie, B. (2016), *gridExtra: Miscellaneous Functions for "Grid" Graphics*, R package version 2.3. Available from `https://CRAN.R-project.org/package=gridExtra`.

Baddeley, A., Rubak, E., and Turner, R. (2015), *Spatial Point Patterns: Methodology and Applications with R*, Boca Raton, FL: Chapman & Hall/CRC.

Bakar, K. S. and Sahu, S. K. (2015), "spTimer: Spatio-temporal Bayesian modelling using R," *Journal of Statistical Software*, 63, 1–32.

Banerjee, S., Carlin, B. P., and Gelfand, A. E. (2015), *Hierarchical Modeling and Analysis for Spatial Data*, 2nd ed., Boca Raton, FL: Chapman & Hall/CRC.

Bates, D. and Maechler, M. (2017), *Matrix: Sparse and Dense Matrix Classes and Methods*, R package version 1.2-14. Available from `https://CRAN.R-project.org/package=Matrix`.

Becker, R. A. and Wilks, A. R. (2017), *maps: Draw Geographical Maps*, R package version 3.3.0. (R version by Brownrigg, R. Enhancements by Minka, T. P. and Deckmyn, A.) Available from `https://CRAN.R-project.org/package=maps`.

Benjamini, Y. and Hochberg, Y. (1995), "Controlling the false discovery rate: A practical and powerful approach to multiple testing," *Journal of the Royal Statistical Society, Series B*, 57, 289–300.

Berliner, L. M. (1996), "Hierarchical Bayesian time series models," in *Maximum Entropy and Bayesian Methods*, eds. Hanson, K. M. and Silver, R. N., Dordecht: Kluwer, Fundamental Theories of Physics, 79, pp. 15–22.

Berliner, L. M., Milliff, R. F., and Wikle, C. K. (2003), "Bayesian hierarchical modeling of air-sea interaction," *Journal of Geophysical Research: Oceans*, 108.

Bivand, R. and Lewin-Koh, N. (2017), *maptools: Tools for Reading and Handling Spatial Objects*, R package version 0.9-3. Available from `https://CRAN.R-project.org/package=maptools`.

Bivand, R. S., Pebesma, E., and Gomez-Rubio, V. (2013), *Applied Spatial Data Analysis with* R, 2nd ed., New York: Springer.

Blangiardo, M. and Cameletti, M. (2015), *Spatial and Spatio-Temporal Bayesian Models with R-INLA*, Hoboken, NJ: John Wiley & Sons.

Box, G. E. P. (1976), "Science and statistics," *Journal of the American Statistical Association*, 71, 791–799.

— (1979), "Robustness in the strategy of scientific model building," in *Robustness in Statistics*, eds. Launer, R. L. and Wilkinson, G. L., New York: Academic Press, pp. 201–236.

Branstator, G., Mai, A., and Baumhefner, D. (1993), "Identification of highly predictable flow elements for spatial filtering of medium- and extended-range numerical forecasts," *Monthly Weather Review*, 121, 1786–1802.

Briggs, W. M. and Levine, R. A. (1997), "Wavelets and field forecast verification," *Monthly Weather Review*, 125, 1329–1341.

Bröcker, J. and Smith, L. A. (2007), "Scoring probabilistic forecasts: The importance of being proper," *Weather and Forecasting*, 22, 382–388.

Brown, B. G., Gilleland, E., and Ebert, E. E. (2012), "Forecasts of spatial fields," in *Forecast Verification: A Practitioner's Guide in Atmospheric Science*, eds. Jolliffe, I. T. and Stephenson, D. B., 2nd ed., Chichester: John Wiley & Sons, pp. 95–117.

Carlin, B. P. and Louis, T. A. (2010), *Bayes and Empirical Bayes Methods for Data Analysis*, Boca Raton, FL: Chapman & Hall/CRC.

Carroll, S. S. and Cressie, N. (1996), "A comparison of geostatistical methodologies used to estimate snow water equivalent," *Water Resources Bulletin*, 32, 267–278.

Carvalho, A. (2016), "An overview of applications of proper scoring rules," *Decision Analysis*, 13, 223–242.

Christakos, G. (2017), *Spatiotemporal Random Fields: Theory and Applications*, 2nd ed., Amsterdam: Elsevier.

Cohen, A. and Jones, R. H. (1969), "Regression on a random field," *Journal of the American Statistical Association*, 64, 1172–1182.

Congdon, P. (2006), "Bayesian model choice based on Monte Carlo estimates of posterior model probabilities," *Computational Statistics & Data Analysis*, 50, 346–357.

Cook, R. D. (1977), "Detection of influential observation in linear regression," *Technometrics*, 19, 15–18.

Cressie, N. (1990), "The origins of kriging," *Mathematical Geology*, 22, 239–252.

— (1993), *Statistics for Spatial Data*, revised ed., New York: John Wiley & Sons.

Cressie, N. and Huang, H.-C. (1999), "Classes of nonseparable, spatio-temporal stationary covariance functions," *Journal of the American Statistical Association*, 94, 1330–1339.

Cressie, N., Shi, T., and Kang, E. L. (2010), "Fixed rank filtering for spatio-temporal data," *Journal of Computational and Graphical Statistics*, 19, 724–745.

Cressie, N. and Wikle, C. K. (2011), *Statistics for Spatio-Temporal Data*, Hoboken, NJ: John Wiley & Sons.

Crujeiras, R. M., Fernández-Casal, R., and González-Manteiga, W. (2010), "Nonparametric test for separability of spatio-temporal processes," *Environmetrics*, 21, 382–399.

Dahl, D. B. (2016), *xtable: Export Tables to LaTeX or HTML*, R package version 1.8-2. Available from https://CRAN.R-project.org/package=xtable.

Delmonico, R. (2017), *The Philosophy of Fractals*, independently published, ISBN 978197343066.

Diggle, P. J. (2013), *Statistical Analysis of Spatial and Spatio-Temporal Point Patterns*, Boca Raton, FL: Chapman & Hall/CRC.

Diggle, P. J. and Ribeiro Jr., P. J. (2007), *Model-Based Geostatistics*, New York: Springer.

Douc, R., Moulines, E., and Stoffer, D. (2014), *Nonlinear Time Series: Theory, Methods and Applications with R Examples*, Boca Raton, FL: Chapman & Hall/CRC.

Ebert, E. E. and McBride, J. L. (2000), "Verification of precipitation in weather systems: Determination of systematic errors," *Journal of Hydrology*, 239, 179–202.

Finley, A. O., Banerjee, S., and Carlin, B. P. (2007), "spBayes: an R package for univariate and multivariate hierarchical point-referenced spatial models," *Journal of Statistical Software*, 19, 1–24.

Fisher, R. A. (1935), *The Design of Experiments*, 8th ed., Edinburgh: Oliver and Boyd.

Gamerman, D. and Lopes, H. F. (2006), *Markov Chain Monte Carlo: Stochastic Simulation for Bayesian Inference*, 2nd ed., Boca Raton, FL: Chapman & Hall/CRC.

Gelfand, A. E. and Ghosh, S. K. (1998), "Model choice: A minimum posterior predictive loss approach," *Biometrika*, 85, 1–11.

Gelman, A., Carlin, J. B., Stern, H. S., Dunson, D. B., Vehtari, A., and Rubin, D. B. (2014), *Bayesian Data Analysis*, 3rd ed., Boca Raton, FL: Chapman & Hall/CRC.

Genton, M. G., Castruccio, S., Crippa, P., Dutta, S., Huser, R., Sun, Y., and Vettori, S. (2015), "Visuanimation in statistics," *Stat*, 4, 81–96.

Gilleland, E. (2018), *SpatialVx: Spatial Forecast Verification*, R package version 0.6-3. Available from `https://CRAN.R-project.org/package=SpatialVx`.

Gilleland, E., Ahijevych, D. A., Brown, B. G., and Ebert, E. E. (2010), "Verifying forecasts spatially," *Bulletin of the American Meteorological Society*, 91, 1365–1376.

Gneiting, T. and Katzfuss, M. (2014), "Probabilistic forecasting," *Annual Review of Statistics and its Application*, 1, 125–151.

Gneiting, T. and Raftery, A. E. (2007), "Strictly proper scoring rules, prediction, and estimation," *Journal of the American Statistical Association*, 102, 359–378.

González, I. and Déjean, S. (2012), *CCA: Canonical Correlation Analysis*, R package version 1.2. Available from `https://CRAN.R-project.org/package=CCA`.

Goodfellow, I., Bengio, Y., Courville, A., and Bengio, Y. (2016), *Deep Learning*, Cambridge, MA: MIT Press.

Goulet, V., Dutang, C., Maechler, M., Firth, D., Shapira, M., and Stadelmann, M. (2017), *expm: Matrix Exponential, Log, "etc"*, R package version 0.999-2. Available from `https://CRAN.R-project.org/package=expm`.

Gräler, B., Pebesma, E., and Heuvelink, G. (2016), "Spatio-temporal interpolation using gstat," *R Journal*, 8, 204–218.

Hanks, E. M., Schliep, E. M., Hooten, M. B., and Hoeting, J. A. (2015), "Restricted spatial regression in practice: Geostatistical models, confounding, and robustness under model misspecification," *Environmetrics*, 26, 243–254.

Harvey, A. C. (1993), *Time Series Models*, 2nd ed., Cambridge, MA: MIT Press.

Haslett, J. (1999), "A simple derivation of deletion diagnostic results for the general linear model with correlated errors," *Journal of the Royal Statistical Society, Series B*, 61, 603–609.

Hastie, T., Tibshirani, R., and Friedman, J. (2009), *The Elements of Statistical Learning*, 2nd ed., New York: Springer.

Henebry, G. M. (1995), "Spatial model error analysis using autocorrelation indices," *Ecological Modelling*, 82, 75–91.

Henry, L. and Wickham, H. (2017), *purrr: Functional Programming Tools*, R package version 0.2.5. Available from `https://CRAN.R-project.org/package=purrr`.

Hlavac, M. (2015), *stargazer: Well-Formatted Regression and Summary Statistics Tables*, Harvard University, Cambridge, MA, R package version 5.2.2. Available from `http://CRAN.R-project.org/package=stargazer`.

Hodges, J. S. and Reich, B. J. (2010), "Adding spatially-correlated errors can mess up the fixed effect you love," *American Statistician*, 64, 325–334.

Hodges, J. S. and Sargent, D. J. (2001), "Counting degrees of freedom in hierarchical and other richly-parameterised models," *Biometrika*, 88, 367–379.

Hoeting, J. A., Madigan, D., Raftery, A. E., and Volinsky, C. T. (1999), "Bayesian model averaging: A tutorial," *Statistical Science*, 382–401.

Hooten, M. B. and Hobbs, N. (2015), "A guide to Bayesian model selection for ecologists," *Ecological Monographs*, 85, 3–28.

Hotelling, H. (1936), "Relations between two sets of variates," *Biometrika*, 28, 321–377.

Hovmöller, E. (1949), "The trough-and-ridge diagram," *Tellus*, 1, 62–66.

Hughes, J. and Haran, M. (2013), "Dimension reduction and alleviation of confounding for spatial generalized linear mixed models," *Journal of the Royal Statistical Society, Series B*, 75, 139–159.

Jaeger, H. (2007), "Echo state network," *Scholarpedia*, 2, 2330.

James, G., Witten, D., Hastie, T., and Tibshirani, R. (2013), *An Introduction to Statistical Learning*, New York: Springer.

Johnson, R. A. and Wichern, D. W. (1992), *Applied Multivariate Statistical Analysis*, 3rd ed., Englewood Cliffs, NJ: Prentice Hall.

Jordan, A., Krueger, F., and Lerch, S. (2017a), *scoringRules: Scoring Rules for Parametric and Simulated Distribution Forecasts*, R package version 0.9.5. Available from `https://CRAN.R-project.org/package=scoringRules`.

Jordan, A., Krüger, F., and Lerch, S. (2017b), "Evaluation of probabilistic forecasts with the scoringRules package," in *EGU General Assembly Conference Abstracts*, vol. 19, p. 3295.

Kahle, D. and Wickham, H. (2013), "ggmap: Spatial Visualization with ggplot2," *R Journal*, 5, 144–161.

Kalman, R. E. (1960), "A new approach to linear filtering and prediction problems," *Journal of Basic Engineering*, 82, 35–45.

Kang, E. L., Liu, D., and Cressie, N. (2009), "Statistical analysis of small-area data based on independence, spatial, non-hierarchical, and hierarchical models," *Computational Statistics & Data Analysis*, 53, 3016–3032.

Kendall, M. G. and Stuart, A. (1969), *The Advanced Theory of Statistics*, vol. 1, 3rd ed., New York: Hafner.

Kornak, J., Irwin, M. E., and Cressie, N. (2006), "Spatial point process models of defensive strategies: Detecting changes," *Statistical Inference for Stochastic Processes*, 9, 31–46.

Krainski, E., Gómez-Rubio, V., Bakka, H., Lenzi, A., Castro-Camilo, D., Simpson, D., Lindgren, F., and Rue, H. (2019), *Advanced Spatial Modeling with Stochastic Partial Differential Equations Using R and INLA*, Boca Raton, FL: Chapman and Hall/CRC.

Kuhnert, P. (2014), "Editorial: Physical-statistical modelling," *Environmetrics*, 25, 201–202.

Kutner, M, H., Nachtsheim, C. J., and Neter, J. (2004), *Applied Multiple Regression Models*, Irwin, OH: McGraw-Hill.

Laird, N. M. and Ware, J. H. (1982), "Random-effects models for longitudinal data," *Biometrics*, 963–974.

Lamigueiro, O. P. (2018), *Displaying Time Series, Spatial, and Space-Time Data with R*, 2nd ed., Boca Raton, FL: Chapman and Hall/CRC.

Laud, P. W. and Ibrahim, J. G. (1995), "Predictive model selection," *Journal of the Royal Statistical Society, Series B*, 247–262.

Le, N. D. and Zidek, J. V. (2006), *Statistical Analysis of Environmental Space-Time Processes*, New York: Springer.

Lindgren, F. and Rue, H. (2015), "Bayesian spatial modelling with R-INLA," *Journal of Statistical Software*, 63, 1–25.

Lindgren, F., Rue, H., and Lindström, J. (2011), "An explicit link between Gaussian fields and Gaussian Markov random fields: The stochastic partial differential equation approach," *Journal of the Royal Statistical Society, Series B*, 73, 423–498.

Lindstrom, J., Szpiro, A., Sampson, P. D., Bergen, S., and Oron, A. P. (2013), *SpatioTemporal: Spatio-Temporal Model Estimation*, R package version 1.1.7. Available from `https://CRAN.R-project.org/package=SpatioTemporal`.

Livezey, R. E. and Chen, W. Y. (1983), "Statistical field significance and its determination by Monte Carlo techniques," *Monthly Weather Review*, 111, 46–59.

Lucchesi, L. R. and Wikle, C. K. (2017), "Visualizing uncertainty in areal data with bivariate choropleth maps, map pixelation, and glyph rotation," *Stat*, 6, 292–302.

Lukoševičius, M. (2012), "A practical guide to applying echo state networks," in *Neural Networks: Tricks of the Trade*, eds. Montavon, G., Orr, B., and Müller, K.-R., 2nd ed., New York: Springer, pp. 659–686.

Lukoševičius, M. and Jaeger, H. (2009), "Reservoir computing approaches to recurrent neural network training," *Computer Science Review*, 3, 127–149.

Lumley, T. (2017), *leaps: Regression Subset Selection*, R package version 3.0, based on Fortran code by A. Miller. Available from `https://CRAN.R-project.org/package=leaps`.

Lütkepohl, H. (2005), *New Introduction to Multiple Time Series Analysis*, Berlin: Springer.

Mateu, J. and Müller, W. G. (2013), *Spatio-Temporal Design: Advances in Efficient Data Acquisition*, Chichester: John Wiley & Sons.

McCullagh, P. and Nelder, J. A. (1989), *Generalized Linear Models*, Cambridge: Cambridge University Press.

McCulloch, C. E. and Searle, S. R. (2001), *Generalized, Linear, and Mixed Models*, New York: John Wiley & Sons.

McDermott, P. L. and Wikle, C. K. (2016), "A model-based approach for analog spatio-temporal dynamic forecasting," *Environmetrics*, 27, 70–82.

— (2017), "An ensemble quadratic echo state network for non-linear spatio-temporal forecasting," *Stat*, 6, 315–330.

McDermott, P. L., Wikle, C. K., and Millspaugh, J. (2018), "A hierarchical spatiotemporal analog forecasting model for count data," *Ecology and Evolution*, 8, 790–800.

Micheas, A. C., Fox, N. I., Lack, S. A., and Wikle, C. K. (2007), "Cell identification and verification of QPF ensembles using shape analysis techniques," *Journal of Hydrology*, 343, 105–116.

Milliff, R. F., Bonazzi, A., Wikle, C. K., Pinardi, N., and Berliner, L. M. (2011), "Ocean ensemble forecasting. Part I: Ensemble Mediterranean winds from a Bayesian hierarchical model," *Quarterly Journal of the Royal Meteorological Society*, 137, 858–878.

Monahan, A. H., Fyfe, J. C., Ambaum, M. H., Stephenson, D. B., and North, G. R. (2009), "Empirical orthogonal functions: The medium is the message," *Journal of Climate*, 22, 6501–6514.

Montero, J.-M., Fernández-Avilés, G., and Mateu, J. (2015), *Spatial and Spatio-Temporal Geostatistical Modeling and Kriging*, Chichester: John Wiley & Sons.

Morrison, P. and Morrison, P. (1982), *Powers of Ten: About the Relative Size of Things in the Universe*, Redding, CT: Scientific American Library, distributed by WH Freeman.

Murphy, A. H. (1993), "What is a good forecast? An essay on the nature of goodness in weather forecasting," *Weather and Forecasting*, 8, 281–293.

Nakazono, Y. (2013), "Strategic behavior of Federal Open Market Committee board members: Evidence from members' forecasts," *Journal of Economic Behavior & Organization*, 93, 62–70.

NCAR – Research Applications Laboratory (2015), *verification: Weather Forecast Verification Utilities*, R package version 1.42. Available from `https://CRAN.R-project.org/package=verification`.

Neuwirth, E. (2014), *RColorBrewer: ColorBrewer Palettes*, R package version 1.1-2. Available from `https://CRAN.R-project.org/package=RColorBrewer`.

Nychka, D., Furrer, R., Paige, J., and Sain, S. (2015), *fields: Tools for spatial data*, R package version 9.6. Available from `www.image.ucar.edu/fields`.

Obled, C. and Creutin, J. D. (1986), "Some developments in the use of empirical orthogonal functions for mapping meteorological fields," *Journal of Climate and Applied Meteorology*, 25, 1189–1204.

O'Hara-Wild, M. (2017), *ggquiver: Quiver Plots for 'ggplot2'*, R package version 0.1.0. Available from `https://CRAN.R-project.org/package=ggquiver`.

Overholser, R. and Xu, R. (2014), "Effective degrees of freedom and its application to conditional AIC for linear mixed-effects models with correlated error structures," *Journal of Multivariate Analysis*, 132, 160–170.

Paradis, E., Claude, J., and Strimmer, K. (2004), "APE: Analyses of phylogenetics and evolution in R language," *Bioinformatics*, 20, 289–290.

Patterson, H. D. and Thompson, R. (1971), "Recovery of inter-block information when block sizes are unequal," *Biometrika*, 58, 545–554.

Pebesma, E. (2012), "spacetime: Spatio-temporal data in R," *Journal of Statistical Software*, 51, 1–30.

Pinheiro, J., Bates, D., DebRoy, S., Sarkar, D., and R Core Team (2017), *nlme: Linear and Nonlinear Mixed Effects Models*, R package version 3.1-131. Available from `https://CRAN.R-project.org/package=nlme`.

Prado, R. and West, M. (2010), *Time Series: Modeling, Computation, and Inference*, Boca Raton, FL: Chapman & Hall/CRC.

Rasmussen, C. E. and Williams, C. K. I. (2006), *Gaussian Processes for Machine Learning*, Cambridge, MA: MIT Press.

R Core Team (2018), R: *A Language and Environment for Statistical Computing*, R Foundation for Statistical Computing, Vienna.

RESSTE Network et al. (2017), "Analyzing spatio-temporal data with R: Everything you always wanted to know – but were afraid to ask," *Journal de la Société Française de Statistique*, 158, 124–158.

Robinson, D. and Hayes, A. (2018), broom: *Convert Statistical Analysis Objects into Tidy Tibbles*, R package version 0.5.0. Available from `https://CRAN.R-project.org/package=broom`.

Rue, H. and Held, L. (2005), *Gaussian Markov Random Fields: Theory and Applications*, Boca Raton, FL: Chapman & Hall/CRC.

Rue, H., Martino, S., and Chopin, N. (2009), "Approximate Bayesian inference for latent Gaussian models by using integrated nested Laplace approximations," *Journal of the Royal Statistical Society, Series B*, 71, 319–392.

Sarkar, D. (2008), *Lattice: Multivariate Data Visualization with R*, New York: Springer.

Schabenberger, O. and Gotway, C. A. (2005), *Statistical Methods for Spatial Data Analysis*, Boca Raton, FL: Chapman & Hall/CRC.

Scheuerer, M. and Hamill, T. M. (2015), "Variogram-based proper scoring rules for probabilistic forecasts of multivariate quantities," *Monthly Weather Review*, 143, 1321–1334.

Schott, J. R. (2017), *Matrix Analysis for Statistics*, 3rd ed., Hoboken, NJ: John Wiley & Sons.

Searle, S. R. (1982), *Matrix Algebra Useful for Statistics*, New York: John Wiley & Sons.

Shaddick, G. and Zidek, J. V. (2015), *Spatio-Temporal Methods in Environmental Epidemiology*, Boca Raton, FL: Chapman & Hall/CRC.

Shen, X., Huang, H.-C., and Cressie, N. (2002), "Nonparametric hypothesis testing for a spatial signal," *Journal of the American Statistical Association*, 97, 1122–1140.

Sherman, M. (2011), *Spatial Statistics and Spatio-Temporal Data: Covariance Functions and Directional Properties*, Chichester: John Wiley & Sons.

Shumway, R. H. and Stoffer, D. S. (1982), "An approach to time series smoothing and forecasting using the EM algorithm," *Journal of Time Series Analysis*, 3, 253–264.

— (2006), *Time Series Analysis and its Applications with R Examples*, 2nd ed., New York: Springer.

Siegert, S. (2017), *SpecsVerification: Forecast Verification Routines for Ensemble Forecasts of Weather and Climate*, R package version 0.5-2. Available from `https://CRAN.R-project.org/package=SpecsVerification`.

Simpson, D., Rue, H., Riebler, A., Martins, T. G., and Sørbye, S. H. (2017), "Penalising model component complexity: A principled, practical approach to constructing priors," *Statistical Science*, 32, 1–28.

Spiegelhalter, D. J., Best, N. G., Carlin, B. P., and van der Linde, A. (2002), "Bayesian measures of model complexity and fit," *Journal of the Royal Statistical Society, Series B*, 64, 583–639.

Stanford, J. L. and Ziemke, J. R. (1994), "Field (MAP) statistics," in *Statistical Methods for Physical Science*, eds. Stanford, J. and Vardeman, S. B., San Diego, CA: Academic Press, pp. 457–479.

Stevens, B., Duan, J., McWilliams, J. C., Münnich, M., and Neelin, J. D. (2002), "Entrainment, Rayleigh friction, and boundary layer winds over the tropical Pacific," *Journal of Climate*, 15, 30–44.

Tobler, W. R. (1970), "A computer movie simulating urban growth in the Detroit region," *Economic Geography*, 46, 234–240.

Ver Hoef, J. M. and Boveng, P. L. (2015), "Iterating on a single model is a viable alternative to multimodel inference," *Journal of Wildlife Management*, 79, 719–729.

Verbeke, G. and Molenberghs, G. (2009), *Linear Mixed Models for Longitudinal Data*, New York: Springer.

Von Storch, H. and Zwiers, F. W. (2002), *Statistical Analysis in Climate Research*, Cambridge: Cambridge University Press.

Waller, L. A. and Gotway, C. A. (2004), *Applied Spatial Statistics for Public Health Data*, Hoboken, NJ: John Wiley & Sons.

Weigel, A. P. (2012), "Ensemble forecasts," in *Forecast Verification: A Practitioner's Guide in Atmospheric Science*, eds. Jolliffe, I. T. and Stephenson, D. B., Chichester: John Wiley & Sons, pp. 141–166.

Wickham, H. (2011), "The split-apply-combine strategy for data analysis," *Journal of Statistical Software*, 40, 1–29.

— (2016), *ggplot2: Elegant Graphics for Data Analysis*, 2nd ed., New York: Springer.

Wickham, H. and Chang, W. (2017), *devtools: Tools to Make Developing R Packages Easier*, R package version 1.13.6. Available from `https://CRAN.R-project.org/package=devtools`.

Wickham, H., Francois, R., Henry, L., and Müller, K. (2017), *dplyr: A Grammar of Data Manipulation*, R package version 0.7.6. Available from `https://CRAN.R-project.org/package=dplyr`.

Wickham, H. and Grolemund, G. (2016), *R for Data Science: Import, Tidy, Transform, Visualize, and Model Data*, Sebastopol, CA: O'Reilly Media.

Wickham, H. and Henry, L. (2017), *tidyr: Easily Tidy Data with "spread()" and "gather()" Functions*, R package version 0.8.1. Available from `https://CRAN.R-project.org/package=tidyr`.

Wikle, C. K. and Hooten, M. B. (2016), "Hierarchical agent-based spatio-temporal dynamic models for discrete-valued data," in *Handbook of Discrete-Valued Time Series*, eds. Davis, R. A., Holan, S. H., Lund, R., and Ravishanker, N., Boca Raton, FL: Chapman & Hall/CRC.

Wilks, D. S. (2011), *Statistical Methods in the Atmospheric Sciences*, 3rd ed., Waltham, MA: Academic Press.

Wood, S. N. (2017), *Generalized Additive Models: An Introduction with R*, 2nd ed., Boca Raton, FL: Chapman & Hall/CRC.

Wu, C.-T., Gumpertz, M. L., and Boos, D. D. (2001), "Comparison of GEE, MINQUE, ML, and REML estimating equations for normally distributed data," *American Statistician*, 55, 125–130.

Xie, Y. (2013), "animation: An R package for creating animations and demonstrating statistical methods," *Journal of Statistical Software*, 53, 1–27.

— (2015), *Dynamic Documents with R and knitr*, 2nd ed., Boca Raton, FL: Chapman & Hall/CRC.

Xu, K., Wikle, C. K., and Fox, N. I. (2005), "A kernel-based spatio-temporal dynamical model for nowcasting weather radar reflectivities," *Journal of the American Statistical Association*, 100, 1133–1144.

Zammit-Mangion, A. (2018a), *FRK: Fixed Rank Kriging*, R package version 0.2.2. Available from `https://CRAN.R-project.org/package=FRK`.

— (2018b), *IDE: Integro-Difference Equation Spatio-Temporal Models*, R package version 0.2.0. Available from `https://CRAN.R-project.org/package=IDE`.

— (2018c), *STRbook: Supplementary Package for Book on ST Modelling with R*, R package version 0.1.0. Available from `https://github.com/andrewzm/STRbook`.

Zammit-Mangion, A. and Huang, H.-C. (2015), *EFDR: Wavelet-Based Enhanced FDR for Signal Detection in Noisy Images*, R package version 0.1.1. Available from `https://CRAN.R-project.org/package=EFDR`.

Zeileis, A. and Hothorn, T. (2002), "Diagnostic checking in regression relationships," *R News*, 2, 7–10.

Zhang, J., Craigmile, P. F., and Cressie, N. (2008), "Loss function approaches to predict a spatial quantile and its exceedance region," *Technometrics*, 50, 216–227.

Subject Index

367

Author Index

R Function Index

Pages refer to the end of the code chunk containing the function. The function may hence appear on the page preceding that listed when the code chunk spans more than one page.

Pages refer to the end of the code chunk containing the function. The function may hence appear
on the page preceding that listed when the code chunk spans more than one page.

Pages refer to the end of the code chunk containing the function. The function may hence appear on the page preceding that listed when the code chunk spans more than one page.

Pages refer to the end of the code chunk containing the function. The function may hence appear
on the page preceding that listed when the code chunk spans more than one page.